ESTIMATING CONSTRUCTION COSTS: A Conceptual Approach

ESTIMATING CONSTRUCTION COSTS: A Conceptual Approach

Keith Collier

RESTON PUBLISHING COMPANY, INC.
A Prentice-Hall Company
Reston, Virginia

Library of Congress Cataloging in Publication Data

Collier, Keith
Estimating construction costs.

Includes index.
1. Building—Estimates. I. Title.
TH435.C719 1984 692'.5 83-24500
ISBN 0-8359-1792-4

10 9 8 7 6 5 4 3 2 1

PRINTED IN THE UNITED STATES OF AMERICA

To Beatrice, with loving thanks.

Contents

Preface

Most people in construction know something about estimating. At the mention of a *conceptual estimate* there are questions because the term is relatively new. The term received official recognition in 1975, when a federal government publication said that construction managers needed more expertise in conceptual estimating.[1] Conceptual estimates are those made before construction drawings are completed, often before they are hardly begun. They are essential in the fairly new method of contracting that employs construction managers instead of general contractors and that travels the fast track to completion rather than the traditional road. Conceptual estimates, as one kind of preliminary estimate, are also used to set construction budgets, to make feasibility studies for developers, and to control construction costs at the most critical stage, during the design.

For years designers tried to predict construction costs, and eventually developers and building owners turned to other ways of avoiding excesses, particularly to construction management (CM) and to target figure contracts which both need estimates before a design is completed.

Not much has been published about conceptual estimates and related topics other than scattered articles on the use of *building elements.* This book seeks to fill that need and to promote better skills in making conceptual estimates by using elements and other means. It examines all kinds of construction estimates, their strengths, their weaknesses, and also their relationships with each other. It shows how to make the best possible preliminary estimates, especially

[1] *The GSA System for Construction Management* (Washington, DC: General Services Administration, Public Buildings Service, 1977).

conceptual estimates. It shows how conceptual estimates are made even before preliminary design sketches and why conceptual estimates have increased in importance since traditional contracting methods have given way to other methods.

New directions and new methods in construction, contracting, and development; the advent of microcomputers into business and of research into the economics of buildings and their construction—all these have changed ideas about the role and practice of cost estimating and what is expected from estimators. On the one hand computers make it possible to estimate more efficiently, more analytically, and in greater detail; on the other hand concepts like *life-cycle costing* and the *investment approach to budgeting construction costs* cause one to question the need for more details about the capital costs of construction while we ignore the often larger costs of maintenance, renovation, financing, and the dramatic effects of violent fluctuations in the marketplace. This book examines these questions and provides some answers.

Primarily this is a practical book for those who want to make better preliminary estimates of construction costs and better building appraisals based on the cost approach. It contains a number of selected examples of preliminary estimates (some conceptual, some not) together with complete drawings for three buildings and a number of sketches and figures. It also contains many estimating exercises, questions, and topics for discussion and for testing. There is a comprehensive glossary developed in my three previous books and further developed for this one, as well as a bibliography. For easy reference the Index is more extensive than some. I hope this book will be read and used not only by those in construction but also by others in related fields such as appraisal and construction financing, because the economics of construction are a common ground for all of their work.

In this book I have for the first time written an introduction. Many introductions remain unread, so I advertise it here as an integral part of the text in which I have tried to show—through personal experience—that estimating is not entirely a science but also an art that often employs scientific methods and techniques, as do all the arts. Estimating is not just a dreary side issue to the more dramatic acts of designing and building; it is an essential part of construction and it can be creative and interesting, as indeed can be any task that is worthwhile, that demands knowledge and skill and that is well done. This is especially true with *conceptual estimates* because they require creative thinking and the fullest use of an estimator's knowledge. I hope this book helps to convince you of that and helps you to make better estimates.

Acknowledgments

Acknowledgment is due to my friend and colleague, Glenn M. Hardie, of the British Columbia Institute of Technology (Building) for reading through the manuscript and for his suggestions; also to Anna Maharajh of BCIT for her help with some of the estimating examples; and to the staff at Reston Publishing Company for their cooperation with my first electronic manuscript. Specific acknowledgments are generally made in the appropriate places in the text where they have greater significance.

Introduction

Some of the estimates used as examples in this book were done in the 1960s; and I have updated their cost figures so as not to confuse the reader. Costs have since increased more than threefold, but the essential things of estimating and construction have not changed. In the 1960s, I was fortunate in having the chance to make preliminary estimates with little fear of trouble if they were not accurate (I was with a government department and did them almost as a favor because they were not really part of my job description.) Perhaps I did them almost lightheartedly. I was not afraid to be experimental. Nevertheless, I found that usually I could get surprisingly good results in a short time. Sometimes an estimate took only a few hours, at other times a few days. Sometimes I had full drawings and specifications and the time to peruse them. Sometimes I had enough time to make only the most cursory survey and to take a few overall measurements. I began to collect cost information from my own estimates and from the schedules of values of current projects; my main resource was an appraisal manual and its periodic updates. During that period, and in that place, the construction market was fairly competitive and stable, but the buildings for which I did the estimates were an unusual lot and included prison blocks, trout hatcheries, and government buildings of uncommon design and in isolated places. It was a grand opportunity to experiment and to learn more about making estimates rapidly and often with minimal information. I explain all this by way of encouragement.

Later I did the same kind of estimates for a consultant's fee, and I took up teaching estimating and construction economics at the British Columbia Institute of Technology (BCIT). Teaching required me to rethink what I had learned about construction, and it gave me cause to think more about conceptual estimates. As a consultant I was surprised to hear people express amazement and curiosity over the accuracy of preliminary estimates; once I was asked if estimating skills could be inherited. I told the truth; there were Gypsies among my ancestors. In those days most people thought of estimating as something done only by contracting companies, and designers were still having problems designing buildings that could be built within their clients' budgets. Some designers have learned to make good estimates, but for many, estimating is either incidental or a necessary evil to be avoided whenever possible.

About that same time, in the early 1960s, the Construction Management (CM) approach to major building projects began to spread. I was involved in some of the earliest CM projects in the Pacific Northwest and did conceptual estimates for them. About the same time I also managed my first target-figure contract; that too required an accurate preliminary estimate. I saw then the beginning of conceptual estimating's growing importance, particularly to new contracting methods.

This book has four parts:

1. *Part One*, which deals with the background of conceptual estimating, with the nature of construction, with its costs, and with the different kinds of estimates, particularly as they relate to *conceptual estimates*. It also introduces the relationship between estimates and those appraisals based on costs.

2. *Part Two*, which contains examples of the application of several kinds of preliminary estimates, each reproduced in synoptic form to show the method without all the detail; also there are estimating exercises.

3. *Part Three*, which deals with estimating precepts and practices and continues to survey the things that estimating and appraisal have in common. It examines new estimating techniques and proposes a new look at and a new methodology for the cost approach to appraisals of value.

4. *Part Four*, which discusses the nature and sources of information and data for estimating and introduces microcomputers as a means of better estimating and construction management.

The *Glossary* originated in my first book, written in 1973, and in each book since I have expanded and improved the Glossary because precise terminology is important to good communication and understanding. Readers are urged to check the meanings of designated words and phrases in the Glossary as it will help them in following the text's comparisons and arguments. For teachers and students the Glossary also makes a useful source of material for short tests. For more demanding studies, longer tests, and examinations there are the questions at the end of almost every chapter. I have put some effort into the *Index* for similar purposes.

It was not unusual for an owner-developer to give his designer a budget amount and to spend more, often without too much concern on the part of the owner at the increased expenditure because he knew that the development's value was increasing all the time. But economics and attitudes have changed, and today an owner-developer usually wants to know not only what the bids for construction will be, he also he wants to know what his development's economic condition will be at varying times during its life. Life-cycle costing is an inevitable outgrowth from estimating, but it requires different kinds of estimates. It is here that the estimating of future costs and the appraisal of present value through costs come together in construction economics, of which conceptual estimating is an important part.

The reader experienced in construction may prefer to pass over Part 1, at least initially, and go directly to Parts 2 and 3. I suggest an initial survey of the whole book, a careful reading of Part 3, and then a study of selected examples in Part 2 probably with periodic returns to Part 3. For students and novices I suggest an initial reading through the text but with considerable skimming over the more stolid parts intended for future reference. After that I suggest following the same order proposed for the experienced reader, preceded by a second reading of Part 1 in order to get the broad view first.

At the suggestion of my teaching colleague, Glenn Hardie of BCIT, I expanded the book's original concept to make it a teaching text; consequently teachers will find much material for discussions, assignments, tests, and examinations. Experience shows that for students with some elementary background in construction and cost estimating the best approach to the subject of construction economics is one with few formal lectures and many assignments and discussions especially discussions that involve comparisons of the results of students' efforts. No two estimates are ever the same; it is instructive to discover why. Ideally this book will be used as a text in a senior course after a course on estimating fundamentals based on my first book,[1] as the two books

[1] *Fundamentals of Construction Estimating and Cost Accounting* (Englewood Cliffs, NJ.: Prentice-Hall, Inc., 1974).

have some things in common and are sequential. However, this book can well serve as the text for a first course in estimating for designers and for others who do not need the detail and exactness found in the first book.

I hope that appraisers will read this book, use it, and gain something from it. It is a lot to ask someone to learn everything necessary to use estimating techniques in appraisals based on proper costs; unless appraisers do, they cannot claim to be using a valid cost approach. At the same time the "cost approach to value" would be greatly enhanced by the use of the estimating techniques explained here; the cost data bases now available in the marketplace certainly facilitate this.

Finally, I hope my references to computers, especially those in Chapter 3, Part 4, will cause some readers to look into their uses, both personal and for business; here I am primarily addressing my own generation, as the younger need no such encouragement. The computer industry, especially that part dealing in personal computers, is in a great rush, and the ordinary customer often does not get the kind of service that he deserves. Some minimal preparation is advisable before entering into this boisterous marketplace, but the intelligent and wary need not fear. There are few good reasons for postponing a personal computer if one has the work for it. The *Bibliography* will lead to some introductory books on computers that may encourage and give useful advice.

Keith Collier
Kaneohe, Hawaii
July 1983

PART ONE

1

Construction and Construction Sites

In Part One *construction*[1] is reviewed as it relates to estimating, for it is the unique nature of construction that gives rise to the need for estimating and in particular makes *conceptual estimating*[G] necessary. It is important not only to understand this but also to keep it in mind when making construction estimates. Simple things are often taken for granted, and sometimes their purpose is forgotten, especially those

[1]Phrases and words, such as *construction*, that are included in the Glossary are indicated by a [G] when they first appear in a chapter or whenever the reader's attention should be drawn toward a precise meaning.

things done repeatedly. Often one overlooks the *why* because of preoccupation with the *how*.

The purpose of Part One is to review the *why* before studying the *how*. But first we need to be clear about our subject: conceptual estimating. For the present, we may say that *conceptual estimating* is that estimating done before a construction design is made in detail, when the design is expressed only in outlines as a concept of the designer, as distinct from that estimating (better known and understood) by a bidder[G] from detailed drawings and specifications for a lump-sum contract.[G] Later, Chapter 4 examines in greater detail the nature of estimates and estimating of several kinds. Here, we shall see that construction is unique and that therefore construction costs also are unique. From this it follows that estimating and cost accounting[G] in construction are fundamentally different from similar activities in other industries, a fact often initially rejected by accountants. Those readers well grounded in construction may choose to skim Part One but are urged not to as they may find some fundamentals worth reviewing and perhaps reestablishing.

The Design of Construction Work

By *construction*[G] we mean the design and production of *improvements*[G] to land, using the term in its legal sense as it is used in appraisals and assessments,[2] but not necessarily in an aesthetic sense. By *work*[G] we mean work as defined in most standard construction contracts.[G] We ignore the relatively recent and artificial separation of design from the production of buildings by architectural institutes as something that is neither necessary nor usual in the general history of engineering and building; in *construction* we also include *design*.[3] The design of structures of all kinds requires the calculation of the effects of several natural forces: in ship design, the forces due to wind and water pressures; in aircraft design, the forces due to air pressures and currents. In building design, the universal force of gravity and variable forces due to winds and earthquakes must be taken into account.

Of course, there is much more to design, but whereas with ships and aircraft the designers are concerned with movement, with buildings the designers are primarily concerned with stability. Ships and aircraft are always recognizable because of the constraints on their designers, but with buildings—and some of them look like ships or aircraft—designers have much more freedom of design because their creations are static. That is not quite true; they do move, but they are also fixed in place. Ships and aircraft are inevitably shaped by the dynamic forces that act on them while they are in motion. Buildings have only one universal characteristic, the verticality of gravity. Typically the external lines of ships and aircraft are streamlined for horizontal motion while those of buildings are mostly vertical because of gravity.

In buildings the walls are vertical because of gravity and because of the kind of space that they enclose. (If structures are occupied by creatures other than human the enclosing planes of structures may be different; for example, curved.) Roofs may be either sloped (to shed water) or more or less horizontal. (Absolutely flat roofs are not practical because they rarely stay flat, and consequently they never drain properly.) Exterior surfaces must resist wind pressures and the entry of moisture and air. Foundations must sit on stable subsoil and transmit the loads of the building (both live and dead loads, due to occupancy and the weight of the building itself) to the ground. They also must be able to resist lateral movement and overturning and possible disturbance or destruction by natural forces such as earthquake, fire, and flood. Other than these, there are few major restrictions to building design, especially when buildings are compared with ships and aircraft. (Most restrictions on buildings are regulatory and manmade.) Buildings and other structures, ships, and aircraft—all these have something in common; they are more or less custom-designed; but because of the freedom in building design and because of the unique relationship of buildings with the land buildings are unique products of a unique industry.

No two buildings are ever absolutely alike, even if they are built side by side and from the same drawings and specifications, because inevitably they sit on two different parcels of land in two different locations. They may be similar, but they are always different. A few lightweight buildings such as some residences are built with only slight contact with the ground (on piers, for example), but most buildings and other structures must be built into the ground and down to a suitable bearing. Also buildings require connections to underground service systems and to utilities. Often the exact design of the underground work, including foundations and connections to services and utilities, cannot be determined until site work exposes the subsurface con-

[2]Both *appraisal* and *assessment* are defined in the Glossary. Generally the former term is used in referring to both.

[3]Recently some regulations have been changed to allow architects to both design and build, but it is not yet a common practice in the English-speaking world.

ditions. Consequently in both estimating and quantity surveying[G] it is common practice to separate the *substructure* from the *superstructure* of a building; the design and measurement of work in a substructure is considered provisional until the exact requirements are uncovered and then suitable contractual adjustments are made. Typically this division between substructure and superstructure is made at the floor slab nearest to the ground surface.[4]

As in buildings so in other structures, one can visualize the same natural forces at work; in all of them one sees the vertical lines of gravity combined with the horizontal lines of utility; the lines of horizontal planes designed to resist the force of gravity and to support what is placed on them. Besides buildings, bridges are a good example of structures that combine both the vertical lines of gravity and the horizontal lines of utility. In one sense all buildings are bridges, and all structures are rooted into the ground each in a unique location. It is this fact above all others that makes every structure unique.

The uniqueness of buildings and similar structures originates with the sites upon which they stand. The next three chapters examine more closely the nature of construction, but construction sites ——that part of real property known as land on which all improvements are constructed——are examined first.

Construction Sites

When all societies were agricultural most buildings were relatively simple and often consisted of materials obtained directly from the land itself: timber, stone, and clay burned in the form of bricks and tiles. (Under everything lies the land, the source of all materials and all wealth.) Not long ago few architects specialized in design; major construction work was usually supervised by master builders and surveyors.[G] Then, surveyors (that is, overseers) were not mapmakers but managers of landed estates and their buildings.[5] Instead of on plans and maps they recorded information about land, its boundaries, dimensions, features, and titles, in verbal descriptions by *metes* and *bounds* (a term later garbled into *leaps and bounds*).

Buildings and the professions connected with them (of architects, engineers, surveyors, managers, and others) all grew out of the land, but now, with intensive urbanization and mammoth buildings, people often overlook the relatively insignificant but all-important parcels of land beneath them. Beneath them all lies the land, and the fundamental relationship between land and buildings persists, even though today materials taken directly from the site are rarely used in construction, and even though the soil removed from around foundations is often replaced by imported material.

No building can escape the influences of its site: the *subsoil conditions* that determine the character of the foundations; *the location* that determines so much of a building's design and the methods of its construction; and even a site's *relationships* with its surroundings which determine when the site is ripe for development and therefore when construction takes place. Legally and practically land and the buildings erected on it (all the improvements, in fact) are one.

Site Locations

All sites are unique in at least one unchanging feature: the location; and location is very important to construction costs. As real estate people say (and so should estimators): location, location, location. Of course, all sites are also more or less different in their individual characteristics, both above and below the surface, and the differences are virtually unlimited. Some of the more important differences among construction sites——set down by way of contrast in two main classifications——follow:

Urban and Suburban Sites

1. Easier access
2. Closer to resources
3. Knowledge and experience of locality available
4. Fewer and smaller risks in construction
5. Lower overhead costs

Rural and Remote Sites

1. More difficult access
2. Further from resources
3. Knowledge and experience of locality sparce
4. More and larger risks in construction
5. Higher overhead costs

These lists are not complete and other contrasting characteristics may come to mind. But the distinc-

[4]The division between *substructure* and *superstructure* may be formalized, as in a typical *contract with quantities*,[G] or it may be made informally by an estimator for his own purposes.

[5]See F. M. L. Thompson, *Chartered Surveyors: The Growth of a Profession* (London: Routledge & Kegan Paul, 1968).

tion is clear. At one extreme is the typical *urban site*, close to a population center if not within it, well known, and readily accessible. At the other extreme is the typical *rural site*, often more or less remote from population centers, more or less unknown, and usually much more difficult to reach. In between lies an almost infinite variety of sites sharing an almost infinite variety of characteristics that affect the costs of construction.

Site Information for Estimating

What information does an estimator[G] need in order to consider properly a site and its effects on construction costs? The scope of such information is broad and absolute definition impossible, but in view of the importance of sites on costs and in view of the carelessness with which they are sometimes handled let us review the most common items of information that should be contained in an estimator's site survey before estimating costs:

1. Postal address and a site's location relative to adjoining roadways, railways, rivers, canals
2. Legal description of the property of the site, with a site plan or a surveyor's drawing of the site
3. Orientation of the site's boundaries by compass bearings
4. Dimensions of the site's boundaries and the site's area
5. Relative elevations of the topography and of any unusual features or existing improvements on the site (including benchmarks)
6. Dimensions of any existing improvements such as buildings, roads, fences, and their distances from the site's boundaries (to locate improvements on site)
7. Locations and types of all existing utility lines and existing services to the site with their elevations, inverts, and descriptions; all items such as parking meters, bus stop signs, and other objects belonging to others installed on or near the site
8. Locations and descriptions of all easements, rights of way, footpaths, roads, streams, gullies, ponds, and similar features on the site
9. Locations of all existing access points to the site, such as crossovers, gates, and any easements and rights of way over other property leading into the site
10. Distances from the site to nearby communities and to other major points such as highways, railways and stations, airports, rivers, and canals.

Some of this needed information should be available on a legal survey plan of the site (a copy is probably held by owner); in the case of a site in an organized area the rest of the information should be available at local government offices and at the local library. For a site in a remote area, it may be necessary to make a survey of the site from the air or by other means. In any event, and for all sites, a personal examination is essential in order to discover information about the following:

1. The site's subsurface conditions (if not already provided by the owner or designer, and which may require confirmation)
2. Existing trees, other vegetation, and all other natural features of substance or consequence
3. Details of existing buildings and any other major improvements on the site
4. Restrictions on access to the site and the causes and frequencies (for example, heavy snowfalls, floods, and weight and height limits for bridges and railway crossings)
5. Availabililty of resources (materials, labor, and equipment) and transportation to remote sites
6. Prevailing weather at different times of year and the probabilities of extreme conditions (such as the annual hurricane or monsoon or the 16-year flood)
7. Legal or cultural restrictions on access to or use of a site or of its contiguous areas.

Adequate information about the subsurface conditions of a site (most important to design and installation of foundations and to all other underground work) can be obtained only by penetrating the surface and by actual examination of the subsoil; this is usually not an easy thing for a bidder to do, especially at a remote site. It is a source of disputes between contractors and owner-designers,[G] particularly because most bidders naturally take the easier way and rely on whatever information about subsurface conditions they receive from the owner-designer, which is invariably incomplete and at best probably obtained by drilling test holes.[6] The most common procedure is to hire a soils engineering consultant to make test holes and to provide a soils' report. The information is limited to the test holes and may not tell the whole story. However, when making a conceptual estimate more or better infor-

[6]For discussion of subsurface conditions, claims, and disputes, see the author's earlier work, *Managing Construction Contracts* (Reston, VA: Reston Publishing Co., 1982).

mation may not be available. Because more information is always needed for design, it makes sense to obtain this information earlier and in time to be of use in a conceptual estimate.

The oversight of existing improvements and natural features on a site may cause delays and problems when the time comes to start construction: for example, a 20-feet deep ravine on an urban site concealed by dense vegetation (who would have believed it was there?) remained undiscovered until a contractor's superintendent arrived one wet Monday morning with two laborers and a truck to begin construction. A laborer bearing a leveling rod stepped backward for a better sighting and disappeared. And how many gas and water lines have been ripped out by impatient bulldozers because "nobody knew it was there"?

Restrictions to access may frustrate the use of certain equipment or impede the delivery of large material components to a site; the effect on construction costs may be serious. Remote sites may necessitate the use of radio communications; and even on an urban site the cost of courier services for the speedy delivery of documents, drawings, and messages (the federal postal service is not always fast enough) can create significant costs.

Weather does not receive enough attention from many estimators, especially the infrequent but more onerous weather that can be predicted with more or less accuracy. If a site is in an area that will be inundated once every 20 years (as records show and local people tell), should not an estimator allow for the costs related to a 5 percent chance of a flood at that site, or at least the costs to provide special insurance coverage? Should not all such things be calculated risks in any estimate?

There are numerous risks in construction that estimators and their employers and owners and their designers are aware of but are overlooked or ignored because either—it is said—there is not sufficient time or information to deal with the risks or (as some estimators are fond of saying) the inclusion of allowances for such risks in an estimate and bid would inevitably make the bid uncompetitive. But reflection shows that this is not good sense: not to calculate foreseen risks and not to allow for them in an estimate is just not good business. The real problem is not that most estimators do not want to allow for risks but that they supect that their competitors will not; as a result, their own estimate and bid will be higher in amount and therefore unsuccessful. Some say that by its nature construction is a gamble and that this cannot or should not be changed. But gambling involves the calculation of risks. Recognizing and calculating allowances for risks in construction does not remove the element of gambling; it simply distributes the risks more equitably between the owner and the contractor.

Perhaps the time has come, now that the boom era of construction is over, for new and different attitudes and approaches for those who estimate and bid construction costs—and for those who pay for them—a time in which all risks in a construction project are discovered and calculated and allowed for in a fair and reasonable manner so that more risks are equitably carried by both parties in construction contracts. In construction there will always be the unpredictable—enough to satisfy most appetites for gambling. But what can be foreseen should be calculated as best as it can and included in the contract. This may require contracting methods different from those in use today. To begin with, after surveying a site, and as part of an estimate, an estimator should prepare a written report on a standard survey form (especially designed for each company) that will act as a checklist for essential information. (The subject of an estimator's responsibilities is taken up in Chapter 4).

Site Work

Division 2 of the *Masterformat*^G^, entitled "Site Work," contains the following "broadscope" headings:[7]

1. Clearing of site
2. Earthwork
3. Piling
4. Caissons
5. Shoring and bracing
6. Site drainage
7. Site utilities
8. Roads and walks
9. Site improvements
10. Lawns and planting
11. Marine work.

Except for *excavating* and *backfilling*,[8] and a few minor and unusual items, all of the work contained in this standard division of work,^G^ *site work*, is not

[7] *Broadscope* and *narrowscope* are descriptive terms used to define the scope of work under standard titles.

[8] A narrowscope heading under *Earthwork*.

directly related to the actual erection of a building but to preliminary preparation and final landscaping and grooming of the site. Often, even excavating and backfilling is not totally related to a building but also to work in site improvements such as walks and roadways. In other words *site works* are precisely just that: primarily work to change and to improve a site, whether or not there is to be a building on it; for example, in the creation of a public park. Clearly this type of construction work is essentially different from any other: one that does not necessarily depend on a building and a building's design for its scope and nature; a class of work that is dependent mostly on the site itself—hence its name.

Site work is primarily work done to improve a site. In estimating construction costs, site work is one of the "problem divisions" of work,[9] one in which either an estimator expects a shortage of information and many risks, or else he is happily ignorant—especially while making a conceptual estimate. This fact should be noted and applied later in this chapter and in other chapters. Now we shall deal with specific aspects of site works in estimates and real-property appraisals. As a construction person, should the reader wonder about his interest in appraisals, believe that there is a meaningful connection.

Given a site intended for development, and given that few sites are entirely ideal, it follows that an amount of site work directly proportional to the extent to which a site falls short of an ideal site is always needed to ameliorate the site in addition to that site work needed for actual construction (for example, for a building's foundations). This identification and separation of two kinds of site work—(1) for site upgrading and (2) for improvements, such as a building, on the site—is important to estimators in conceptual estimating (and in other estimating), and it is also important to real-property appraisers.

Estimators making conceptual estimates need to separate these two kinds of site work because they need to relate the specific costs of site works to specific parts of buildings in order to create specific cost information that can be used for estimating the costs of similar work in other buildings, as described later. Similarly appraisers and assessors need to know how much of the total costs of site works is attributable to the value of a site (as land) as distinct from the buildings thereon (as improvements). In fact we can be even more analytical and consider a third kind of site works. For the present we shall leave this topic with these three classifications in mind:

[9]As the junior estimator said, an excavation is the only thing that gets bigger when one takes something away from it.

1. Site work required to improve a site, to upgrade it from a less than ideal site to a site suitable for an intended project
2. Site work as a part of major improvements (such as buildings), including excavation, backfilling, and filling for foundations and building services
3. Site work for minor site improvements (other than buildings) such as walkways and roads.

As with all classifications, these are not absolute but they will prove useful.

Site Utilities

Site utilities, or *site services* as they are often called, are another part of site work that requires special analysis and definition for estimators and appraisers; the definition is different from that found in Masterformat and in most contract documents.[G] There are at least three useful classifications:

1. Off-site services existing outside a site's boundaries to which new service lines to a building will be connected
2. On-site services to be installed inside a site's boundaries and up to an imaginary perimeter line around a building (often defined as 3 feet or 1 meter from the building's exterior surface)
3. Building services to be installed within a building and extending outside up to the imaginary perimeter line.

As to the precise division of services, local custom may vary as may the scope of the work included in *site work.* (For example, services' piping may be included in *plumbing.*) But some such division of this work is necessary in estimating. Costs of off-site services (utilities) are usually quoted by authorities as lump-sums,[G] while the costs of on-site sevices and building services are measured and priced by an estimator. In making most estimates, especially preliminary estimates,[G] site work often appears to the estimator as a confusion of indefinite work that is partially concealed (below ground), entirely difficult to measure and price (because it mostly involves only labor and equipment and few tangible materials), and, all too often replete with errors of omission.

Thus far, we have related different classes of site work to their causes and origins so as to have a clearer understanding of what site work is and what it involves and to illustrate the uniqueness of building sites and therefore the uniqueness of each construction project.

Aspects of Estimating

Estimating and the *cost approach* to the appraisal of real property values (which is based on estimating) is essentially a process of analysis and synthesis; of taking construction work apart (mentally), and of putting it together again in a different format of words and figures. Construction drawings and specifications are analyzed and translated by an estimator into quantities of *items of work*[G] and then dollars representing the *total costs of the work*[G]. This process is partly scientific (and partly an art) and therefore it uses classifications. We have just identified in this chapter some new classifications for site works to improve estimating techniques.

At first meeting, and to many, especially the student who has aspirations to be a designer or a construction superintendent, estimating may seem to be a dull and tedious subject, especially because it starts with the *measurement of work.* The student cannot at first appreciate that estimating is mostly imaginative and that conceptual estimating (to which he is usually introduced later) is almost entirely imaginative; that is, conceptual. At first the student cannot understand that an estimate is derived not only from design information[G] but also from concepts of design and construction gained through experience and stored in the estimator's mind. There are times in estimating, such as during the measurement of repetitive items like doors and windows, when one becomes bored; but so it is with all trades and all professions. Many have found estimating to be a challenging and creative occupation. Nobody can deny that estimating the costs of site work is challenging; and it is creative because it requires an estimator to visualize a site with all its natural characteristics, to speculate on its latent features, and to conceive of the site as a finished plot: graded, landscaped, and designed in its topography to accommodate the buildings erected on it. The following chapters endeavor to convince the reader of the pleasures and value of an estimate well done. There is little pleasure in anything not done well, and great pleasure and satisfaction in most things when they are done well. The objective is not only to understand conceptual estimating but also to do it well and to enjoy it.

Questions and Topics for Discussion

1. List in order of importance the main constraints in building design and describe them.

2. Explain the common characteristics of buildings, what they are, and why they exist in their typical form.

3. Explain why the construction industry is properly described as unique among industries.

4. Describe and compare two main classes of sites that establish a wide range of differences among sites generally.

5. Design a pro forma for a site-survey report for use in a general contractor's office. In a preamble state assumptions as to the nature of the company and the kinds of work that it usually performs.

6. Select a vacant site and make a written, verbal survey of the site for estimating purposes.

7. Define the different kinds of *site work* that are significant to the estimator and the appraiser. Explain why they are so.

8. Make a list of and describe the *site services* and *utilities* commonly found in a mixed commercial and residential area; classify services generally in a way significant to an estimator.

9. Define the following terms: (a) *land,* (b) *improvements,* (c) *fixtures,* (d) *site works,* and (e) *a unique industry.*

10. Write a report on the reliable sources of weather information (both long- and short-term) in your locality that are available to a construction company and describe how a company can obtain information about weather for estimating and for planning and scheduling purposes.

11. Write a report to an estimator on the specific sources of information of all kinds concerning construction sites (real property) in your area that are available to construction companies.

2

Construction Work

Construction work[G] has already been defined and it means far more than the ordinary term implies. Here we examine its meaning and content more closely. We shall examine the nature of *construction costs* in Chapter 3. The immediate objective is to find classifications of construction work that are useful in estimating, which means seeking classifications of work that have meaning in the field of construction costs. We are not looking for classifications according to function or architectural style, although in some ways they are related to costs. Rather, we are looking for classifications that arise from construction economics and from technology. The purpose is to understand construction work better from the economic and technical point of view and thereby to better understand construction costs and how they can be estimated.

All construction work may be divided into two main classifications:

1. *Building construction work*, which creates a sheltered environment for human society

2. *Engineering construction work*, which embraces all other kinds of construction.

Because we are primarily concerned with *building construction work*, it is expedient to include all other work in a single, second class. In reality other classes for other purposes require separate identities; for example, industrial construction. But let us proceed by comparing these two main classes of work in the light of estimating construction costs.

Building Construction Work

Chapter 1 introduced the subject of design of building construction work and talked about its constraints and main characteristics and about the relative freedom of designers of buildings compared to that of designers of ships and aircraft. Now let us go further and examine some other characteristics of buildings and see how this may help in estimating their costs.

The Primary Characteristics of Buildings

Buildings designed as shelters against the natural environment consist of enclosing planes and of conditioned spaces, of walls and decks enclosing lighted and heated, or cooled, rooms. Most of the costs of construction are spent on the enclosure and on access, but the rest (typically called the mechanical and electrical services, or M & E work) may in some buildings (such as hospitals) cost up to half of the total costs of construction. But even M & E work in a building bears some relationship through function and design to a building's enclosures. Heating and cooling needs are in part related to the surface areas of the enclosures; lighting is largely related to floor area. Even more esoteric construction work such as that for acoustical and decorative purposes is largely related to inner surface areas.

Enclosure means weatherproofness to prevent the entry of the natural elements. Window vents control the entry of unconditioned air, and other kinds of *valves* control the entry and exit of other things and people. Extractors are required for the expulsion of waste matter (air, smoke, sewage). Elevators are required to move people and things vertically. Partition walls are required in most buildings. For estimating purposes all these and all other *building elements*[G] may be put into one of three main classes:

1. Elements related to building area
2. Elements related to building perimeter
3. Elements related only to use and occupancy.

We might consider another class (at least one more): those elements related to a building's volume; but as most building types have more or less standard ceiling heights (and when it comes to costs we are concerned with classes of buildings according to type of construction and use), usually we can effectively relate the elements related to volume and their costs to a building's area. From among those elements already mentioned the most important are those of mechanical and electrical (M & E). But building volume cannot be ignored entirely.

Elements Related to Building Area

These elements are in a majority because buildings as defined are designed for human use and occupancy; when we say *area* of course we primarily mean *floor area*. The main elements related to building area are:

1. Floors
2. Roofs
3. Partitions
4. Finishes (installed on the other elements)
5. Mechanical (heating, cooling, air conditioning)
6. Electrical (lighting, heating, power).

These elements require definition; the Appendix provides a standard list of elements. There are other lists, but the one in the Appendix is better than most. Until there is a North American standard list of elements, it will suffice. Let us review each of these area elements:

Floors are often separated into (1) lowest floor and (2) upper floors because of their physical difference. The lowest floor element usually consists of a complete strata of items on the ground and up to the floor surface, but the floor finish is usually separated. Upper floors include their supports, usually columns and beams.

Roofs, like upper floors, include their supports as well as any overhang. They may include the roof finish. Usually, finishes are separated from structure, and so is roofing.

Partitions are not directly related to area as they are linear, but in many kinds of buildings (a good example, apartments) there is a fairly constant relationship between the total length of partitions on a floor and the floor area because of typical design requirements.

Finishes on the preceding elements are usually separated because they are highly variable, even among buildings of the same class.

Mechanical heating and cooling, air-conditioning, and electrical work are classed as area elements for the reasons already given. The standard list of elements in the Appendix shows different classes of elements from those listed above. The standard list was first developed by quantity surveyors[1] and it is widely used. But it does not convey the fundamental facts about elements that we are searching for here, particularly their origin and nature. Nor does the standard list indicate the possible flexibility of their applications in conceptual estimating, and that is one of our objectives.

Elements Related to Building Perimeter

These elements contain a lesser proportion of the total work in most buildings, but the first element listed is among the most important, in part because it is among the most variable:

1. Exterior cladding
2. Perimeter foundations
3. Perimeter eaves and parapets.

Exterior cladding always has a big effect on costs; its area and its costs are obviously related to both perimeter and to building height, and perimeter is a major characteristic of plan-shape. A square-shaped plan is the most economical for a building's perimeter (apart from a circular plan, and circular-buildings are neither economical nor popular). A long, narrow rectangular-shape is the least economical because of the high proportion of perimeter to area, as shown in Figure 2-1.

Perimeter foundations are the main foundations in many buildings (but not in curtain-wall clad buildings). The standard list in the Appendix includes the whole of *Substructure* in one "area element," but that is not necessarily the best way. Most of a substructure's costs may be in the perimeter foundations and may therefore bear little relationship to the costs of the substructure of another building of different shape or one of a different construction and enclosure. (Perimeter foundations may include perimeter drains.)

Perimeter eaves and parapets also may have no relationship to a building's area. They may be extensive, expensive, or nonexistent.

The simple relationship of these first two classes of elements to a building's design and particularly to its area, perimeter, and plan-shape should be apparent.

[1]Members of the Royal Institution of Chartered Surveyors and the Canadian Institute of Quantity Surveyors.

Elements Related Only to Use and Occupancy

In this group are all the other elements, those that rationally cannot be related directly to either *building area* or *building perimeter*:

1. Conveying systems
2. Plumbing, drainage, and similar systems
3. Chimney stacks and similar items
4. Special fixtures, equipment, and other installations.

Conveying systems include stairs, escalators, elevators, and all other kinds of lifting and conveying systems.

Plumbing and similar systems are included because they bear no relationship to a building's plan or shape and not much to its area (except through population), unlike most other mechanical work.

Isolated items, such as stacks, must be included. They bear a relationship only to use and occupancy; one that cannot easily be quantified.[2]

Special fixtures, equipment, and other installations are similar in nature and relation to the other isolated items. They include, for example, equipment related to a special use and occupancy, special-purpose rooms (darkrooms, coolers) and any number of special installations that are limited and isolated within a building; that is, they have little or no relationship to the building as a whole, only to its special functions. Reflection and research show that there are many latent relationships among a building's elements, among its shape, area, perimeter, and other major characteristics, but in estimating costs there may not be time to pursue them all. (There is much need here for research.)

Estimates are like suits, they must be cut according to the available cloth; the time and money available from the client-owner and the information and data available to the estimator. By describing the elements related to area, perimeter, and occupancy, we are revealing a view of buildings unfamiliar to many in construction. Essentially it is the designer's view, and it shows a building in simplified and holistic form as an arrangement of horizontal and vertical planes enclosing spaces and items (the third class of elements); not as a complex conglomeration of hundreds of items of work, as a building appears to an estimator in a construction company, but as a complete entity which is then perceived as an arrangement of the main parts, or elements. This view

[2]This may not be true in *industrial construction* where a structure may be entirely utilitarian in design and function.

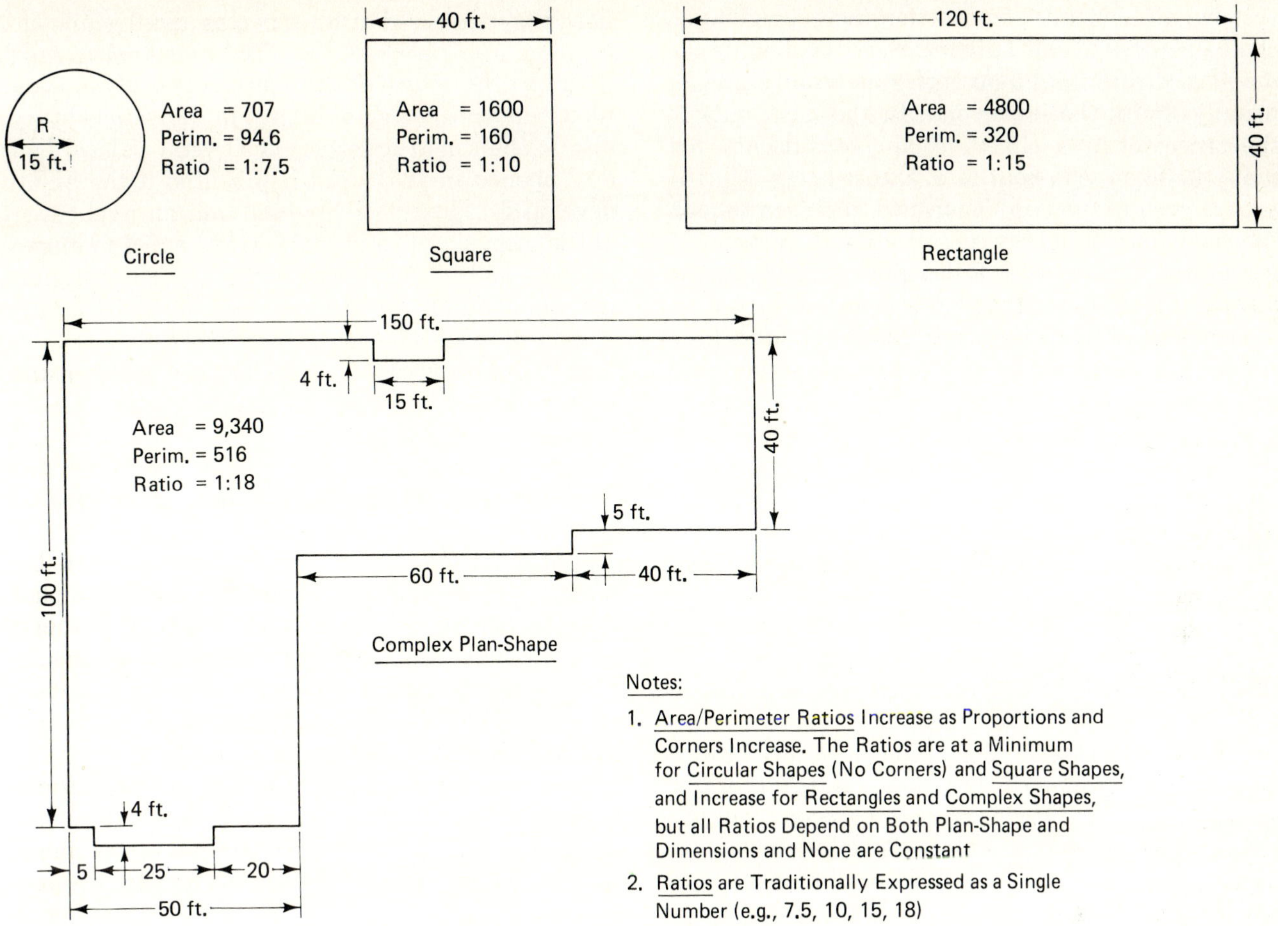

FIGURE 2.1 Plans and Perimeters

of a building made up of elements makes certain kinds of conceptual estimates possible; the kind most useful to designers and to owners in the early days of a project when it is necessary to delineate a design and its costs with a broad economic brush.

Building Services Systems

Earlier this chapter referred to the large proportion of a building's costs taken up by mechanical and electrical work, and this fact needs some elaboration and emphasis. Look at these representative percentages of costs quoted from a perennial source:[3]

Class of building (by function)	*Median costs of mechanical & electrical work (as a percentage of total building costs)*
Apartments (high-rise)	25.30%
Offices (high-rise)	25.90%
Schools (elementary)	29.20%
Hospitals	40.30%

[3]*Building Construction Cost Data 1982* (Kingston, MA: Robert Snow Means Co., 1982).

The significance of these figures and their effect on building costs generally is increased if we add to them the proportion of total costs applicable to a *building's structure.* Without a structural frame there is no building and there are no services. In most buildings, especially in those kinds of buildings listed, the proportion of total costs paid for the skeletal structure is about 20 to 25 percent (and even more if it is of precast concrete). By adding structural costs to costs of mechanical and electrical (M & E) about 50 to 65 percent of total costs are accounted for. If the structure is enclosed, 70 to 90 percent of the total costs are probably accounted for. These figures are only generally illustrative, but they do show some important facts about building costs: that M & E costs often are one of the biggest group of costs, and often the most variable of costs, depending on building type; that M & E costs are often rivaled by structural costs in magnitude; and that after allowing for cladding there is often not much left for anything else.

A building's structure is fundamental and utilitarian and it holds little room for variations that will significantly affect costs. An enclosure is essential, and although there is a variety of enclosure systems, in economic construction there is not much choice as design and building code constraints demand a certain minimum enclosure while a budget usually limits the upper end. In the past, outside the offices of M & E engineering consultants, not many knew much about M & E costs. It is still the area of greatest general ignorance, and even though the economics of energy conservation have drawn a lot of attention to M & E designs and costs in the last few years, much more knowledge and education are needed. M & E systems in buildings are where the greatest technical changes will take place, and of course this will affect the costs of M & E systems and of buildings. In a building nothing appears more confusing than the M & E installations before they are concealed by other work. To measure and price such complex work from bidding documents[G] appears as a formidable task. To do so in any meaningful way from preliminary design sketches seems quite impossible. Yet it can be done.

The first requirement of estimating is to be able to see the trees in spite of the tangled forest. Hidden in that thicket of ducts, tubes, conduits, and wires are the trees of the M & E systems. In any M & E system or subsystem the tree trunk is the *main line* or *stack*; there are branches radiating out from the trunk and extending to the tree's extremities, and at the ends of the minor branches are the fruits and the flowers of the tree, each according to its kind: colored fixtures supported by plumbing and drainage systems, round air registers on the branches of another system, and glittering light fixtures on another. All these trees have the same general format: a main trunk, major and minor branches, and terminal units. (The measurement of M & E work is discussed in detail in Chapter 3, Part 3.)

Standardization and Prefabrication in Construction

In discussing the nature of construction from the point of view of the estimator we cannot ignore the standardization and the prefabrication of construction components.[G] Both are quietly but steadily evolving, and both are effecting changes in contracting methods as well as in construction. Let us first examine *standardization* because it is the more common.

Standardization in Construction

The Glossary makes useful distinctions among the meanings of *materials,*[G] *products,*[G] and *components.*[G] The idea of standard building products is not at all new; they appeared at least two hundred years ago. Giedion, in his book, *Space, Time and Architecture,*[4] refers to the industrialized production of iron and its use in bridges and buildings in the 18th century; also to the construction of the Crystal Palace, London, opened in 1851, that had behind it the highly developed industry of England and an application of the most simple and rational system of manufacturing—that of serial production. . . . The design of the whole building was planned around the largest standard sheet of glass, which was only four feet long.[5] Giedion goes on to say that the building was unprecedented, that it contained some 800,000 square feet of area (on one level), and that it was erected in six months. This was achieved through "serial production" as Giedion calls it, that is, by the mass production of standard products. The revolutionary design of the Crystal Palace was based on a standard product already extant at the time; the four-foot sheet of glass. The Crystal Palace and innumerable later buildings were built from standard products; now we take such construction for granted. The Crystal Palace was also built with *standard components*: with iron lattice girders, with cast-iron columns, and with wooden frames for the glass. It was constructed mostly of *prefabricated components* but, unlike the standard sheets of glass, they were specially designed for the Crystal Palace project.

Today in a few places we can occasionally see innovations in serial production; for example, in the use of standard building components of high complexity for the construction of several similar buildings under one authority; such as a school board. Briefly, and in general terms, the concept calls for the planning of the construction of a number of buildings of similar use and design over a specific period. Although the buildings are designed for the same purpose (say, education), local needs are for buildings of different sizes; different sites require different substructures and different arrangements of service connections. But beyond those special requirements, the buildings can be more or less standardized through the use of large and complex standard components for such building elements as exterior and interior walls, floors, and roofs, each complete with all surface finishes and with the M & E services built-in.

To attain this advanced level of standardization it is necessary to limit its application; for example, to a school district in which one authority can plan, implement, and control the standardization. But planning must include not only the buildings but

[4](Cambridge: Harvard University Press, 1954) 167ff.

[5]Ibid., 249.

also the production of the standard components at a certain minimum volume to make it economical for the manufacturers. One obstruction to wider use of such standard components, apart from general acceptance, is the initial cost of planning, designing, and setting up the production of the components. Nevertheless, it has been done (in at least two school districts); there are no impossible obstacles, only difficulties arising from acceptance, implementation, and economics. We can see, for example, a similar but much less complex degree of standardization in standard wood roof trusses designed for residential construction. Because a wood truss is not complicated and because the spans and loads needed for houses are not highly variable, the trusses can be truly standard; that is, they are not custom designed but made according to standard drawings when an order is placed. In the school developments cited, the manufacturers of the standard components were brought into the projects and their planning from the outset; indeed they were essential participants. To catch and hold and bind their commercial interest it was necessary that they be guaranteed a certain volume of work in order for them to agree to take part and to carry the initial setup costs.

In a free market it usually takes a manufacturer to initiate steps toward the standardization of components, and such steps are naturally taken with great caution. Despite the general dearth of research and development in construction and its contracting methods, we may expect continued movement toward more standardization of building components.

Prefabrication in Construction

Prefabrication and standardization go together. Standardization as a means to economy in construction leads directly to what Giedion called "serial production," the beginning of mass production as we know it. Standardization also leads to prefabrication because prefabrication involves production in a factory instead of at a construction site. But we say *prefabrication* in reference to the factory production of *components* rather than of *products* as we have defined and used those terms here. Products are assembled into components, and when this is done off the site and in a factory, we say that components are *prefabricated.* Most prefabricated components are serial products—for economy, several or many are produced together—but not necessarily. For example, the steeple for a church may be prefabricated and lifted and placed in position by a crane or by a helicopter. The implications of prefabrication and standardization for construction costs and economics are obvious.

The economics of mass production and standardization are generally understood, and standard products are normal in building construction. Most architects now use nonstandard products only when they believe that their design needs them for aesthetic reasons. It is always difficult to balance costs against aesthetic values, especially for a designer. Even in designing speculative dwellings some designers try, for example, to include nonstandard windows and doors in their search for a better appearance; but owner-developers do not always see eye to eye with their designers because custom-made windows and doors can cost several times more than standard items.

The practice and economics of prefabrication can be complicated.

Case study

My first experience with prefabricated architectural concrete involved narrow precast concrete panels designed as sculptured and decorative structural columns. My specifications allowed the contractor to fabricate the panels on site, but the contract's requirements were based on panels fabricated by a precast concrete manufacturer in a plant. For fabrication at the site, the contractor was required to produce initial samples of the panels for approval before proceeding to general production. The specified standard of finish was high. Because of forms distorted by steam curing the contractor was not able to attain the specified standard and approval until he had experimented with several batches of panels. The result was a loss for the contractor of several tens of thousands of dollars (in today's dollars) because he lacked the expertise of a precast concrete specialist.

The primary advantages of prefabrication are serial production under the controlled conditions of a factory. Even if there is no serial production, it may be cheaper to fabricate in a factory and to transport a complex component to the site for installation in one operation, as opposed to piecemeal installation or site fabrication. There lies the question, Is it cheaper or not to prefabricate? Some of the important things to consider follow:

1. Are conditions at the site tolerably suitable for site fabrication, and if not, what will it cost to create tolerable conditions if they can be created?

2. What is the nature of the component to be produced; is it of a common or uncommon type, or is it unique?

3. How many of the component are required?

4. Does the contractor have the needed expertise to produce the component, or must he contract with a subcontractor or supplier for all or part?

5. Is the needed expertise to produce the component available locally?

6. Is there a more or less competitive market locally in which to obtain the component?

7. Does an initial physical setup (of forms, jig, and such) already exist in one or more factories, or will the component require an entirely new setup?

8. Is a high level of control required by production such that can be attained only in a factory (for example, for special curing), or is production control at a site feasible?

9. Is transportation from factory to site expensive or prohibitive?

10. Is special equipment needed for installation of the component at the site; if so, is it economically available to the contractor?

11. Does fabrication or erection of the component require workers of a particular union or workers of more than one union?

12. Are such workers readily available to the contractor?

13. Is the contractor likely to need to produce similar components for other projects in the foreseeable future?

These are some of the questions that may require answers in order for a contractor to make a reasonable decision whether to prefabricate or to have a component prefabricated by others, assuming that the contract gives him the choice. For some components the answer quickly becomes clear, while with others a contractor may still be wondering whether he made the better decision years later. For an estimator making a conceptual estimate, a decision may be even more elusive. As an employee or agent of an owner-developer,[G] a cost consultant may also have to consider the time and the timing of construction relative to a volatile real estate market.

If the nature, multiplicity, and economics of a component lead directly to the idea of prefabrication the key questions remaining probably are: Are transportation and handling and erection of the prefabricated component economically feasible? And is there a practical alternative to prefabrication? If the component represents a significant part of total costs, do not suppose that the costs in a conceptual estimate cannot be so accurate that it does not matter whether the estimator assumes the components are prefabricated. The difference, for example, between precast (prefabricated) and cast-in-place concrete structural systems for apartment buildings is significant. On the other hand one must observe that for most buyers of apartment condominiums it probably makes little difference which type of concrete structure is used. And how does an appraiser respond if that proposition is correct? Presumably, if precast concrete is used there are good economic reasons, such as a shorter construction period and therefore a lesser job overhead cost[G] which in turn results in a building improvement equal in cost and value to a similar building but with a cast-in-place concrete structure. However, it is even more complicated than that.

Other Economic Considerations

There was a time when if you spoke of a building's costs, it was assumed that you meant only the capital costs of construction, but that time of innocence has passed. Now, it is not enough to be able to estimate with a 10 percent accuracy the median bid for a building's construction contract; now the enlightened owner wants to know what it will cost him to own and rent the building over the next fifty years—and what the rate of inflation will be in the year 2001. Arthur C. Clarke may be able to tell him, but not you or I. If you undertake this kind of prediction—called *life-cycle costing*[G]—you had better stick to present-day dollars; but some cost consultants like to think of themselves as seers.

Life-Cycle Costing

As the name implies, life-cycle costing requires us to look at a building after it is built and occupied and to consider its future use in terms of costs expressed in a specific currency of a value related to a specific date, usually within the next year or so, because it is not normally within an estimator's expertise (or anybody else's) to predict accurately the value of a currency far into the future. You may then wonder why anybody should bother at all. If we cannot predict what inflation will be in five years' time, what makes us think that we can predict other facts that touch on the future economics of real property? We may have some sympathy with that point of view, but those in development and investment in real property need something on which to base their decisions; a well-made study of life-cycle costing (LCC) is better than most other things—especially if it confirms an investor's hunch.

And as for inflation, the argument for using only today's (or next month's) dollars also says that because inflation is a general phenomenon, all economic relationships are equally affected and will remain the same, which is also the reason we may claim some value for LCC: because every investor is involved in a world of economic relationships, it is the relative economic position that is significant.

Life-cycle costing may be used to shed light on the possible long-term effects of:

1. Starting and finishing a construction project at different times;
2. Selecting one design rather than another;
3. Selecting among alternative construction methods and materials with different economic lives and with different maintenance requirements and costs; and
4. Choosing from among different development proposals involving differing capital costs, construction periods, economic lives, and operating costs.

In cost estimating we consider the effects of *time* only during the construction period and during the following warranty period because that affects overhead costs, especially financing costs, and cash flow. In life-cycle costing we consider *time* and its effects over a building's projected life, even up to and including its demolition. Life-cycle costing is useful therefore in a number of ways, all involving a building's life and use. But some things are difficult to quantify and to express in terms of cash, and that is an important limitation of a technique essentially based on (1) the cash value of things and benefits and (2) the present worth of future benefits and the present worth of future dollars (afghanis, zlotys, or what have you).

For buildings the fundamentals of life-cycle costing may be expressed as follows:

1. There is a time element in the value of money, such that one dollar *invested today* is worth more than the promised receipt of a dollar sometime (anytime) in the future because the invested dollar earns interest, but the interest rate must be higher than the rate of inflation.
2. It is necessary to convert all future costs and benefits to a common base—to their present worth, or their worth at some specified date—expressed in a currency (e.g., U. S. dollars).
3. Building components (with some exceptions) have a shorter economic life than the building as a whole, and most components require maintenance during the building's life while others must be replaced once or several times during a building's life.
4. Future costs-in-use of a new building include (a) maintenance costs, (b) replacement costs, and (c) operating costs.
5. Income from passive sources may offset some costs. (See the following.)

To be clear about the costs involved in a building and its life, let us review them.

Capital Costs of Building Construction

1. Land and development costs (whether these need to be included depends on the nature of the cost study undertaken)
2. Design fees and other fees for consultants, agents, professional advisors, and local government, and payments related to development permits and such like
3. Costs of financing a project, including any related points and or fees
4. Costs of construction
5. Costs of installed equipment.

Maintenance and Replacement Costs

1. Periodic cleaning and preventive maintenance costs
2. Costs of periodic redecorating and repairs
3. Costs of replacements of fixtures and components.

These classifications are not hard and fast and they do overlap.

Operating Costs

1. Staff wages and related overhead costs
2. Electric power, fuel, and other utilities
3. Insurance premiums, taxes, and similar costs
4. Periodic cleaning costs, if not included in maintenance.

Income from Passive Sources

This is income other than that for which the building is operated (e.g., rented) and includes income obtained from appreciation in value of the property, subsidies, and other incidental sources.

Operating Income

This is the income expected from the investment and obtained by the operation.

In the past it was difficult to keep track of and to categorize all income and expenditure in many commercial operations, let alone try to predict them over a period of years. Microcomputers with software designed for business uses make it easier, thus opening the door to easier and more effective life-cycle costing.

Classifications of Construction

For our use in discussing costs I have put all construction into two main classes, with many possible subclasses:

1. Building construction, and
2. Engineering construction.

As our interest is primarily in the former, the latter class is used here as the residual class containing everything that is not building construction.

Building Construction

Many classification formats are used for different purposes. Here is one for use with cost data:

Occupancy	Residential, commercial, public, government, industrial
Height	Up to 3 stories, 3 to 6, 6 to 12, 12 to 24, 24 to 30, 30 plus
Construction	Wood frame (light) (heavy), wood and steel frame mixed, fireproof (up to 12), fireproof (over 12 stories)

Most buildings can be classified using these three characteristics, with extra classes and subclasses added as needed.

Engineering Construction

Within this superclass are all other kinds of construction: highways, bridges, dams, sea defenses, industrial plants, public works for such as utilities, etc. The list is almost endless.

Differences among the many kinds of engineering construction are greater than among buildings, which is one reason why engineering construction, unlike building construction, rarely uses the stipulated-sum contract.[G] Nevertheless, estimating is just as much needed in engineering construction but it does require some different techniques.

Questions and Topics for Discussion

1. Define *building construction* as distinct from all other kinds of construction by describing fundamental characteristics.

2. Describe those parts of a building other than the *enclosures* and describe their uses, functions, and their design requirements.

3. Describe three major classes of building elements by their relationships to a building's primary characteristics. Can you think of any other classes of building elements significant enough to possibly equal the three classes cited in the text? If so, describe them. If not, find and describe three major classes of elements in a different kind of construction (e.g., engineering, industrial).

4. Explain in detail why the most commonly used and referred-to building elements are those related to building area; explain why at least one other major class of elements and its fundamental relationship to building design cannot be ignored.

5. Explain why among buildings the vertical elements of a building are much more open to variation than the horizontal elements. If you disagree with the premise of this question, explain why in detail.

6. Explain fully the significance of a building's shape to its costs of construction.

7. Explain how in industrial construction the elements unrelated to *enclosure* may be quantified for future reference and use as data in estimating. Explain why this is not usually possible in building construction.

8. Describe diagrammatically and verbally the basic physical format of M & E systems in buildings and the format's significance to estimating costs.

9. Define the following terms: (a) *materials*, (b) *products*, and (c) *components*. Explain the historical significance of these three related terms.

10. Describe the potential effect of the widespread prefabrication of standard building components on construction economics.

11. Describe the requirements for economical prefabrication of building components.

12. Explain fully the monetary principle that underlies *life-cycle costing.*

13. Explain fully the technical principles (other than the monetary principles) underlying *life-cycle costing.*

3

Construction Costs

One of the best places to find clarification of construction costs is in the typical standard form of the *cost-plus-fee contract*,[G] because, for this kind of contract in which the owner pays the costs as they are incurred, costs need to be defined clearly, as opposed to the more commonplace *stipulated-sum contract*[G], in which the owner simply pays an agreed stipulated sum, or contract amount. In the cost-plus-fee contract one finds definitions for two main classes of costs:

1. *Reimbursable costs*,[G] those that the *contractor*[G] has to account for fully to the *owner-designer*[G] and that we have defined as *direct costs*.[G]

2. *Nonreimbursable costs*,[G] those that are paid for by the owner as a *fee* to the contractor, and that, according to a specific contract, are either calculated as a stipulated percentage of the reimbursable costs or are already stipulated in the contract as a fixed fee; those costs that we have defined as *indirect costs*.[G]

There are only two basic kinds of construction contracts (all others are hybrids): (1) the cost-plus-fee contract and (2) the stipulated-sum contract, popularly called the lump-sum contract. In the former, only the fee is agreed to before the work begins; in the latter, the entire contract sum. In the cost-plus-fee contract, the indirect costs[G] (operating overhead[G] and profit[G]) are settled in advance as the fee, and the owner pays all other costs, usually at the end of the month, as they are incurred. These direct costs are classified as:

1. Material costs,
2. Labor costs,
3. Plant and equipment costs,
4. Small tool costs, and
5. Job overhead costs.[1]

Let us enlarge on the definitions of these costs.

Material Costs

Material costs[G] appear as the most easily identified costs as they arise from tangible items: lumber, concrete, hardware, and such like; yet still there are problems in estimating these costs. For example, I got my first estimating job with a contracting company because I knew, it was said, that sawing wood produced sawdust; something that the previous estimator seemed not to understand, according to the manager. He was, of course, talking of *waste,*[G] something that an estimator must assume invariably occurs even with products[G] such as plumbing fixtures. Also, some items do not fit neatly into our classifications, for example, concrete formwork; an item in construction worthy of a thesis for a Ph.D. candidate because it exemplifies so much of construction economics, as we shall see.

There are many kinds of *concrete formwork*, so let us consider just two classifications:

1. Formwork fabricated at the site by carpenters
2. Formwork prefabricated elsewhere.

Formwork made at the site usually consists of form plywood, yard lumber, and hardware (both carpentry hardware and formwork hardware). It requires labor for fabrication, for erection, and for stripping, removal, and repairs before its reuse. It also requires, in many projects, equipment to lift it into place. Nevertheless, formwork is not normally incorporated into the building work; it is put up, used, and taken down. Should the plywood, lumber, and hardware it consists of then be classed as materials and their costs as material costs for cost accounting and estimating purposes? No, they should not be classed as materials because they have not been incorporated into the work. If at this point you want to say, So what? and to suspect me of being a pedantic academician, let me hasten to defend myself by drawing your wavering attention to the other kind of formwork mentioned.

Formwork prefabricated elsewhere is usually of steel, or of another metal alloy, sometimes of plastic or of plywood and lumber. It makes no difference which, because—and this is the point—it is not incorporated into the work; it simply does a job and is then taken away, just like any piece of equipment. For that is what we should see formwork as—equipment[G]—even when it is made from plywood and lumber by carpenters at the site. Only a portion of the formwork costs should be allocated to the project on whose site the forms are made, and wood forms should be treated for estimating and cost accounting purposes just the same as prefabricated forms rented from someone else, used, and later returned.

Ah! some readers will say, but what about forms in housing construction that are made from plywood and lumber and later used in the actual framing and cladding of the houses? A common enough thing in frame construction. Certainly those methods complicate estimating and cost accounting. (Just how much plywood and lumber are wasted while in use as forms, and how much retrieved for incorporation into the actual construction work?) But these methods should not, cannot, change the principles involved, the principles of estimating and cost accounting, which, if ignored, will cause in the minds and accounts of a contractor and his staff confusion and ignorance about the true profit (or loss) of the project involved.

Clearly, then, forms are forms whatever form they take, and they must be treated (in both estimating and cost accounting) as equipment[G] and not as materials[G] and labor.[G] Consequently we cannot say categorically that, for estimating purposes, all plywood and lumber are *construction materials.* (You may be regarded as eccentric, however, if you point to a pile of form ply and tell somebody to "Get that equipment off the site.") This incidentally shows the importance of defining technical terms such as those in the Glossary.

Waste[G] can also cause problems. Waste, as an expense, is the cost of material not incorporated into the work. It has many causes, from normal usage to theft, and it is highly variable, which is why estimators cannot ignore it or treat it lightly. By estimators, waste must be assumed to be always potentially present; a cost accountant should be

[1]These five classes of costs are defined in the Glossary.

skeptical if it appears nonexistent in an account. For some items for which waste is an exception, such as the plumbing fixtures already mentioned, an estimator should allow for the *risk of waste* unless the risk is obviated by insurance. (Waste and laps[G] are not the same.)

Labor Costs

There are both *direct labor costs*[G] and *indirect labor costs.*[G] The latter depend on the former, on the laws and regulations that give rise to them, and on wage agreements. These all vary according to time and place, otherwise they are a matter of fact. Direct labor costs are more interesting as they depend on variables that are of the essence of construction: labor productivity and management efficiency. The two are closely related, and the only way that an estimator can quantify and measure their effect and their results is through cost accounting. In *work study*[G] the researcher uses techniques not suitable for the estimator, who must predict the results of future work, something much harder than analyzing actual work in progress.[2]

There is wrong thinking among some estimators about labor costs, as some think of wages only as labor costs and they treat indirect labor costs[G] as a job overhead cost.[G] They may do this because they prefer to add indirect labor costs as one total figure to the total net labor costs on the summary sheet. But its location in an estimate does not define and identify a construction cost; rather it is its origin that identifies a cost.

The definition of direct costs[G] requires that costs be allocated to those *items of work*[G] or the project as a whole that gives rise to the costs, and therein lies the distinction between direct and indirect costs.[3] A bricklayer lays bricks at a site, and his labor costs are clearly part of the project's direct costs. A mechanic overhauls and lubricates a pump in the yard; to what may his labor costs be allocated? Should all projects be charged with part of his labor costs, or only those that used the pump, or should it be only the project on which the pump was last used?

Let not the conceptual estimator who estimates for reasons other than for making a profit from performing construction work think that such questions are of no concern to him. For unless he can answer such questions, he will not be able to answer correctly other questions about construction costs that more obviously bear on his conceptual estimates.

Plant and Equipment Costs

Whole books have been written about the singular topic of plant and equipment costs[G], it is so complex. In brief, the main general points to be considered in estimating are:

1. Rental terms for equipment (and plant),[4]
2. Efficiency of equipment, and
3. Productivity of equipment.

Even if equipment is not to be rented, and even if a contractor intends to use his own, for estimating and cost-accounting purposes the equipment should be treated in every way as though it were rented. (Some prefer to set up their own rental company to hold their equipment and to rent it to themselves.) This keeps everyone's thinking, as well as well as their accounting, more objective.

Rental terms may include different rates for operating time and idle time,[G] specified responsibilities for transportation to and from a site, for setup ready for use and dismantling later, repairs, maintenance, fuel, lubricants, other consumable items, replacement parts, insurance, taxes, and the operator. All terms should be written into a lease or agreement.

Efficiency means selection and use of the economically right piece of equipment, and a skilled operator, for a specific purpose. It has a direct bearing on the next item.

Productivity depends on management, assuming that we have an efficient piece of equipment. Poor management may cause excessive idle time and unnecessarily high costs. Some projects' costs are practically all equipment costs (for example, many civil engineering projects), and good productivity of equipment is essential to the contractor's success. Even in building projects with a relatively small amount of equipment use, through delays and increased overheads the low productivity of equipment may increase equipment costs beyond the costs allocated to the equipment in the project estimate.

In building projects, the usual way to estimate

[2]See my book, *Managing Construction Contracts,* for more on this topic, including a brief description of some work study techniques used.

[3]Note that there are direct and indirect costs, generally, and direct and indirect *labor costs,* each as defined in the Glossary.

[4]Plant is no longer common on building sites. It is more common on certain kinds of engineering projects (e.g., crushers and mixers, for highway work). For brevity's sake we refer only to equipment.

equipment costs is to assess the amount of time that a particular piece of equipment is required on site and then to apply the following costs:

1. Mobilization costs (initiation, preparation, transportation to site, set-up ready for use)

2. Operating costs (at a periodic rate; per hour, day, week, month)

3. Idling costs (at a periodic rate; per hour, day, week)

4. Demoblization costs (shutdown, dismantling, and transportation to owner's yard).

The synthesis of costs for a major item of equipment must usually follow this format in any kind of estimate, bid or conceptual. There is no other effective way because the initial and terminal costs bear no relationship to the operating and idling costs.

For some kinds of plant and equipment, particularly heavy-duty items with more or less continuous usage in engineering projects (rather than building projects), costs can be realistically related to the production of plant or equipment (e.g., tons per hour, cubic yards per hour), especially for conceptual or other preliminary estimates. In other projects involving much equipment to be used over long periods (months, years) it may suffice to estimate a major portion of equipment costs by "writing off" the entire cost of new equipment to the one project. But this still requires estimation of the other costs for repairs, maintenance, consumables, and operator.

Small Tools

Small tools[G] are usually hand tools and small standing tools and machines such as saws and grinders. Economically, they are exactly the same as plant and equipment, and in principle the same costs are incurred; the only principal difference is the order of magnitude. Therefore, in estimating costs, the logical way to decide whether to treat an item as a tool or as piece of equipment is to consider the relative costs involved: If they are significant, treat the item as a piece of equipment. Otherwise, small tools are usually lumped together and their costs are based on a percentage of the labor costs.

Logically, more accurate costs for small tools can be obtained through cost accounting if tools for each trade are costed separately, and that should apply at least in bid estimates. With the increased use of microcomputers there is no reason why this and many similar refinements in estimating should not be applied in all but the most crude preliminary estimates—and it should no longer be expedient to make crude preliminary estimates, as we shall show. It may be tiresome to repeat it, but microcomputers make it possible to handle much more information much more effectively and much more economically. Estimating of all kinds should therefore advance immediately—but not unless the necessary principles and precepts of estimating and cost accounting (as outlined here) are followed.

Job Overhead Costs

Many estimators claim that job overhead costs[G] are the most difficult of all costs to estimate accurately. The probable reason is that these costs are almost all directly related to a project's duration, and typically projects often run into extended time for a thousand different reasons. It is perhaps the most typical undesirable characteristic of a construction project that it will overrun its *contract time*[G] and therefore its *contract sum.*[G]

If a project is to be performed with a stipulated-sum contract because it has all the necessary characteristics (adequate information about the site and the works are primary), then it should be possible to estimate accurately its job overhead costs. If not, then it probably should be performed not with a stipulated-sum contract but with a contract of another kind.

The breakdown of job overhead costs for a major building project was as follows:

Job Overhead Items	*Estimated Costs*	*Percentages (approx.)*
Supervision and staff at site	$244,000	24.50
Fees, permits, and utilities	$200,000	20.05
Rental and use of equipment[5]	$163,000	16.50
Bonds and insurances	$87,000	8.70
Rubbish removal & cleaning	$72,000	7.20
Temporary services & facilities	$72,000	7.20
Site office and equipment	$37,000	3.70
Small tools	$30,000	3.00
Security	$27,000	2.70
Drawings and documents	$19,000	1.90
Pumping	$15,000	1.50
Climatic conditions	$10,000	1.00
Finish protections	$8,000	0.80
Hoardings and barricades	$7,000	0.70
Testing	$5,000	0.50
Total	$996,000	100.00*

**Rounding off percentages causes a minor discrepancy in addition.*

[5]In this project, equipment costs were treated as a job overhead because of the general nature of the equipment used; chiefly large mobile cranes and hoists.

For another project the items of job overhead would be different. Although the preceding example contains common items, because it was for a precast concrete building development the project's duration (18 months) was less than that for a poured concrete project; as a result supervision was a smaller proportion of the total of job overhead costs. As mentioned, construction projects are unique, and therefore few examples can be called typical. It is therefore necessary to be able to explain the relative magnitude of items in the light of each project's peculiar characteristics and conditions. That is fundamental to estimating.

The total estimated cost for the project from the above preceding example was about $30 million; the job overheads listed amounted to about 3 percent, which is less than on many building projects, again, because the project was designed in precast concrete (at a premium on structural costs) to minimize the project's duration. Both higher construction costs and a shorter contract period for precast concrete caused the costs of supervision to be proportionately less.

In the project cited, fees, permits, and utilities had a high cost because of local conditions and the high payments required by local government and public utility companies. Rental and use of equipment might have been put into a separate category, as stated, and the total figures would have been different; supervision would have been proportionately higher, for example. Thus, in this last example we see both total job overheads and the main component—supervision—proportionately lower than in many projects. In other examples (in Part 2) job overhead costs have different contents and proportions.

The point made by the example and the related comments is that in most instances it is not possible to estimate the job overhead costs accurately for a project without a detailed analysis and estimation of each major overhead item unless one can make good comparisons among similar projects done under similar conditions. As such similarity is unusual, we may accept the point made as a general rule for estimating.

Indirect Costs

With another look at the typical cost-plus-fee contract for an understanding of construction costs, and at that standard contract's typical division of total costs into direct (reimbursable) costs and indirect (nonreimbursable) costs, reflection on the nature and procedures of a cost-plus contract shows us the fundamental differences between direct and indirect costs and their practical significance. *Direct costs* are classed as reimbursable costs[G] (according to the meaning in the standard contract) because they are directly attributable to the project in hand, they can be verified by the owner-designer, and they are reimbursed to the contractor as they are incurred. *Indirect costs*[G] are the opposite; they are not directly attributable to that specific job (indeed, most are incurred whether a particular project exists for the contracting company or not), they cannot be verified by the owner-designer (as they are incurred other than at the site), and they are paid only as an agreed fee. They are nonreimbursable costs[G] (according to AIA's standard contract).

An owner-designer[G] agrees to pay indirect costs and knows that they probably consist of:

1. Operating overhead costs[G] (popularly known as head office overheads),
2. Profit[G], and
3. Contingency sums[G] (if the contract is for a stipulated sum).

In cost-plus-fee contracts they are paid as a fee; in lump-sum contracts they are part of the contract sum. The owner-designer probably has only a rough idea of the amount that he will pay for each of these indirect costs, and often the contractor's idea may not be much more precise. How do contracting companies estimate indirect costs?

Operating Overhead Costs

Operating overhead costs[G] are the costs of being in business and of obtaining work, costs that are in part incurred independently of projects performed and of their direct costs. The only way that a contracting company can pay these indirect costs (without actually putting money into the company themselves, which at times they may have to do) is to include them as part of the overhead costs of the projects that they do perform. If they do no work, they not only make no profit, even more critically they cannot pay their operating overhead costs: the office rent, staff salaries, postage, and such. But that is only the beginning; the problem grows in trying to gather these overhead costs by including them in estimates and bids for work, because this enlarges the amounts of bids, which in turn makes them less competitive and less likely to be accepted.

You might think that the solution is to obtain work (in order to make a profit and to pay the overheads) by not getting involved in the crassness of competitive bidding. Surely, some may ask, is it possible to find enough work without all this pushing and shoving among competitors? A few contracting companies can find work in other ways, but the amount of such work available (like all good things) is limited, and most companies must get into the marketplace and compete. Some get work through business connections; through friends and colleagues in other companies that buy construction work; through favoritism bought by political donations; by all kinds of means—legal, illegal, ethical, and unethical—often simply by a good reputation; or by just being in the right place at the right time. Some of the most successful contracting companies obtain most of their work without bidding and through a good reputation, but they are a minority and usually they must demonstrate the fairness of their prices by comparisons and periodically by bidding for work to show their clients that they are still competitive.

Most construction work, however, is obtained by competitive bidding, which means estimating; in turn that means including allowances for operating overhead costs in the estimates and bids. How to calculate the markup[G] to be added (the reasonable share of overhead for each job) and how to remain competitive is a constant problem. Fortunately one's competitors have the same problem, so that reduces its import and it becomes a question of the differences among competitors' markups complicated by the fact that a larger volume of work reduces the amount of operating overhead that needs to be collected from each project, and vice versa. Yet, an estimator cannot simply add a minimal amount for overhead; otherwise his company will not be collecting enough from the interim contract payments received each month to pay for the ongoing operating costs. It is one of the biggest problems in the management of construction. This dilemma is highlighted by the fact that it is not unknown for a contracting company to do work with no markup for indirect costs in its estimate and bid, and therefore with no markup in the contract sum, simply in order to create some cash flow through the company; in fact, borrowing from one project to pay for another and to pay some of the overheads to keep the wolf from the door for a few more months until conditions improve.

If calculating operating overhead is problematic for a contracting company that knows all the facts only too well, how will an estimator making a preliminary estimate[G] manage to estimate overhead costs realistically?

In estimating overhead and profit, the first thing to consider is the current construction market. Is there a buyer's market or a seller's market in the locality of the project? And to what extent is it one or the other? The greatest problem for the estimator[G] of a preliminary estimate arises from the extremely active seller's market in which costs are inflated. Then markup amounts may have little or no connection with actual costs. When costs are deflated in a buyer's market, the same estimator's job is much easier because it is his goal to show the owner, his client, or his employer, a total estimated cost that will not be exceeded by the lowest bidder.[6] For this estimator the best of times is during the stable market when costs are fairly constant and he can calculate the ranges of markup being included in bids by making valid comparisons among projects and their costs.

Buying construction work is not like buying groceries; most people can delay buying construction work when the economic conditions are not suitable (as during the winter of 1982–1983); add to that the fact that construction is cyclical and dependent on seasons, labor-management wage agreements, and supply and demand, each overreacting and trying to attain equilibrium. The sea of construction economics is constantly on the move, with rising and falling tides, ebbs and flows, swells and storms, and sometimes with stultifying calms in which company ships reduce their crews and let their sails flap in the weak breezes of economic doldrums.

Consequently, estimating costs in preliminary estimates is partly science and partly art. Indirect costs are the residual amounts after deducting direct costs (fairly easily estimated) from the contract sum. But this is a disconcerting procedure when the residual amounts are either (1) nonexistent, or (2) far greater than any normal markup for indirect costs for the investigating cost consultant-estimator to identify the normal limits. Often it is of the greatest aid to an estimator struggling with a preliminary estimate in a confusing market to have the confidence of colleagues in contracting companies.

Profit as an Indirect Cost

Whatever else it may be called, profit is not without honor. It is a valid motive for a contracting company and it is a cost to an owner in a construction contract. Therefore it must be assumed and allowed for in preliminary estimates, no matter that it may on occasion be omitted from a bid estimate. There is something to be said for estimators in contracting

[6]By *lowest bidder* we mean *lowest valid bidder*, because despite its attractions an erroneously low bid ultimately may be a source of grief for the owner who accepts it.

companies not deciding on the amount to be added for profit, because profit is not a cost to the contractor, and its determination should be a management decision. But the estimator for a cost consultant, a construction manager, a quantity surveyor, or a construction economist must often estimate profit amounts for preliminary estimates.

During 1980 in western parts of Canada and of the United States some trade work[G] such as framing and drywalling was costing almost twice what it had cost two years earlier, and only a small part of the increase in costs could be truly attributed to general inflation; such are the costs in a seller's market. Sometimes costs inflate so rapidly due to a sudden surge in demand and a shortage of skilled workers that a preliminary estimate made barely a year earlier is much below the current market price. There are things an estimator should do to protect himself and his client or employer in such cases which are discussed in Part 3.

Contingency Allowances as Indirect Costs

In a stipulated-sum contract the owner agrees to pay the stipulated contract sum for the work shown and described in the contract documents. This agreement is without regard to the actual costs of construction, whatever they prove to be; the owner is not directly concerned with them. The contractor has estimated the costs and has committed himself to do the work for the contract sum. Therefore, if the contractor has allowed in his estimate and bid for contingencies that subsequently do not arise, assuming that all else remains as expected, the contractor benefits by in effect receiving additional profit. At the same time the converse is true should unexpected and unallowed contingencies arise. This is the nature of the traditional stipulated-sum (or lump-sum) contract.

But how can an estimator making a preliminary estimate of costs make any reasonable provision for contingencies[G] before the site has even been drilled for test holes let alone excavated? The answer is, he cannot. If an owner so wishes (and some do), the estimator may include in his preliminary estimate allowances for certain assumed and described contingencies (this looks better than an allowances for unspecified contingencies), or he may specify in the verbal report accompanying his preliminary estimate that no contingency sums[G] have been included. Either way an estimator should always make clear exactly what his estimate does and does not include. Many misunderstandings about preliminary estimates would be avoided if estimators wrote descriptive reports with their estimates. (More on that aspect of preliminary estimates appears in Chapter 2, Part 3.)

In certain circumstances some contingency allowances are prudent and called for in any kind of estimate. In many areas, for example, rock blasting is often needed for building foundations, even though test holes may have exposed no rock below the surface. Make a suitable contingency allowance. In other kinds of work other kinds of contingencies are common. If it is a matter of reasonable supposition that a certain requirement and cost will be incurred, even if the only evidence is from previous projects in that area, an estimator may give more weight to his contingency allowances by basing them on assumed quantities (of, say, rock to be blasted out) priced at current unit prices.[G] This however may only look more scientific than the contingency allowance simply expressed as a lump-sum but probably calculated in the same way. It is an axiom of estimating to set down in any estimate all assumptions on which estimated costs are based, or to specify that an amount is purely a guess made to avoid the omission of any allowance at all. Honesty is always the best policy, in estimating as elsewhere, as it dispels illusions, avoids delusions of accuracy, and helps to avoid misunderstandings on the part of those who do not fully understand estimating.[7]

Finally, in considering contingencies and their allowances in estimates, we should not be misled by the contents of the last paragraph into thinking that costs must always be related to a measured quantity of work. Indirect costs[G] by definition are quite unrelated to actual quantities of work, unlike most direct costs. Rather, it is of the essential nature of indirect costs, and especially contingency allowances, that they are related to *risk*. Therefore, it is a good estimating technique in all kinds of estimates to attempt to measure a risk and thereby assess the costs related to it. For example, if in an estimate we wish to put a fine point on a contingency, instead of simply making an allowance to provide for a specific event (such as the occurrence of a frost requiring protection of exposed ground in excavations), we may consider the *probability* of the contingency arising at all. We may in effect say that in the light of certain information there is a 20 percent chance of a frost occurring; therefore we shall allow for that risk by including in our estimate 20 percent of the total estimated costs that will be incurred in providing remedial action.

[7]Surprisingly this group often includes professionals with whom one would naturally assume such an understanding exists, including some who regularly deal in construction finances.

Costs in General

It is a common view that all construction costs are more or less directly related to the quantities of specified work, although the last people to swallow that whole are quantity surveyors[G]—despite their archaic title. That is more or less true of most *direct costs* in a stable and well-balanced market, but that kind of market is not the general rule much of the time. Therefore, we must say that *often* construction costs are *not* quantifiable and sometimes not even reasonable, from one point of view or another. Consequently estimating is not entirely subject to mathematical formulation and to rational analysis; it is not entirely scientific, and in part it depends on an estimator's feelings or intutition. There is more than one definition of *construction costs* depending on the payor's or the payee's position, contractual and otherwise, and depending on the timing of a project's duration. Let us review some differing definitions of costs, including:

1. Conceptual costs,
2. Contract-bid costs,
3. Contract-amount costs,
4. Final-account costs, and
5. Actual costs.

Conceptual costs are those estimated in a conceptual estimate; they are not simply vague shadows of real costs but are accepted for working purposes as real facts by a number of persons connected with a project, including the owner (for budgeting and financing and decision making), the designer (for similar purposes to those of the owner and as a basis for fees) and the construction manager (for management purposes, including planning, scheduling, and planning cash flow needs).

Contract-bid costs are those that make up the total of a bid for work as arrived at in the preceeding bid estimate.

Contract-amount costs may not be the same as contract-bid costs because of subsequent negotiations and resulting adjustments made by the contracting parties.

Final-account costs may be different from contract-amount costs (and frequently are) because of modifications[G] and changes[G] in the work of the contract according to the contract's provisions. These are the contract costs finally paid by the owner, and from his viewpoint these are the actual costs of construction.

Actual costs, however, are not so easily defined, even if we ignore complications in costs arising from contractual ramifications, such as interest lost or gained (depending on whose position is taken) on holdback money[G] (retention money), and costs incurred during a contract's warranty period. As noted, what is a cost to an owner is not always a cost to his contractor. Consideration of operating overhead costs and their allocation further complicates the question: What are the actual costs of any construction project? To some significant extent, construction costs are what we make of them; to a not insignificant extent they are subjective and variable, and to some extent they may be indeterminable. Whatever they are, they are not cut and dried or simply the results of bookkeeping and accounting.

If there is any doubt about this description of the nature of construction costs, and if for some it sounds like a writer's attempt to create a strawman ready for burning, or simply verbosity, consider once again the nature of *formwork* in concrete-building construction as one good example of where costs become less than hard and fast.

In summary let us review some of the main points in this chapter that cause us to say that construction costs are almost always an approximation, rather like the platonic shadow of an unseen reality:

1. *Site and location.* Change the site and costs will vary; move a building on its site and costs will vary.

2. *Time and duration of construction.* At a different time and season, during a different wage agreement, in a different construction market, and in a different economic climate, construction costs will vary.

3. *Design of work.* Costs will vary not only according to the design but also according to the designer, the use of standard and prefabricated components, legal requirements, inspection, and spontaneous on-site decisions by the owner-designer and their employees.

4. *Market conditions.* Costs will vary with supply and demand, with the cost of credit, with the need for cash flow, and with the efforts of contracting companies to maintain that cash flow.

5. *Construction contracts and contractual arrangements.* The type of contract, the choice of traditional or CM arrangement of contracts,[G] the use of sequential design and production procedures, or the use of phased (fast-track) construction methods will make a difference.

6. *Laws and regulations.* Federal laws and requirements for federal projects and similar requirements of other levels of government, standards of construction required by building codes incorpo-

rated into local laws, and restrictions on the use of new materials and methods of construction by obsolete regulations will cause costs to vary.

7. *Union or nonunion projects.* This includes the often higher costs of union projects, sometimes providing greater efficiency and labor stability; the often lower costs of nonunion (merit shop) projects; and losses due to disputes, lockouts, and strikes.

All these and many other things too numerous to mention affect construction costs. But the point has already been made: Construction costs are elusive and are not simply subjects for general accounting or even for cost accounting.

In Part 2 are examples of estimates that illustrate relative costs; in Part 3 costs are expressed in different ways and further discussed in the light of different kinds of estimates; and in Part 4 costs as information and data are examined in detail.

Questions and Topics for Discussion

1. In which type of standard form of contract can we find good definitions of construction costs? Define *costs* according to that standard form and according to this text.

2. Define the following terms as used in this text: (a) *nonreimbursable costs* and (b) *reimbursable costs.* Explain how the former term can be applied in a contract; is it truly descriptive and precise? Suggest two equivalent but more descriptive terms.

3. Define and compare the following in the context of construction costs: (a) *plant,* (b) *equipment,* and (c) *small tools.*

4. Among the different classes of construction costs, which do you believe is the most difficult to estimate? Explain your answer in detail.

5. State and explain the text's axiom concerning the allocation of costs in estimating and cost accounting.

6. Discuss formwork for cast-in-place concrete, its costs, and its economics in order to illustrate fundamental principles and definitions of construction costs.

7. For each of the classifications of costs of materials, labor, and equipment, state and explain at least one major difficulty in estimating those costs and discuss the ways of overcoming each of the difficulties.

8. Describe the origins and controlling facts of job overhead costs; explain what determines their amounts and their relative proportions.

9. Explain in detail precisely why there are *direct* and *indirect* costs in construction.

10. Describe the effects of different construction market conditions on building construction costs.

11. Describe in detail the relationship between *risk* and *costs* in construction.

12. Briefly describe at least four different definitions of the total costs of construction relative to payor and payee, and to a construction contract.

13. Which contractual arrangement brings about the most accurate and most equitable payment of costs for construction work? Explain your answer by detailed arguments.

4

Construction Estimates

We have mentioned and briefly described two kinds of estimates in construction:[1] (1) *bid estimates*[G] made by contracting companies in obtaining work and (2) *conceptual estimates*[G] made by others (including cost consultants,[G] quantity surveyors,[G] and designers[G]) for various purposes of an owner[G] in construction, including feasibility studies[G] at the outset of a project and cost controls[G] during a project's design phase.[G] In this chapter, we examine estimating and different kinds of estimates more closely before we study examples of actual estimates in Part 2. In Part 3 we shall examine and discuss the precepts and practices used in different kinds of estimates for different purposes.

All estimates are made to predict future costs, but different kinds of estimates are made at different times and for different reasons. The two main classes of estimates were just mentioned: (1) estimates for contractors' bids and (2) estimates for an owner's purposes. But, as with all classifications of types, there is always at least one more that lies somewhere between: in this instance those estimates that are partly conceptual and yet are made in order to obtain work or to

[1]The astute reader will have noted mention of *preliminary estimates*[G] in Chapter 3. This term is explained later.

manage it. We shall examine these in Part 3. For the present let us compare the two main classes that we have identified.

Bid Estimates

Because in most construction projects, so far as an owner is concerned, the costs are largely determined before work starts by an estimator making a bid estimate (which leads to a bid, or offer and, with its acceptance, to a contract), we may realistically typify bid estimates as the only true estimates and all other kinds of estimates as simulations. When an owner's agent or employee makes a conceptual estimate, what is that person doing? That agent or employee of the owner is attempting to anticipate the subsequent bid estimate that will lead to the successful bid in order to know the total costs of the work—before it is bid for by contracting companies and before a construction contract is made. For this reason we may say that *bid estimates* are the *original model estimates* and that *conceptual estimates* are but *preliminary outlines*.[2] Further, we may say that to understand conceptual estimates and how they are made it is necessary to understand bid estimates and how they are made. Here lies the stumbling block for many estimators. It has been observed that estimators with contracting backgrounds, accustomed to making bid estimates, often have difficulties with conceptual estimates, and that is a problem for many contracting companies that go into Construction Management (CM).[G]

Bid estimates, when properly made, consist of:

1. Items of work,[G]
2. Job overhead items,[G]
3. Operating overhead costs,[G] and
4. Profit.[G]

Each of these must be properly identified and priced—the first two (the direct costs) by appropriate unit prices[G] applied to the measured quantities of work or by properly analyzed lump-sum prices; the last two (the indirect costs) usually by percentage markup[G] amounts added to the direct costs. As in the last chapter, in bid estimates every job overhead cost should be identified and analyzed, and the need for any contingency sums[G] should be fully described.

[2]That is not to say that all bid estimates are better estimates than all conceptual estimates, as that depends on the particular estimators and estimates involved.

In other words the contents of a properly made bid estimate are fully detailed and described, and it is the product of the best in estimating practice.

The Two Main Activities in Bid-Estimating

The two main and distinctly different activities in bid-estimating are:

1. Measurement of the work, and
2. Pricing.

Measurement of work requires the translation of drawings and specifications into items and quantities of work and items of job overheads adequately described to permit accurate pricing. This requires in turn the use of some *standard method of measurement*[G] and of description, whether it be national, local, or individual (i.e., belonging to one contracting company). Without some standard method, estimating is not possible except in the crudest manner; similarly cost accounting requires a standard vocabulary of construction work.

Pricing requires more skill, knowledge, and experience than measurement and it is, therefore, often done by a senior estimator who may also supervise and check the measurement by a junior estimator for general accuracy and for the specific accuracy of specific items of work by comparing their quantities. (In construction there are many fairly constant relationships among many common items, such as between structural concrete and reinforcing steel, that make fast but effective checking possible.)

There is much more on both measurement and pricing practices later in Part 3.

Bid-Estimating Procedures

With an example of a bid estimate by a contracting company as a *primary bidder*[G] for a major building project to be performed by means of a stipulated-sum contract, the estimating and bidding procedure might go somewhat as follows—although many alternative procedures are possible.

Bidding documents[G] are obtained from the owner-designer.[G] Because a primary bidder may need to have more than one person at a time working on a major project's estimate, he may need more than one set of copies of the bidding documents. If documents are not distributed to trade contracting companies[G] as sub-bidders[G] through a bid depository, or through local plan rooms,[G] a primary bidder may need several additional sets of documents to ensure that his preferred trade contracting companies have an opportunity to make and submit sub-bids.[G]

Owner-designers often try to save money by limiting the numbers of sets of documents that they make available to bidders. This may be false economics because bids made in haste without adequate information often contain errors of serious omissions and magnitude that, may make a bid useless or conversely, if of duplication, overpriced. Some may think of a low bid due to errors of omission as a windfall benefit to an owner, but more likely it will lead to contractual troubles because there are many ways a losing contractor may select to make up his losses on a project.

Allocation of estimating effort is done by the senior estimator in charge of a project's estimate. After perusing the bidding documents he may pass them to one or more junior estimators to measure specific work. If, for example, the project is a reinforced concrete building, one estimator may measure the concrete structure, another the finishes, and another the site works and the substructure. It may be necessary to provide certain trade contracting companies (trades) with bidding documents to estimate the costs of such work as mechanical and electrical (M & E). This may mean splitting up and sharing sets of documents, and this creates the risk of errors.

Sub-bids from trades should be specifically requested by primary bidders.[G] Typically, general solicitations are advertised, specific solicitations are mailed. Primary bidders may also register their names with any designated bid depository, although the rules vary. Sometimes, especially in a busy market when trades have more than enough work, a primary bidder may be concerned that he may not receive sufficient (or any) sub-bids for important trade work, such as that of M & E and the major finishing trades. A primary bidder can rely only on his requests and his reputation as a fair general contractor[G] to receive sub-bids. For some important trades a primary bidder himself may measure and price the trade work to ensure that at least he can complete his bid for the project by the time and date for the receipt of bids set by the owner-designer. If his bid proves successful and results in a contract, the primary bidder, as the contractor, may again seek sub-bids for that trade work that are lower than his own estimated costs. Of course, there are risks in this practice: He may not be able to find lower sub-bids. If sub-bids that are lower than a primary bidder's own estimates for trade work arrive before his prime bid is submitted, the bid can be reduced accordingly. However, if sub-bids subsequently prove higher than the estimates of the primary bidder (now the contractor) he must work energetically to find trade companies that will negotiate a lower price, or the contractor will be faced with a loss before he begins work.

Toward the latter part of the bidding period and when the junior estimators have practically completed their measurement of the work (the *takeoff of quantities*[G]) the senior estimator will begin to price the estimate. He may price only the major items, leaving the pricing of minor items to the juniors because, as the old axiom has it, 80 percent of the costs are in 20 percent of the items, which is to say that in an estimate a much larger proportion of the total costs of work is contained within a much smaller proportion of the total number of items.

It is common practice to make up a chart, or spreadsheet, for the sub-bids that are to be received from trades, so that all sub-bids received for each trade or for each section of trade work may be compared and, if necessary, analyzed and adjusted. Thus, ideally, every sub-bid for a trade is for exactly the same work so that all comparisons are valid and only the truly lowest bid for each trade is selected for inclusion in the primary bid.

The final summary of a primary bid is often prepared well in advance of completion of the estimate to serve as a working checklist and as a visible goal to be attained. If it appears that sub-bids or other cost figures are not forthcoming in time, alternative arrangements are made (such as for trade estimates by the primary bidder) to ensure there will be valid cost figures for the estimate. These are entered in the summary as provisional figures to be replaced by more accurate or more competitive figures should they become available in time. Apart from ensuring completion of a valid and timely bid, such early but tentative completion of an estimate's summary serves to provide the estimator with a reasonably accurate total cost for the work early enough to be used as the basis for estimating some other important costs, particularly job overhead costs,[G] such as the costs of a performance bond and the premium costs for builder's risk insurance. The completed estimate summary, even if premature and not precise, also provides information for preliminary *planning and scheduling*[G] of the work in order to determine the project's probable duration, something on which, as we have seen, several important costs (chiefly job overhead costs) depend.

Preliminary Check Estimates

It is not unusual for an estimator in a contracting company to make a preliminary estimate[G] before making a bid estimate.[3] This preliminary estimate may be nothing more than a perusal of the documents, then a period of thoughtful silence while lips are pursed and the mind searches for comparisons, and then a statement as to the probable costs of the work. Or it may actually involve written figures, some minimal calculation, and finally a total cost figure scrawled on a notepad and ringed with a flourish.

In some instances an estimator may undertake to make the rough equivalent of a conceptual estimate, employing one or more of the popular shortcut methods of estimating, such as one based on a unit price per square foot (or square meter) of floor area. Sometimes such preliminary estimates are surprisingly accurate when compared to the bid estimates that follow. Other estimators prefer not to make any preliminary estimate for fear of creating a preconception of total costs, with the risk of influencing their bid. Such *preliminary estimates* as these are not the same as *conceptual estimates* because the estimator invariably has the bidding documents and all of the design information needed to make a detailed bid estimate—which he eventually intends to do. But the practice, potential or real, does indicate a relationship between conceptual estimates and bid estimates on which we shall enlarge later.

Bid-Estimating Standards

Standards of estimating vary greatly in the construction industry; they range from excellent to useless. To review cost estimates of failed contracting companies for receivers in bankruptcy or for accountants conducting autopsies is to see some of the worst of estimates; probably the best estimates are never seen by outsiders. To teach estimating at night school can also be revealing as some students bring from their companies estimates made with dubious estimating techniques to show and tell "how it is done in actual practice."

Estimating in many contracting companies is like specification writing in many design firms: traditionally, neither technique has been formally taught—and many believe that they cannot be taught except on the job. Consequently estimating techniques and estimating standards vary greatly, and inefficient methods are perpetuated because they are rarely questioned, but instead are handed down from the older and wiser to the novices, who, if in a classroom and backed up by 29 other students, might question what they are taught.

In North America no professional institution appears to have had much influence on estimating standards and techniques. There is a *standard method of measurement*[G] published in Canada with some influence but no equivalent publication in the United States. In both these countries specification writers have effective organizations, but building estimators belong to a variety of bodies (or to none), none of which are primarily concerned with estimating building costs. The American Association of Cost Engineers is perhaps the most effective body in the estimating field, but its traditional emphasis is on engineering works and industrial construction, not on building construction.

A strong, national body of building construction estimators could promote the development and use of estimating standards, beginning with a national standard method of measurement; it could also promote estimating education and the exchange and formalization of estimating knowledge and techniques. It could set down standards of professional practice for estimators similar to those for other professions, and it could create recognized standards of quality for estimating and estimators. All of these are needed.

Standard methods of measurement are needed for each major class of construction work: building, civil engineering, and industrial construction. Standards for estimating and cost accounting also are needed, beginning with classifications of costs and recommended formats for estimate summaries.

Standards of estimating practice are needed to deal with:

1. Measurement of work, including standard procedures for dealing with such phenomena as waste,[G] laps,[G] and shrinkage and swell.[G]

2. Estimate layouts and formats, including a thesaurus of recommended descriptions for items, standard methods of dimensioning, and standards for cost allocations.

3. Recommendations to owner-designers and cost consultants on the uses of estimates and contract documents to minimize the risks for all concerned in estimating, bidding, and contracting.

4. The extent of responsibility of estimators for making estimates and for establishing construction costs. (Estimators should be required only to estimate construction costs relative to specific dates and locations; actual bid amounts should be the responsibility of management; and estimators should be required to qualify their estimates, stating all their

[3]In ordinary parlance one would probably say, "...for a contractor's estimator to make...," but I have tried for precision in terminology; until there is a contract, there is no contractor.

assumptions and all limitations, especially the limits of estimated costs due to time and location, much the same as appraisers are required to do by their professional institutes.)

5. Ethics for estimators, related to such things as accuracy, purposes and uses of estimates, ownership of estimates, copyrights, and confidentiality of estimates and information.

6. Standards of training, education, and competence for estimators, and the use of appropriate professional designations.

All these things are similar to the concerns of existing institutes of other professions in construction and it is unfortunate for the construction industry and for the public at large that no professional body exists solely for estimators and estimating in the same way as they exist to deal with accounting, design, specifications, and quantity surveying.

Conceptual Estimates

A professional estimating body such as that outlined ideally would extend its concern to include all kinds of construction estimates (and cost accounts) made for all kinds of purposes, including those described as *conceptual estimates*;[G] those made before a design is completed for budgeting and cost control or for the purposes of a construction manager[G] in a CM project.[G]

Traditionally, before calling for bids for a stipulated-sum contract, a designer[G] (particularly an architect, because we are looking mostly at building construction) usually has, if required to do so, given his client, the owner, a preliminary estimate of the total costs of the proposed work of a project. But generally as a profession, architects have not excelled in estimating construction costs—with some notable exceptions. In most English-speaking countries, at least for most major projects, it is customary for an owner to employ a professional *quantity surveyor*[G] to manage project costs and finances before and during construction, and many building projects in those countries are contracted for by using *contracts with quantities*[G] prepared by the owner's quantity surveyor who works closely with the designer.

Consequently in most English-speaking countries construction costs are the concern of a profession devoted to construction economics and management, but this is not the case in many other countries, including the United States. This situation is, however, changing, and there are now some professional quantity surveyors in almost every free enterprise country, especially those of western Europe and in the Middle East; places where the quantity surveying profession has become more or less active but not yet homegrown. Further, the quantity surveying profession has been recognized and established in Canada for several decades (though not to the extent of using contracts with quantities), and recently quantity surveyors have formed a professional organization in the United States.[4]

In the past, most preliminary estimates by designers were not detailed conceptual estimates; they were made chiefly by applying unit costs to a building's gross area. The development of the use of *building elements*[G] in preliminary estimates originated in Britain in the 1950s and was introduced to North America by quantity surveyors. Conceptual estimates were not, however, unknown in North America before that time. Some were made by informal elemental methods and others by a method that persists and is, for many, still the most effective way: the use of approximate quantities of work and unit prices in a kind of shadow bid estimate.

As Construction Management (CM) developed as an alternative method of contracting in the 1960s, the use of conceptual and other kinds of preliminary estimates also developed because the basis of CM is a good preliminary estimate.[5] Also, as already indicated, *conceptual estimates* do have a place in *bid-estimating*—in bids for work in contracts *other than* traditional stipulated-sum contracts—and in making contracts that are essentially of the cost-plus type, frequently with a stipulated maximum cost (maximum-cost-plus-fee contracts[G])—a hybrid form of contract from "stipulated-sum" and "cost-plus-fee" parents that exists for the reason that a detailed bid estimate and a traditional stipulated-sum contract properly cannot exist, namely, a lack of sufficient design information.

The Initial Conceptual Estimate

A conceptual estimate can be made before any drawings or even rough sketches are made, even before a designer is appointed. Instead of using preliminary sketches, a schedule of the owner's building-space requirements can be used by an estimator to make a conceptual estimate of costs, using his

[4]See the Glossary and Bibliography for further information on quantity surveying.

[5]The distinction between *preliminary* and *conceptual* estimates begins to have significance for us.

own or the owner's concepts of the building as to its general form, specific materials, and its construction methods. Because of the importance of the area-perimeter ratio[G] of buildings and its fundamental effect on costs, some kind of probable building plan is desirable and often not difficult to devise given certain facts about the site and a proposed building's function and general style. But even these things can be expressed in estimating terms; drawings are not essential at this stage. (In fact it is probable that traditional construction drawings will eventually be replaced by alphanumerical series of references that can be put into a computer and transformed into diagrams on portable computer screens at the site.)

This totally conceptual estimate is most effective for speculative shelter construction, commercial buildings, and institutional buildings of high utility, such as schools. It is less effective for those buildings in which aesthetics or symbolism is an important consideration, such as monumental buildings, churches, and other buildings that are not entirely functional and symbolize establishments such as banks, departments of government, and religious sects.

Clearly the statement that we are making here is about *economic design.* If a building's design is to be economical because of economic constraints (of market, politics, or other causes), then it is not difficult for a person familiar with the economics of construction to conceptualize an economic design as a basis for a conceptual estimate. But if you ask me to make a conceptual estimate for a building to accommodate 300 worshippers, first I shall have to ask you about the particular religion and the particular sect before I can conceive of the kind of building that might be required.

If both an owner's mind and the mind of his *conceptual estimator*[G] (a somewhat inaccurate but nevertheless useful term) are thinking in terms of an economic design, then they have that much in common and the estimator can be helpful from the outset and even before design drawings exist. But if no common understanding exists because of the absence of specified economic constraints on design, then mutual agreement must first be reached on the nature of the building's design, perhaps through the development of preliminary design sketches. This brings us to the subject of *information*[G] in estimating.

For estimating there are two kinds of information: (1) *design information*[G] and (2) *experiential information.*[G] The former is normally expressed and conveyed by means of bidding documents[G] in both their final and earlier forms (e.g., preliminary sketches and outline specifications). Experiential information exists in the minds and records of the estimators, the bidders, and of the contractors who eventually perform the work. Both kinds of information are essential to estimating and contracting, and one complements the other. Together they ideally comprise the whole of the information about a project and its work.[6]

In a conceptual estimate almost all the information may be experiential information, originating with the estimator. There may be more or less design information, depending on the circumstances prevailing at different times and stages in the progress of a construction project (from design to production). In any estimate, what the estimator lacks in design information he makes up for with experiential information; for a traditional stipulated-sum building contract the owner-designer must provide specific design information or otherwise accept what the contractor (initially the estimator) assumes and injects into his proposal.[G]

A conceptual estimate, like any other estimate, can be effective only to the extent that design information is provided by the owner-designer and can be supplemented by the estimator with experiential information. An initial conceptual estimate's primary purpose usually is to help the owner set a project's budget. But because a budget usually has many constraints, often including that of commercial feasibility (will it show a profit?), an initial estimate usually does not need much design information to be useful.

The Intermediate Conceptual Estimate

The three descriptive titles used here—preliminary, intermediate, and final conceptual estimate—are not used to identify three specifically different estimates that always exist in a series, but rather the titles are used to illustrate that there is more than one stage in a conceptual estimate each related to a stage in the development of the design information. We began with a conceptual estimate possibly made without drawings, that is, without graphic design information; we move now to the intermediate stage and to the more typical conceptual estimate; to the estimate made from preliminary design drawings and outline specifications, one possibly in the form of sketches, the other in note form, and both expressing only those primary character-

[6]The idea of these two kinds of information and the effects of their presence or absence on *risk* and on contracting procedures is explained in detail in my earlier work, *Construction Contracts* (Reston, VA: Reston Publishing Co., 1979). It is discussed here and in Part 4 of this text.

istics that will make this building different from every other building. At this stage, only this needs to be expressed, and the rest of the design information will be assumed by the estimator from his own experience.

If the reader finds this difficult to grasp—that an estimator can assume so much—let him reflect on what we are discussing: To make any estimate, a certain amount of information is required about the site and about the work of the proposed project. This information consists of two basic kinds: (1) *design information* produced by the owner-designer according to the owner's construction requirements and the site's conditions and (2) the rest of the information, which is supplied by the estimator from his experience and which is therefore called *experiential information.* Each kind of information complements the other so that ideally they together constitute all of the information needed to estimate the costs of the work and subsequently to perform it. But in practice the ideal is never totally achieved, and so practical means are found to overcome this lack of information.

At this point it is necessary to review the relationship between *risk* and *information*, and it is shown graphically in the Figure 4-1.[7] The practical effect of this in contracting is that if, for whatever reason (urgency, unresolved leases, indefinite building requirements, indecision by an owner) an owner-designer is unable to provide sufficient design information to bidders in order to calculate the costs and risks of construction in an estimate, so that they are able to bid and make acceptable proposals to the owner, then the owner-designer[G] cannot realistically expect to make an equitable and mutually desirable *stipulated-sum contract.* Instead the owner-designer should seek some other form of contract: in North America probably a cost-plus-fee or maximum cost-plus-fee type of contract.[8] Because there is never total and complete information and knowledge there are always risks, even in the most researched and best-designed projects. Good contracting should minimize the recognizable risks (by insurances, bonds, and other means), and the remaining risks should be carried by the party most able to control them and to minimize their effects. For a contractor the burden of carrying risks should be offset by potential gain through profit. The greater the risk, the greater should be the potential profit. For an owner (and this should not be overlooked), bearing a risk in construction should likewise be offset by potential gain, in this instance, a possible lower total price for the work.

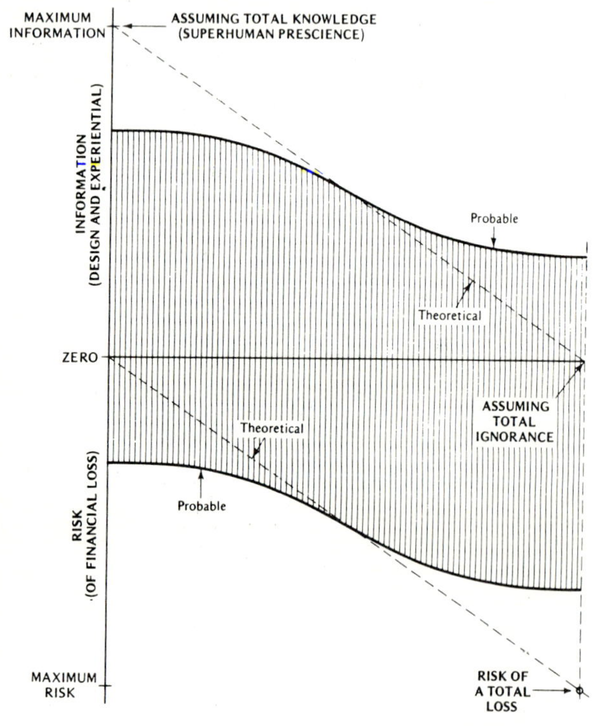

Risk and Information in Bidding and Contracts

FIGURE 4.1 Risk and Information

[7]Reproduced from *Construction Contracts.*

In preparing to bid, an estimator seeks to calculate the costs, the risks, and the potential profit. The conceptual estimator seeks to anticipate the bid estimator. In addition, the conceptual estimator must substitute experiential information for missing design information while assessing what he believes will be acceptable by the owner-designer. The conceptual estimator is less concerned that he is able to anticipate what the owner-designer will require (and eventually specify in the bidding documents), for example, a specific kind of floor covering. Instead he is concerned about being able to include in his estimate an amount of money that is close to equaling the subsequent costs of floor coverings required and specified. This is the general aim of a conceptual estimator for all parts of the work. It is this monetary and economic view of work that makes conceptual estimating possible. I cannot predict what kind of floor covering you might choose for your living room, but given some information about your income, life style, and the kind of house in which you live, I could make a reasonable estimate of how much you might want to spend on covering your living room floor, and an even better estimate of what I think you should spend.

[8]Elsewhere it might be a *unit-price contract,*[G] but generally these are used only in engineering construction in North America. (*Cost-plus-fee* and *maximum cost-plus-fee contacts* are described in the Glossary.)

If you consider a building of normal utility and all its various parts (foundations, structure, cladding, finishes, services, etc.), as already seen in economic terms, most of the parts of a building are chiefly utilitarian; if we know the fundamental design criteria (soil-bearing capacity, loads on floors, weather-resistance requirements, and the needs for lighting, heating, air conditioning, for example), it is possible to estimate the costs of these utilitarian parts with reasonable accuracy. The real problems in conceptual estimating come with those parts of the work that are not utilitarian; the prestigious entranceways, the fancy finishes, and the decoration. Unless an estimator knows specifically about such things, he may not be able to anticipate them or to allow sufficient money for them in his estimate. Consequently an experienced conceptual estimator usually knows what he may assume about a project and what questions he must ask.

It is not enough to anticipate probable costs in a conceptual estimate, it is also necessary to control the costs during the design phase[G] to ensure that the design team stays within the budget set by the owner. Cost control[G] is a primary function of an owner's cost consultant; usually the same person that makes the conceptual estimate. To be effective, cost control must begin before the design in helping to make a realistic budget. Control must then be consistently applied throughout the design phase and into the production phase of construction. This requires, as already seen, more than one preliminary estimate.

The Final Conceptual Estimate

You have undoubtedly realized that timing is important in conceptual estimating and cost control, and some owner-consultant agreements stipulate the stages in design at which estimates shall be revised to check probable construction costs against the budget. There comes a point in designing a building at which the design is no longer only a concept; it exists more or less complete in graphic and verbal form in the almost completed documents; the drawings and the project manual.[G]

Ideally at some point before this the conceptual estimate is replaced by another kind of preliminary estimate; the best kind is based on *approximate quantities of work*.[G] This kind of estimate is examined in Part 3; in a project it may be feasible when the documents are about 50 percent completed; that is, when the main design concepts and details have been set down. (It is not possible to quantify completion of documents accurately, especially as we are concerned with expression of design concepts rather than actual production of drawings and specifications.) Before this, the conceptual estimate is revised and updated and, if necessary, redone according to the design expressed so far on paper. To the extent that an estimate stays ahead of a burgeoning design in all its detail, it is a conceptual estimate.

The Contents of Conceptual Estimates

The contents of a particular estimate tell us the way that it was made and give us a good idea about its effectiveness. Because of the uniqueness of building projects we cannot generalize, but it is probable that the means and therefore the contents of the initial, intermediate, and final conceptual estimates will vary because they are probably based on varying amounts of design information. Estimated costs of buildings may be based on the following:

1. Units of occupancy (per seat, bed, room, etc.)
2. Building volume (per cubic foot or meter)
3. Building area (per square foot or meter)
4. Building elements (described next)
5. Approximate quantities of trade work
6. Accurate quantities of trade work.

The costs of engineering works may be based on different units, for example:

1. Units of production (per ton, barrel, etc., for industrial construction),
2. Units of length (per mile or kilometer for pipelines, highways),
3. Units of area (per square foot or meter for highways, paving),
4. Units of volume (per cubic yard or meter for dams, earthworks),
5. Approximate quantities of trade work, and
6. Accurate quantities of trade work.[9]

Building occupancy is suitable for use only in initial conceptual estimates. It is popular with many owners who naturally relate construction costs to operating costs (of a theater, restaurant, hospital) in

[9]In engineering works it is not usual for either estimates or contracts to be based on precise quantities because it is of the nature of most engineering work to be subject to adjustments as required by the site conditions. Consequently the stipulated-sum contract is not commonly used for engineering works.

making feasibility studies. For many buildings, costs per unit of occupancy must be predictable within a small range, otherwise the project will not be economically feasible. But it is much better to make even an initial estimate on a more detailed basis (such as those described next) and then to express the costs at so much a unit of occupancy. This is more objective and helps to eliminate preconceived ideas about costs.

Building volume has fallen into disrepute in the last 50 years as a basis of estimating costs, perhaps because modern buildings are simpler in form and other units are more easily measured and used. However, many porticos and cupolas and much decoration make cubic-foot costs ridiculous. Reportedly, one architect had great success using the current cost of prime beefsteak as the basic cost per cubic foot for his buildings; which is not as silly as it sounds. The costs of certain products such as beef steaks and suits seem to consistently follow the general level of living costs. The disadvantage of the volume unit is that it does not accurately express either building capacity or the extent of building work (which is chiefly in the enclosure).

Building area is perhaps the currently most popular basis of estimating building costs, and it is heard from almost everyone, from developers to laymen. This popular method has its uses. For example, it makes sense to speak of office buildings as costing so much a square foot when they are leased by the square foot, but only if one remembers that there is more than one kind of square foot: one for construction costs, another for rental costs and each requires proper definition.

Minor height differences among buildings may make little differences among their costs per square foot of area, but other features may have a significant effect on unit area costs, for example:

1. The area-perimeter ratio and the total enclosure;
2. The plan shape, especially the number of corners and breaks;
3. Irregular roof lines, especially the numbers of steps, hips, and valleys;
4. The foundations, especially the depths and widths of footings;
5. The numbers of major isolated elements[G] inside (stairways, elevators);
6. The variations in quality of finishes and fixtures; and
7. The numbers of plumbing and other fixtures.

The list is practically endless, but the point is made: Many features in buildings are not properly accounted for by pricing a building's area.

Building elements were discussed briefly in Chapter 2 where we talked about three main groups of elements: (1) those related to area, (2) those related to perimeter, and (3) those related to neither; the isolated elements related only to occupancy. We also referred to a standard list of building elements in the Appendix. (You may have noticed that the several bases for building costs are reviewed here in order of their general effectiveness, with the least effective and the least detailed listed first.)

Approximate quantities of trade work can be measured with surprising accuracy and detail by an experienced estimator even from primitive sketches, provided overall dimensions are given. The ability to measure what is not shown is always an indication of any estimator's skill, as it takes much less experience to measure accurately from detailed drawings. It is not unusual for professional quantity surveyors to measure in detail (according to a national standard method of measurement) from small-scale drawings (1/100) for *bills of quantities*[G] in *contracts with quantities*,[G] and later for the designer to complete his drawing details from the information in the bills of quantities. Many designers prefer preliminary estimates to be based on *building elements*, as this format facilitates the quick calculation of costs of alternative designs containing similar elements but different overall dimensions. Many others, especially construction managers, prefer estimates based on approximate quantities to those based on elements because quantities of trade work show the total costs of work for the different trades and their proportionate values, and it is with those that a construction manager must deal. It is not difficult to rearrange trade quantities[G] into elements, but the reverse is more difficult.

Accurate quantities of *trade work*[G] are the normal content of bid estimates, but in a conceptual estimate, or any preliminary estimate, some of the work may be measured accurately and in detail as in a bid estimate. (Sometimes it is as easy or easier to measure accurately than it is to be approximate.) From initial to final estimate, during the design phase of a project, the normal progression is from less to more detail and from lesser to greater accuracy in both measurement and pricing.

Ideally, were it affordable just before calling for bids, most owner-designers would prefer to see a final preliminary estimate based on accurate quantities of trade work instead of one based on approximate quantities, on elements, or on a combination of these. A preliminary estimate based on trade work is so much more useful for checking and comparing and negotiating bids. But many owners are not prepared to pay for such effort and detail. Others do not know that it can be done. Some would say that the traditional method of competitive bidding for a stipulated-sum contract precludes the need for a detailed preliminary estimate, and sometimes that is true—but not always. It depends on how complex the work and the contract are, and on the construction market.

Estimating Concepts

The following ideas about estimating and estimators are not the results of formal research; they are opinions based on years of work and observation.

A concept of economic utility. Cost analyses show that there are relationships between *costs* and *utility* in construction. For example, given a design requirement for a certain live load (in say, pounds per square foot) and a certain floor span, the costs per unit of area of floor decks of different construction (wood, concrete, and combinations of materials) often vary only a little, but peculiar circumstances (e.g., a local scarcity of a material) may affect costs more. In other words there is a certain cost for a *specific function* in a building (e.g., supporting a specific load on a deck of a specific span) that varies little, providing alternative designs are efficient. This principle appears to extend beyond obvious examples of efficient design and utility (as in a structure) to other less obvious cases, for example, waterproofing a basement.

The implications of the economics of the costs in use[G] of a building element, or of a combination of elements (a system), cannot be ignored in this context. A clay-tile roof will last longer than one of asphalt, but it will cost more initially. However, with many building systems, such as a structural deck or an underground membrane, costs in use are not of major significance. The differences in costs among systems of similar function are often, it appears, so slight as to be well within the normal margin of variation expected for all estimated costs.

A concept of the distribution of construction costs. It is apparent that a far *greater proportion* of total building costs arise from a far *smaller proportion* of the total number of items of work in detailed bid estimates. (This may not be true in other estimates because of the measurement of elements or of combined items.) Some call this the *80/20 rule;* others call it the *90/10 rule.* The implication is the same: that of the concept just expressed. In this area there appears to be much need for research into the distribution and proportions of building costs and into the implications for estimating. One sometimes suspects that much estimating work is just what many students and novices feel about it—so much drudgery; that some of it could be avoided without loss of accuracy if we knew more about the probabilities of what we are doing. Unfortunately, mathematicians to whom I have spoken about this topic were generally uninterested.[10]

A concept of accuracy in estimating through the assessment of multiple parts. Often in estimating it seems more important to put greater effort into analysis through more detailed measurement than to put more effort into the analysis and synthesis of unit prices. This concept may be seen as contradicting the previous one, but I make no apology for that because all of these are only incomplete opinions until tested by research and I offer them only in that form. Apparent contradictions often are not contradictions—only contra-dictions. It would be too facile to say that the least accurate estimates are always those made by the more simple methods (unit costs per place, per unit of area), and it would not be absolutely correct to say that the most accurate estimates are always those made by detailed measurement. The overbearing problem in estimating (like the problem of predestination in religion) is that of the ultimate incertitude—in estimating, that of the real and proper costs for any construction. There is no absolute truth in estimating construction costs.

A concept of a constant of estimating effort required for efficacy. Although mentioned elsewhere, this concept bears repeating with the other concepts. It was originally expressed as $M+P=E_c$, (where M equal the effort required to measure work; P equal the effort required to price measured quantities of work; and E_c is the constant total effort required to produce an effective estimate). If the measurement effort (M) is less, the pricing effort (P) must make up for M by being more, and vice versa. Practically, this means (and nothing more) that, for example, in an estimate based on building area and one unit price, because the measurement effort is small, a meaningful estimate requires great effort in synthesizing the single unit price. (That can be done by aggregating unit prices related to area for all the building's elements, or even for all the items of work if that were practical, which it usually is not.)

[10]See however an article by Efron and Morris, "Stein's Paradox in Statistics," *Scientific American*, 236:5 (May 1977) and my quotation and remarks about it in *Managing Construction Contracts*.

The preceding equation is only symbolic, and there are no profound facts to be uncovered by rearranging it as there are in the more famous $E = MC^2$. It expresses only a relationship between the two main functions of estimating and their results: that indeed it may be possible to make a good conceptual estimate by measuring only the floor areas of a building and by applying to the total area a unit price that has been carefully built up from many other unit prices.[11]

Questions and Topics for Discussion

1. Identify and describe the differences between a *preliminary estimate* and a *conceptual estimate.*

2. Identify and describe the relationship between a *bid estimate* and a *conceptual estimate.*

3. Explain in detail the relationship in all estimates between *measurement* and *pricing.*

4. Explain in detail why in bid estimating the junior estimator usually does the measurement of work while the senior estimator usually does the pricing.

5. Identify, describe and explain all uses of a conceptual estimate.

6. Explain the relationship between *cost* and *utility* of building elements and building systems.

7. Explain the relationship between *risk* and *information* in contracting.

8. Why does the text refer to initial, intermediate, and final conceptual estimates for a project? Explain in detail.

9. Explain the means and purposes of *cost controls* in the design phase of a construction project.

10. Describe the relationship between *experiential information* and *design information* in construction.

11. Define the following terms: (a) *trade work,* (b) *economic utility,* and (c) *an accurate estimate.*

12. Identify and describe the several bases for estimating costs in conceptual estimates and in other kinds of preliminary estimates, including their merits and deficiencies.

[11]But only for some buildings and only by giving adequate effort and attention to the pricing.

5

Appraisals of Construction Work

If you are not an appraiser and if you are not acquainted with the basic principles of appraisal you may wonder about this chapter and what appraisal and estimating have in common. A partial answer lies in the property tax assessment notice that you as a property owner received at the beginning of the year (with due apologies to all you renters). The assessed value of your buildings was probably arrived at by an appraisal based on the *cost approach*[G] to value. If you are an appraiser you may wonder if a book on estimating can be of any help to you. After all, *appraisals*, you may say, are one thing and *cost estimates* are another; in any event, you may add, the cost approach to appraisal is limited in its applications. Yes, it is true that its applications are limited, but should they be so limited?

The cost approach has been underrated by appraisal textbooks, and one reason for underrating the cost approach to value has been a lack of expertise among appraisers—in part due to a lack of useful and effective cost techniques. Most of the contents of this book are of use to an appraiser who seeks to determine value by the cost approach or who is simply interested in a different view of replacement and reproduction costs. Further, the need for a better cost approach to value is increasing and appraisers need new techniques in costing and in calculating depreciation.

In some respects appraisers have contributed more in the field of conceptual estimating than most others in North America through the development and publication of cost manuals and data about building costs. In the 1960s, in making my earliest conceptual estimates, one of my first tools was an appraisal manual[1] and I found it very useful because little else was available at that time. Since then, a few good cost data books have been published, but there is still room for comprehensive cost manuals to serve both appraisers and conceptual estimators, as they have so much in common both in needs and in objectives.[2]

Estimating is the forward view of a proposed construction project and its costs, while *appraisal* is the backward view of properties that already exist, and of their conditions and their costs. Essentially the substance is the same for both—the costs of construction work.[3] The estimator works from drawings and mental images while the appraiser sees what is extant: the already constructed—and depreciating—building. Assessment of the loss in value due to the passage of time—that is, depreciation[G]—is the building appraiser's bugbear, and here again common estimating techniques are useful.

The declarations of appraisal textbooks about the inefficacy of the cost approach to value, about the need for speed in appraisals (as if estimating were a leisurely process), and about the inability of appraisers to absorb and deal with all the information needed for the cost approach says more about the textbook writers' inabilities to deal with the subject than about appraisal and cost estimating. As for the need for the cost approach by appraisers, let me quote from a recently published paper on financing and investment in real property:

"[Today] the market must, therefore, be using a different basis [from that of not many years ago] with which to buy real estate investments.

The average purchaser today assumes that:

(a) buildings appreciate in value
(b) land value will improve in the long run
(c) rents will increase
(d) governments [will] repay their debts with cheaper dollars
(e) there are tax advantages to holding real estate.

Consequently he [the purchaser] feels that any property should be considered if it can be bought at less than *today's replacement value*, providing it is not functionally obsolescent".[4]

The key to wise buying is knowing the *replacement value*,[G] especially because it appears that as assets many buildings are undervalued. Yet among many appraisers *replacement value* (or *reproduction value*)[5] is almost a proscribed term, and when its determination cannot be avoided many will consult a building contractor; often not the most knowledgeable person on the costs of buildings generally, believe it or not.

My purpose is to argue (and later to demonstrate) that appraisers can make effective and better cost appraisals; when they do, the cost approach to value will acquire greater utility and prestige. As for estimators, they may gain in satisfaction and knowledge in knowing that a large part of the appraisal profession shares with them a common interest in construction costs and economics and a common need for information and data; also, that they too can gain from an increase in information and data generated by a new interest in the cost approach to value and from the development of cost valuation techniques that also can be used to make certain kinds of estimates.

The Cost Approach to Value

The appraiser and the estimator also share a common problem in assessing costs: one of limited time and limited information. Often the latter is due to the former, which in this context is usually due to an economic constraint. Appraisal manuals are a source of data for conceptual estimators, but manuals are limited in the number of useful examples, illustrated models, and other data. Estimators can

[1]Boeckh's Appraisal Manual, fifth edition. Unfortunately now out of print but replaced by later editions. See the Bibliography.

[2]The publications of the Snow Means Co. may be filling that need, and they are discussed later, particularly in Chapter 5 of Part Three.

[3]Appraisal includes much more, but our discussion is limited to the cost approach to the value of buildings.

[4]From an article by John Hawkins, "We Have Heard It Said That Five Percent is the Natural Interest of Money," — Macaulay, 1830, *Input*, April 1982. This is the journal of The Real Estate Institute of British Columbia. (*Author's italics in quotation.*)

[5]*Replacement value* and *reproduction value* are sometimes interchangeable; when they are not, the fact is made clear.

make useful contributions of information and data to appraisers interested in construction costs and ultimately benefit from the results.

Most appraisal manuals employ only one basic approach to building costs: that of *building area times a unit price*; refinements and variations if any are usually found only in the composition of the unit prices. Typically within a building class in an appraisal manual there may be adjustments available for variations in the area-perimeter ratio, for different exterior claddings, and for different interior finishes and fixtures. Some manuals contain unit costs for *building components* (similar to *elements*[G]), but almost always they are mathematically related directly to building area. One uses the given unit cost for a model, or one modifies it for major differences between the model and a subject building (and fine adjustments are usually not possible), or one composes a unit price from those given for building components. It is always essentially the same method, with or without the refinements.

Of course, the almost infinite variety of buildings limits all manuals in their coverage of information. Often, important regional types of buildings are poorly represented, if at all, thus making it almost impossible to appraise some common buildings from an appraisal manual. (For example, the commonplace wood-framed, stucco-plastered, three-story apartment building of the West Coast, usually built speculatively and by nonunion companies, is still poorly represented in most manuals.)

In many manuals the critical subject of *depreciation* is dealt with only superficially by tables of typical rates of depreciation for buildings of a few major classes. Again, one has the impression that the compilers of these manuals either knew nothing more that they could add about depreciation or thought that the readers would not be able to use anything more complicated. Some local assessment authorities' manuals are better equipped to deal with depreciation than most national appraisal manuals.

The use of *building elements* as a means to costing by appraisers is notable by its absence from appraisal manuals; yet one need not be an estimator to measure and price the simple areas of elements. The use of several elements is much superior to the use of only one: the building's floor. All things considered, the general use of the cost approach to the depreciated value of buildings by appraisers is far less effective than it could be. Admittedly a greater knowledge of construction costs among appraisers who use the cost approach is needed, and this can be fostered by using better sources of better information and better construction costing techniques.

What has been said so far applies also to real property tax assessors and to their assessments, which are in fact a kind of mass-produced appraisal of value, and assessors too need better techniques than the universal building area times a unit price to arrive at replacement costs, even if they are only for taxation purposes.

Many appraisal and assessment manuals are already obsolete; their replacement—the portable microcomputer and its software—is waiting to be programmed and fed with data to be ready for field use. The only conceivable objection to the use by appraisers (and assessors) of better appraisal techniques and larger data bases is the time involved in collecting and handling the data. Microcomputers largely overcome this problem, leaving only the question of the costs and the economics of microcomputers and their use in the field. This point I do not propose to argue, because time will soon show that a $2,000 portable computer (including the software) is an economic proposition for anyone who must collect and use information and data. Fundamentally, both appraisal and estimating are the processing of information, and microcomputers are economic and efficient information processors.

However, one does not need a computer just to multiply together an area and a unit price to arrive at an estimated cost. Only if there is something worthwhile to put into a computer is it worthwhile to own and to use one, and that is our subject: the origins of and the ways and means of handling and using cost information for conceptual estimates and appraisals. If in the methods advocated here the amounts of information appear larger than before, if the methods are a little more complicated, remember that now we have the technology to apply them quickly and accurately. The old methods are no longer valid; no longer can we justify old methods by claiming that appraisals and assessments must be done quickly and cheaply, that appraisers cannot be expected to deal with large amounts of cost data, that they cannot be estimators, or that assessments and appraisals do not justify the great detail found in some construction estimates. The combination of conceptual estimating and appraisal techniques together with the computer removes all of these and similar objections to the cost approach to value.

Definition of the Cost Approach

This is not the place to probe into appraisal theory. The appraiser already knows it, and anyone else who is interested is better off reading several appraisal texts, because they do not all agree. Few will argue about these few simple facts on appraisal that follow as an introduction and background for the cost approach to value.

There are three classical approaches to the appraisal of real property values, and in simple outline they are:

1. *The market approach,* whereby a subject property is compared to a number of similar properties recently sold in the same locality and adjustments are made for differences among them to arrive at a *comparable market value* for the subject property;

2. *The income approach,* whereby the gross income from the subject property is reduced to a net income by deducting all expenses, and the resultant *net income* is then *capitalized* at a current rate of investment recovery and return to demonstrate the value of the investment (the property); and

3. *The cost approach,* whereby the *land portion* of a subject property is valued by comparisons made in the current market and the *improvements*[G] (buildings) are valued by estimating their reproduction costs as if new and by deducting their estimated physical *depreciation.*[G]

Some buildings are rarely sold, and so their types cannot be valued by a market comparison; there is no market for some kinds of buildings. Other kinds of buildings do not produce an income. For some properties, and for many purposes, the cost approach is the only way to a valuation. There are many reasons for making appraisals and some of them are directly related to a building's reproduction or replacement cost. They include fire and other insurances, purchase, exchange, mortgage, settlement of estates, and the recognition of good real estate buys, among others.

The cost approach requires estimation of the cost of a new building and the deduction of the calculated physical depreciation to arrive at the current depreciated value. (You may have noticed that there are similarities between life-cycle costing, introduced in Chapter 2, Part 1, and appraisal.) Both the new cost and the value of depreciation to be deducted are best calculated by the use of *building elements,*[G] something still generally foreign to appraisals but common in conceptual estimating.

Building elements (as defined here) are readily understood by persons who have more than a layman's knowledge of buildings but who are not necessarily building technicians. Elements are easily measured and priced. Adjustments for differences in the makeup and specifications of elements are readily made with the proper information. It is impossible to assess accurately the degree of physical depreciation (the loss in value due to age, wear and tear) in a building without considerable analysis. Simply to look at a building and to say that it is wholly 25 percent depreciated in value is almost ludicrous. The only way to measure physical depreciation with any degree of accuracy is to consider each element separately, because all elements depreciate at different rates and because each element represents a different proportion of the total costs.

We have already cited a few purposes of appraisals by the cost approach, and these are well known. Others that are less well known exist, in addition to the one cited at the beginning of this chapter: to judge the worth of a property as an investment. An increasing portion of the nation's construction expenditures is going for the renovation of existing buildings to preserve and convert them to more economic uses, or simply to make them worthwhile for resale or as an investment. The valuation of these renovated buildings requires the expert use of the cost approach, even if the income approach is also used. Even in the usually preferred income approach the adjustments needed in making comparisons often can be made only by using the estimated costs of construction work.

Because of recent changes in the economics of investing in real estate, the use of the cost approach to value in conjunction with one or more of the other approaches is more necessary than ever. Some appraisers claim that ultimately an appraiser should look to only one of the three approaches to value for his ultimate opinion of value, that is, for his appraisal. But is this valid? Should not all three approaches be used wherever possible, and should not the final opinion as to value be based on all three results? Especially now, in view of the apparent shift in the critical facts affecting a decision to purchase real estate in the minds of potential buyers.

Criticisms of the Cost Approach to Value

The following typical criticisms have been gleaned from texts on appraisal theory, and responses have been added.

Few appraisers are qualified to estimate construction costs. This may be true, but it is our intention to help to change this by offering to those appraisers who wish to be so qualified improved means of estimating costs that have been proven effective by conceptual estimators. The advances in conceptual estimating techniques and in collecting and handling information (especially by computers) remove most of the barriers to the appraiser's use of the cost approach.

The cost approach to value has not been upheld by the courts. First, we should ask, Why is this so? It appears that there are two kinds of reasons: (1) those related to the nature of value and (2) those related to the proper application of the cost approach. Among the first kind of reasons we find, for example, judicious criticism that the cost approach is not a good indication of value *unless* the improvements valued are uniquely suited to the needs of the owner. Presumably this means that only in a new building especially designed for an owner does *cost* come close to *value*, assuming that an owner builds only what he alone specifically desires, nothing more or less. But what about the owner who builds or purchases an investment? And cannot that observation be turned aside by appraising value on the basis of the *replacement cost*[G] rather than of *reproduction cost*[G]?(Here the distinction between the two words is critical.)

As previously explained, in conceptual estimating the estimator (like the appraiser with the cost approach) is concerned ultimately with *costs* in dollars, not with specific building materials, elements, or systems (e.g., a floor deck as a floor deck, not as a specific construction of wood or concrete or whatever). Similarly an appraiser is concerned with *value*, which directly relates to function, which in turn, as we have often said, relates directly to cost.

The cost approach takes no account of functional obsolescence. This is true, but a competent appraiser can. He can do so by using the cost approach (in its new form, using elemental costs) to value needed renovations and improvements (to arrive at a negative value to be deducted) or by costing for the replacement of specific items with other items of equal value instead of their reproduction.

It is not possible to calculate the effects on value of physical depreciation. Certainly it is not by the methods commonly used by many appraisers. But if depreciation is assessed separately for each building element, and if the effect of each of these on the whole is calculated, one can arrive at a supportable estimate of depreciation. Let us remember that an appraiser is trying to predict the market in which ultimately all values are established. How sophisticated is the average purchaser in assessing depreciation when he purchases a building? Should we not admit that in fact the market is imperfect in its judgments and insist, therefore, that it is fruitless to attempt to find and apply academic or pedantic judgments?

Who establishes value in the marketplace? It is the average buyer and the average seller interacting at arm's length and with all the open-mindedness and all the information about the property in transaction and all the awareness that are described in appraisal texts and in written judgments. But is there a real person such as an "average buyer" or an "average seller"? Or are they mythological characters created by theoreticians? Is it meaningful to prefer the appraisal approach by market comparison exclusively to any other method when all properties and persons are unique and when virtually all market comparisons require adjustments that are theoretical in content and made using the same techniques that are otherwise considered inferior for making whole appraisals?

The cost approach will often only provide the upper limit of value, rather than actual market value. If this is so, then surely it is something to be sought. How satisfying to a potential buyer to receive an appraisal that shows three different values arrived at by three different methods (providing the differences among them are not too great) and to be aware that the result of the cost approach to value is the upper limit of value and then later to buy the property at a price below that upper limit.

Should an appraiser instead offer only one figure as his sole opinion as to value, as though this were absolutely the "actual value" and all others simply less accurate estimates of value? Rather, should not an appraiser tell his client everything that he can about a subject property, including the varying results of all the approaches taken to discover "value," and not delude anyone by pretending to be able to opine a single, precise, absolute value? In fact it is normal for an appraisal report to explain that an appraised value is an opinion and to show more than one value arrived at by more than one method. But it is also common practice with some to select one of these values as the most representative figure of value. If only one value figure must be presented, why not a weighted average of all the values found?

In some countries it is normal to appraise most residential property by the income approach, yet in North America the appraisal of residences is usually made by the cost approach. Why is that? Often the information is not available to use any other approach because in many places few houses are rented. In many places the rents for houses do not reflect their market value. Why is that? Few houses are built or owned to be rented, and most owners who rent do so expecting a negative cash flow but also appreciation in value that makes it worthwhile to hold on to the property.

Thus it appears that many of the academic and authoritarian criticisms of the cost approach to value are based on weaknesses that arise from defects in its techniques, application, and use rather than on real defects in principle. If the cost aproach were used better, it probably would be more effective and therefore it would be seen and judged as better. Present cost techniques (area times unit price) often provide litle information of worth and little confidence because of their obvious simplicity, lack of precision, and lack of sophistication. Better techniques can provide much more reliable information that is of use either to a potential purchaser or to another as a support and a comparison for valuations by other methods. This topic is taken up again in Part 3.

Questions and Topics for Discussion

1. Interview your local real property tax assessor and write a report (of not less than 1,000 words) on his use of the cost approach to value in residential and commercial property assessments.

2. Interview a practicing professional appraiser on the advantages and shortcomings in the cost approach to value and write a report on your findings.

3. Write an argument in favor of the use of the cost approach (the income approach, the market approach) to residential (commercial, institutional, industrial, educational) property values.

4. Define the following terms used in this chapter: (a) *appraisal*, (b) *assessment*, (c) *replacement cost*, (d) *reproduction cost*, and (e) *depreciation*.

5. Obtain a recent edition of a textbook on appraisal theory and practice, and read what it says about the cost approach. Write an essay (of not less than 1,500 words) on this subject to argue for or against the main points made in the textbook.

6. Compare the actions, and the economic motivations construed therefrom, of the average house buyer 30 years ago with those of the average housebuyer today. Explain your comparisons and results.

7. Define the following terms used in this chapter: (a) the *income approach to value of real property* and (b) the *market comparison aproach to value*.

8. Define the following terms used in this chapter: (a) *building element*, (b) *loss in value*, (c) *obsolescence*, (d) *elemental cost*, and (e) *improvement to land*.

9. Briefly describe the primary objections raised to the cost approach to value as compared to the other valuation approaches.

10. Write well-reasoned arguments in favor of the cost approach that are contrary to the traditional objections.

11. Select a personal computer program available in stores that could be used for real property assessments based on the cost approach (the income approach, the market comparison approach). Describe the program and how it might be used.

12. List and discuss the main shortcomings of any specific appraisal manual containing data for the cost approach. Recommend improvements that might be made to that specific manual.

13. Find original arguments in favor of greater use of the cost approach and write an essay (of not less than 1,000 words) explaining and supporting your arguments.

14. Find a judicial report on the subject of appraisal methods and write a synopsis of the report's findings.

PART TWO

1

The Residence

The format of Part Two is different from the other three parts of the book in that it contains estimating examples with commentaries and estimating exercises. The purpose is to illustrate different estimating methods and practices, especially those described elsewhere in the book and to discuss and apply them.

All the examples are of actual estimates—except the last, which is a descriptive illustration of a different approach to budgets and construction costs, and except the first example, which is derived from an illustrative estimate in my earlier book on estimating;[1] its purpose is explained later. Some of the estimates shown here were first made in the 1960s, and in each estimate the original figures have been multiplied by a factor to represent more up-to-date costs. For example, the costs in one estimate made in 1975 have been multiplied by a factor of 1.67, which brings the costs close to those of 1982 (the latest costs available in 1983). But I have no intention to provide any actual *cost information,* only *information about estimating.* All costs in this book are for illustration or demonstration purposes only.

For uniformity's sake, and to make them easier to read, I have transcribed most of the estimate examples onto one kind of estimating sheet.[2] Some were originally made on other kinds of paper, but these general estimate sheets are particularly useful for preliminary esti-

[1]*Fundamentals of Construction Estimating and Cost Accounting,* the predecessor of this book.

[2]Produced by the Frank R. Walker Co., Publishers, Chicago, whose classic estimating book is listed in the Bibliography.

mates because they are designed for measurement and pricing together. Most of the original sheets are too disheveled to be reproduced, and their transcription has enabled me to edit them and to eliminate calculations that are not needed here. Otherwise, the examples are true transcripts.

This charming residence was designed by E. Kuckein, architect, of Vancouver, British Columbia, for himself. I used it as a subject for estimating examples in my earlier book on estimating, in which I presented several examples of detailed measurement of quantities of construction work. It is my purpose to show that *approximate quantities* are a simple and effective means for preliminary estimates, and this can best be done by comparing the precise and approximate measurements for the same work. But before we look at a fairly detailed preliminary estimate of the residence's costs, let us examine an example of the most common kind of estimate, one based on the total area.

An Estimate for the Residence Based on Area

The limitations of the building area method of estimating costs were described in Part 1. The primary limitation is that the area method often produces misleading results, even in estimates based on

RESIDENCE (1983 COSTS)

Class: AVERAGE (+) 2-STORY

Gross Floor Area = (43-4 x 18-8) + (22-0 x 17-4) = 1,190 SF

Basic Cost: (interpolate: 2200 – 2600 SF) x 2

BASE COST (2400 SF) = $31.50/SF x 2,380 SF = $74,970

MODIFICATIONS:

DEDUCTIONS:–

(1) for built-in GARAGE (2-car) (p. 60)	–1,885	
(2) for Flat Roof, (say) 1,350 sf x 0.25 @ $2.00/sf	675	
(3) for Open LR. CEILING, (say) 220 sf @ $4.00/sf	880	

ADDITIONS:–

(1) WINDOW WALLS; 1/4" Float (gross) area = 720 sf		
Stucco-ed Extr. Walls (#4.5-110) @ $6.10/sf		
Glazed Walls (similar to: #4.7-110) @ $13.76/sf		
720 sf @ $7.66/sf =		5,515
(2) PROJECT'G WING-WALLS; frame & stucco.		
[2(3-0 + 2-6 +) x 19-3] 212 sf @ $6.10/sf =		1,295
(3) INTR. CHMNY & F'PLACE (AVERAGE)		2,145
(4) ROOF PROJECTIONS: 134 sf @ $5.00/sf		670
(5) SCREEN WALL 462 sf @ $6.10/sf		2,820
(6) 1 - FULL BATHROOM (EXTRA) (p. 30)		2,400
(7) 1 - HALF " " "		1,300
(8) 1 - FRIG. + 1 - RANGE ($695 + 1,130)		1,825
(9) Walnut Plywood Panelling 360 sf @ $5.00		1,800
(10) Better than Average Quality Millwork etc.		1,200
(Carry forward)	$ –3,440	+95,940

RESIDENCE (cont.)	(1983 COSTS)	
(brought forward)	$ − 3,440	$ +95,940
(11) 2-Vanity Cabinets @ $350/each		700
(12) SITE IMPROVEMENTS:		
3'x3' P.C.C. paving slabs 71 No. @ $16.00/ea		1,135
2-POOLS (shallow) 630 sf @ $1.00		630
Pea Gravel & Pebbles (allow)		1,500
POOL SLAB 5'x5' "		150
	−3,440	100,055
		−3,440
TOTAL COSTS: (1983 COSTS)		$ 96,615

[$96,615 / 2,380 sf = $40.60/sf]

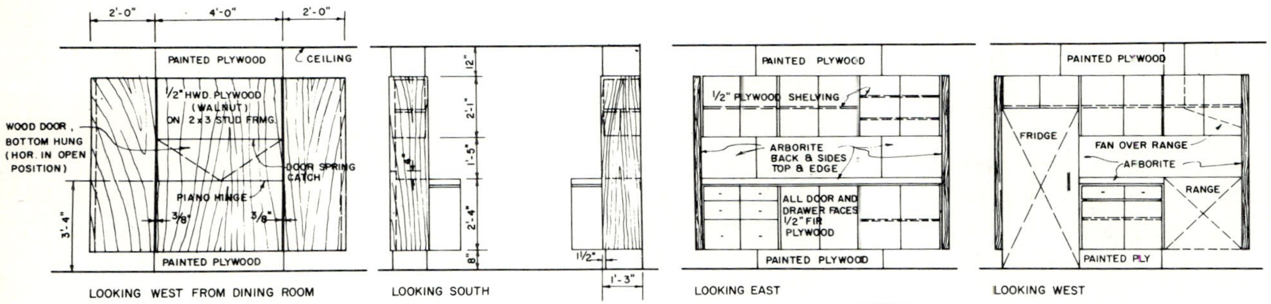

KITCHEN ELEVATIONS

NOTES

1. ALL WOOD & PLYWOOD D FIR UNLESS OTHERWISE INDICATED
2. ALL EXPOSED SURFACES PAINTED TWO COATS UNLESS OTHERWISE INDICATED
3. HWD PLYWOOD SURFACES, OILED & RUBBED FINISH

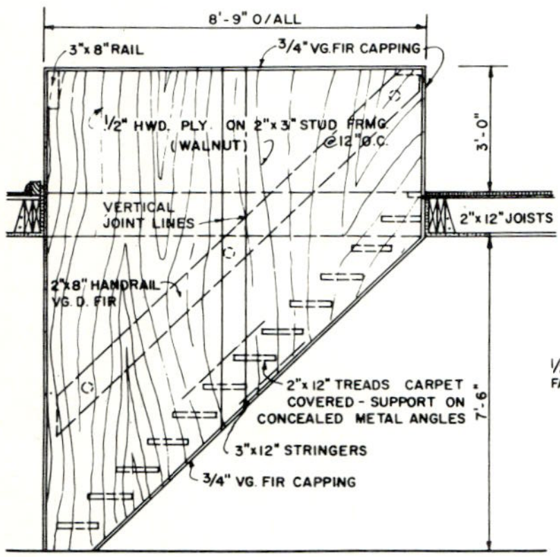

DETAIL OF STAIR

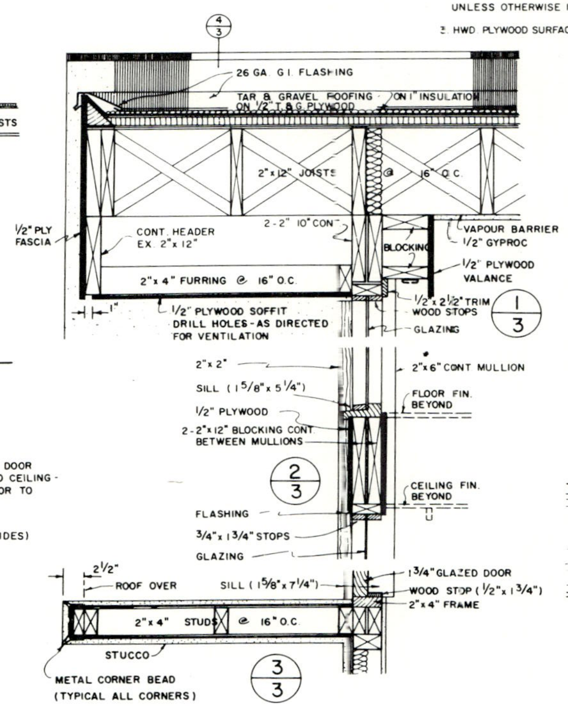

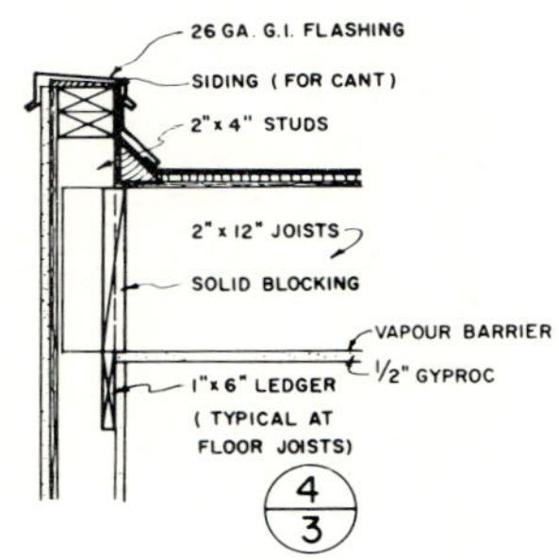

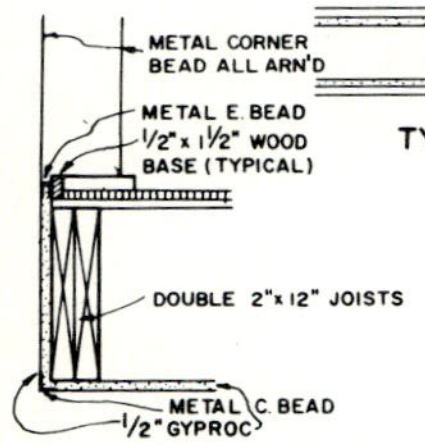

DETAIL OF UPPER FLOOR EDGE

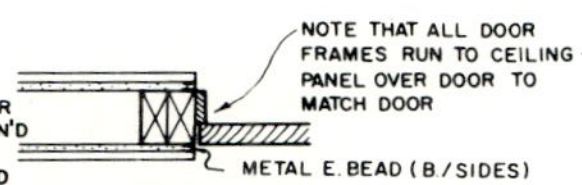

TYP. INT. DOOR JAMB

TYPICAL WALL & ROOF DETAILS

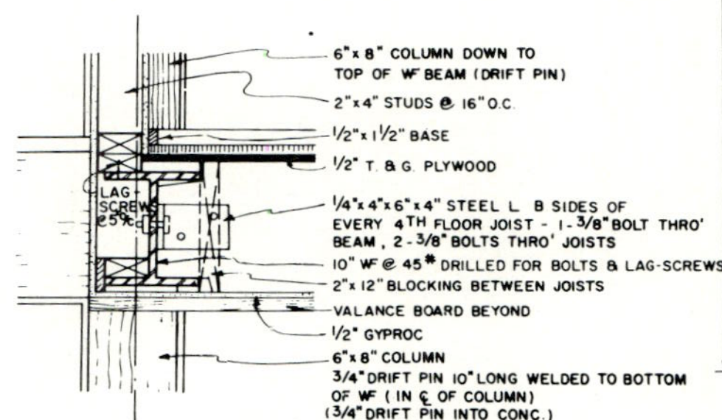

DETAIL OF STEEL BEAM TO COLUMN CONNECTION

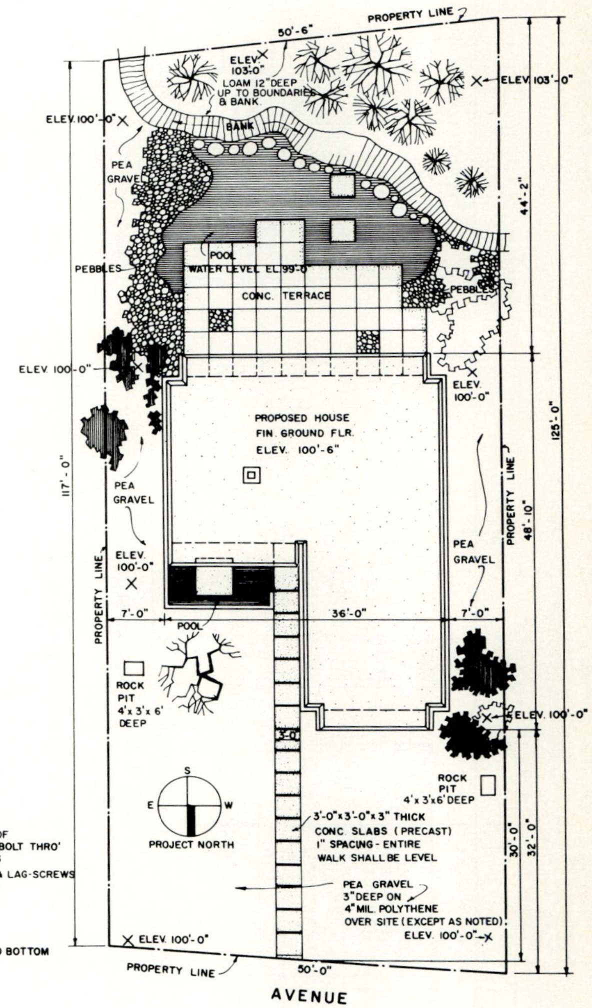

SITE PLAN

RESIDENCE (1)

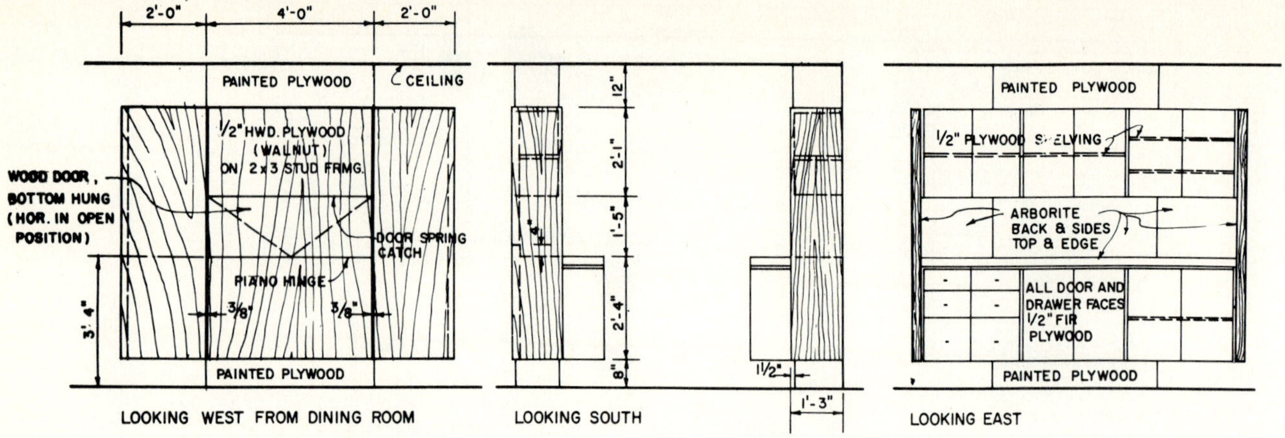

KITCHEN ELEVATIONS

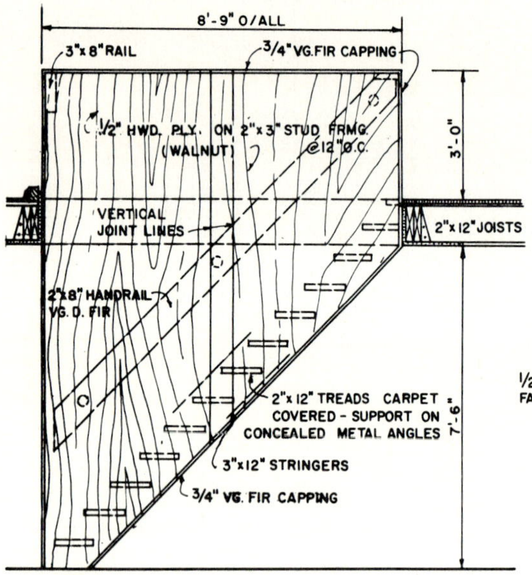

DETAIL OF STAIR

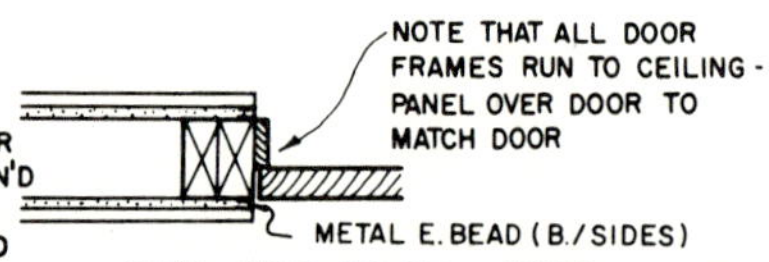

TYP. INT. DOOR JAMB

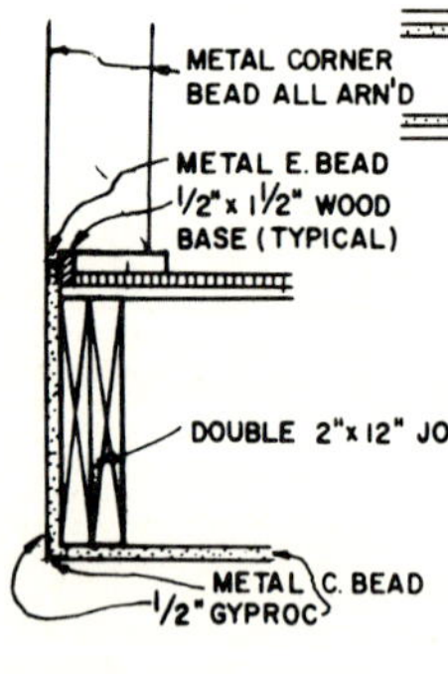

DETAIL OF UPPER FLOOR EDGE

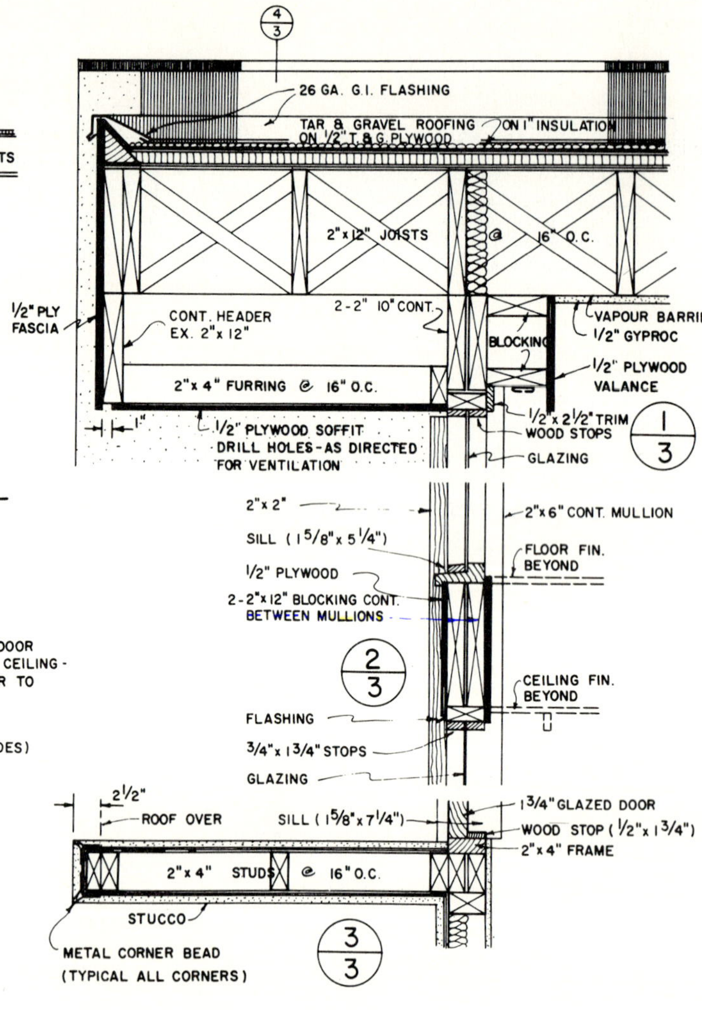

TYPICAL WALL & ROOF DETAILS

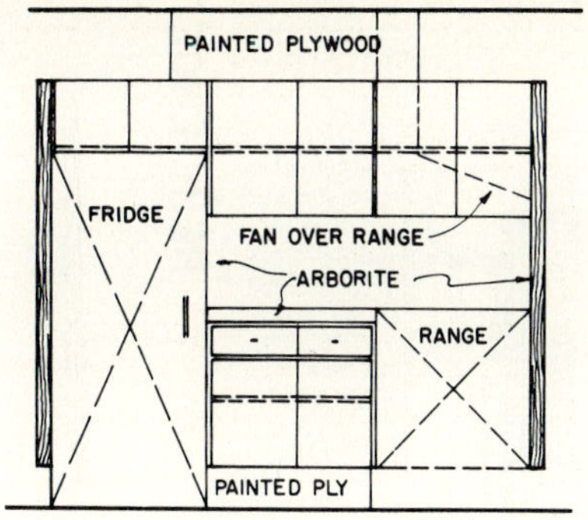

LOOKING WEST

NOTES

1. ALL WOOD & PLYWOOD D. FIR UNLESS OTHERWISE INDICATED
2. ALL EXPOSED SURFACES PAINTED TWO COATS UNLESS OTHERWISE INDICATED
3. HWD. PLYWOOD SURFACES, OILED & RUBBED FINISH

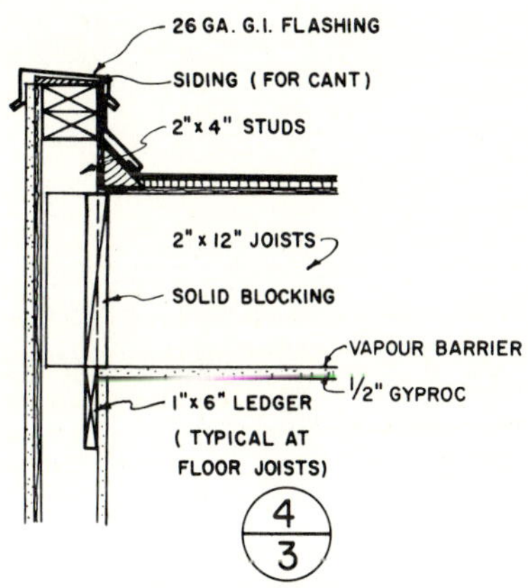

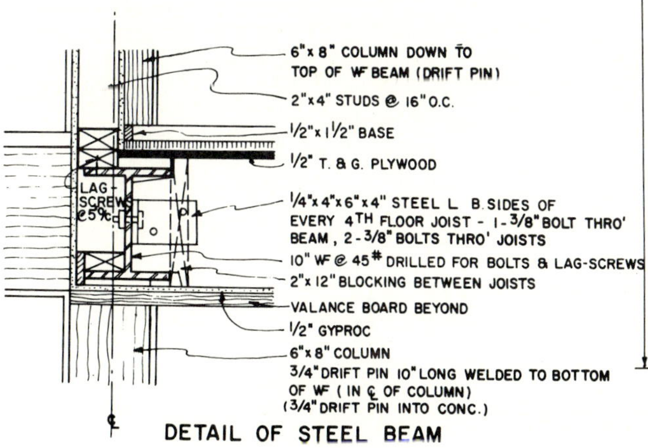

DETAIL OF STEEL BEAM TO COLUMN CONNECTION

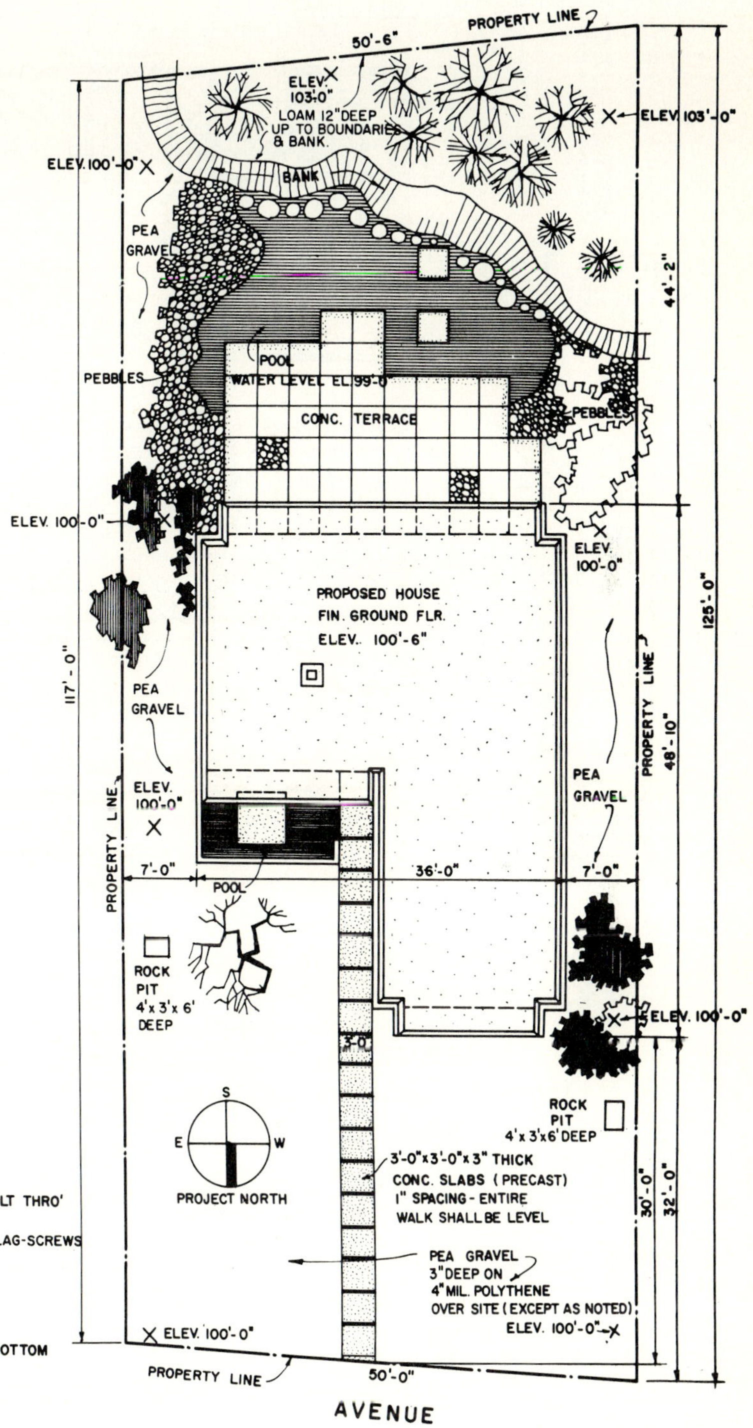

SITE PLAN

RESIDENCE 1

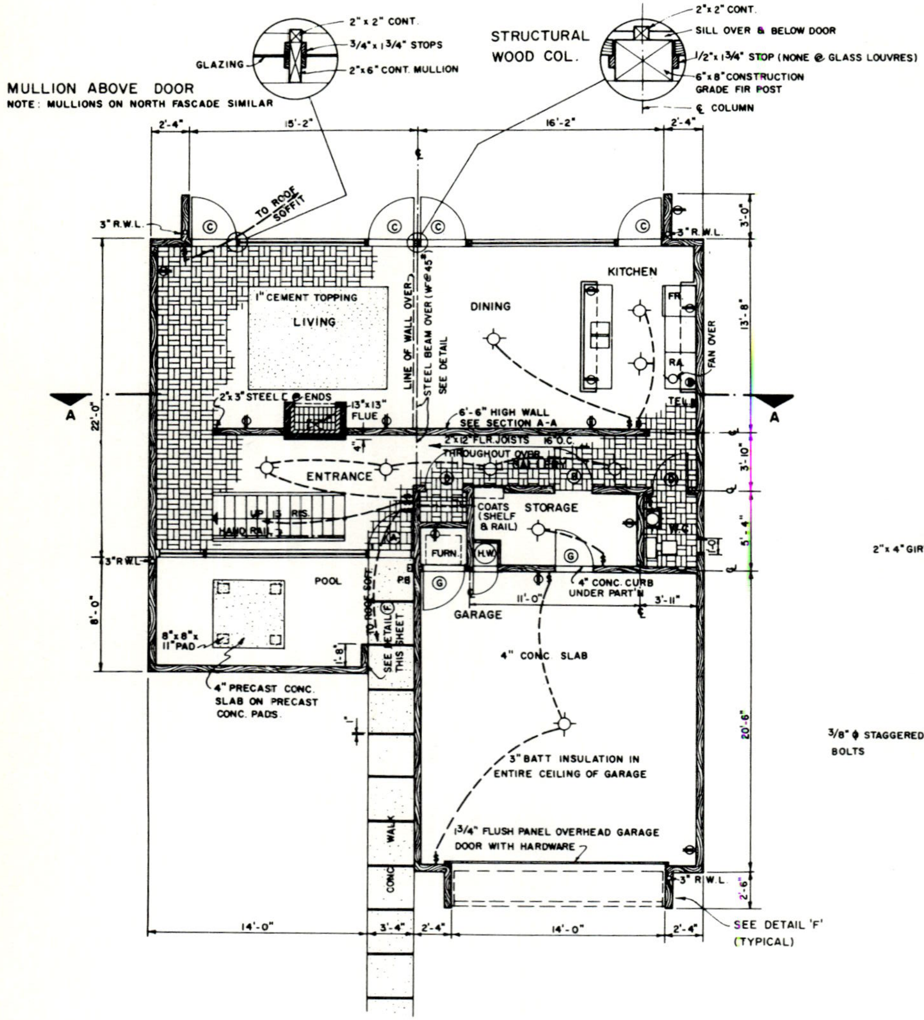

GROUND FLOOR PLAN

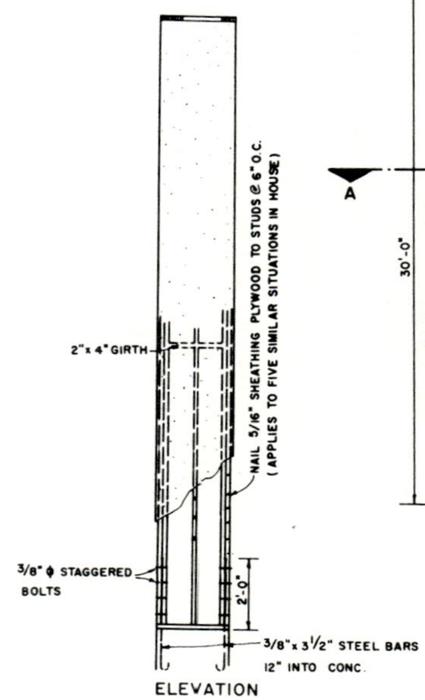

DETAIL 'F' TYPICAL FOR ALL PROJECTING WALLS (1' 8", 2' 6", & 3' 0" WIDE) ON PLAN

DOOR	SIZE	THICKNESS	TYPE	DOOR	SIZE	THICKNESS	TYPE
A	3'-0" x 7'-4"	1 3/4"	SOLID	E	2'-0" x 6'-8"	1 3/8"	HOLLOW CORE
B	2'-8" x 6'-8"	1 3/8"	HOLLOW CORE	F	3'-0" x 6'-8"	1 3/8"	HOLLOW CORE
C	3'-0" x 7'-4"	1 3/4"	GLAZED	G	2'-10" x 6'-8"	1 3/4"	SOLID CORE
D	2'-6" x 6'-8"	1 3/8"	HOLLOW CORE	H.W.	1'-6" x 6'-8"	1 3/4"	SOLID CORE

DOOR SCHEDULE

NOTE THAT DOOR TO H.W. TANK SHALL HAVE 4" ϕ SCREENED COMB. AIR INTAKE INTO TANK ENCLOSURE - DOOR 1 3/4" SOLID

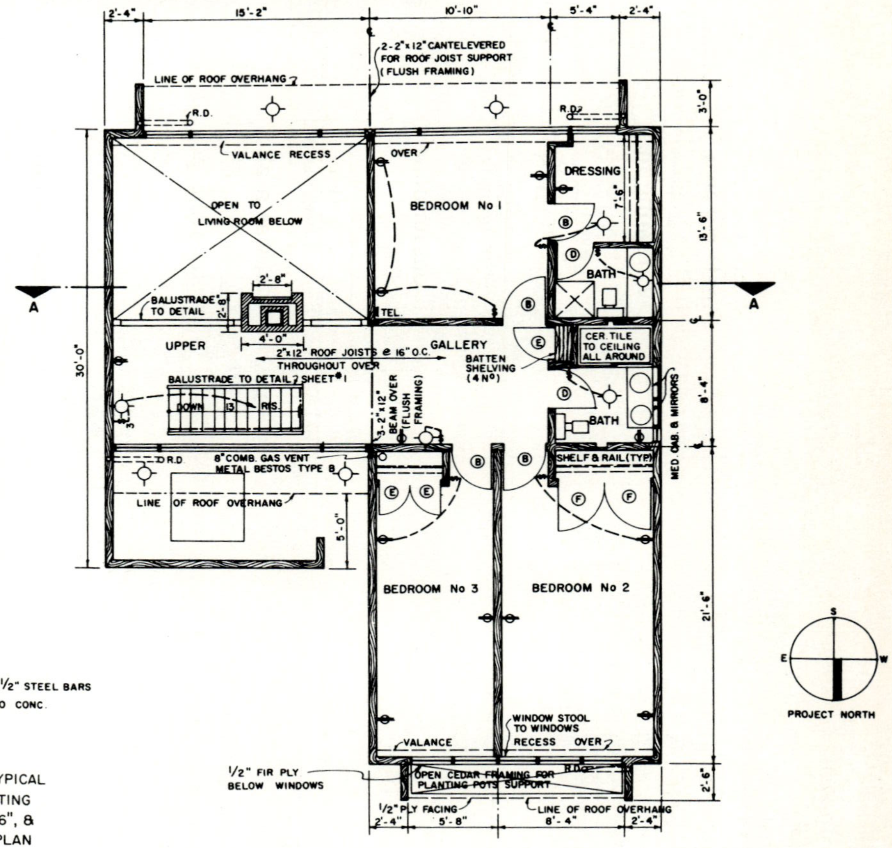

UPPER FLOOR PLAN

RESIDENCE (2)

cost information derived from similar buildings; "similar" for this purpose must include similar shapes, similar area-perimeter ratios, and similar building density,[G] as well as similar design and construction. But the area method is the mainstay of most appraisal manuals and many other cost books whose other compounding weakness usually is that they contain too few building types and models to provide a broad enough range of cost data from which to choose the correct unit cost of a building as a foundation for an estimate.

To attempt an estimate based on building area for this residence, I went to one of the best general sources of cost information, a publication of the Robert Snow Means Company,[3] and in the residential section of that company's manual I searched for a model that had some resemblance to the residence of this example. The residential section of the manual is divided into four main classes of quality: economy, average, custom, and luxury. (Models in the manual are also categorized according to number of stories: one, one plus finished attic, one and one-half, two, two and one-half, three, bilevel, and trilevel.) Selecting the quality class is the stickler. *Average* is described as "built from standard designer plans," which Mr. Kuckein's house is not. It does not look like an average house. Yet it cannot be classed (according to the manual) as one of custom quality, which is described as "usually built from designer's plans"—which it is—because the architect deliberately designed the house to be of commonplace materials and as economic as possible. In other words Mr. Kuckein set out to build an extraordinary house using ordinary construction materials and methods. He succeeded and demonstrated several important things about house design and building, none of which are generally observed by most house builders. I had no choice—if I were to attempt to estimate the costs of Mr. Kuckein's house with a manual—but to class it as average and then to make adjustments in cost for the non-average features.

Turning to the reproductions of my estimate sheets, the house is there described as average (+), meaning that there many nonaverage items included. The gross floor area is calculated as 2,380 square feet on two floors. The base costs for the average two-story houses are on page 30 of the manual; interpolating, I produced a base cost unit price of $31.50 per square foot, giving a total base cost of $74,970. Following that are the modifications, and most are based on costs from the manual including the deduction for a built-in garage. I could find no given adjustments for houses with flat roofs, so I had to estimate my own, and I could see little saving over a pitched roof except in the lesser area of actual roofing. The framing material is not much less for a flat roof, although the labor costs are less. I considered that the overall framing costs would be about the same as for an ordinary house with a pitch roof because the Kuckein drawings show balloon-framing with ledger boards. With framing, one cannot be more precise.

The additions for nonaverage items are numerous, and I will not discuss them all. The estimate sheets are clear and can easily be followed. Glazed walls are uncommon, and the closest item that I could find in the manual (on page 268) in section 3, unit-in-place costs, was a glazed "picture unit" (picture window), size 2 feet 8 inches x 6 feet 8 inches costing $245, or about $7.66 per square foot. (I compared this unit cost to the detailed costs in the other estimate that I had done for the residence and found the costs similar.) I was also obliged to add estimated amounts for the residence's walnut paneling and for better-than-average millwork throughout. The total estimated costs are shown as about $96,600, or, about $40.60 per square foot.

An Estimate Based on Approximate Quantities

As mentioned, this example of an estimate is the only one in the book not taken from an actual estimate. It is used to illustrate just how accurate approximate quantities can be and how in preliminary estimates we may cut corners without any great loss in accuracy—provided that we know what we are doing. By comparing our approximate quantities in this example with the fully detailed and accurate quantities originally measured we can see the evidence for this fact. In addition, we have two preliminary estimates for this residence, and that enables us to make other comparisons. Before we proceed to examine these estimates, let me say again that estimators are just now on the threshold of new ways of estimating. Indeed, some are already there, but from what I see most have only transferred old principles and practices to new systems, and it appears there has been little done in the development of more effective estimating for construction work. Much measurement is still rough and misleading. In the past, one excuse for this was the unwieldy bulk of the written information created by detailed

[3] *Means Square Foot Costs, 1983* (Kingston, MA).

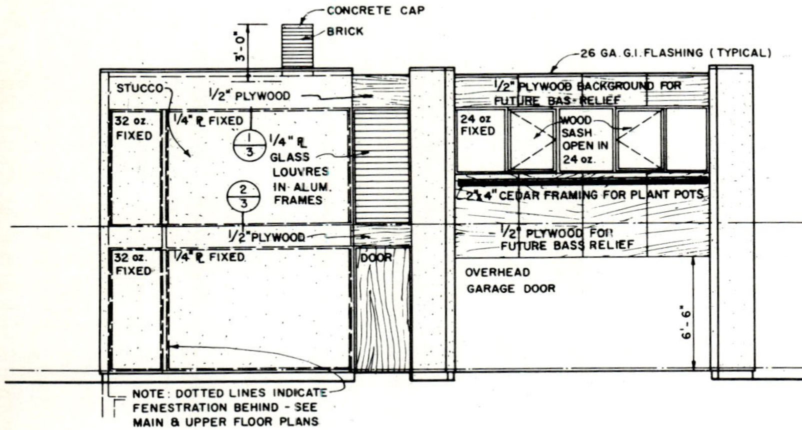

NORTH ELEVATION

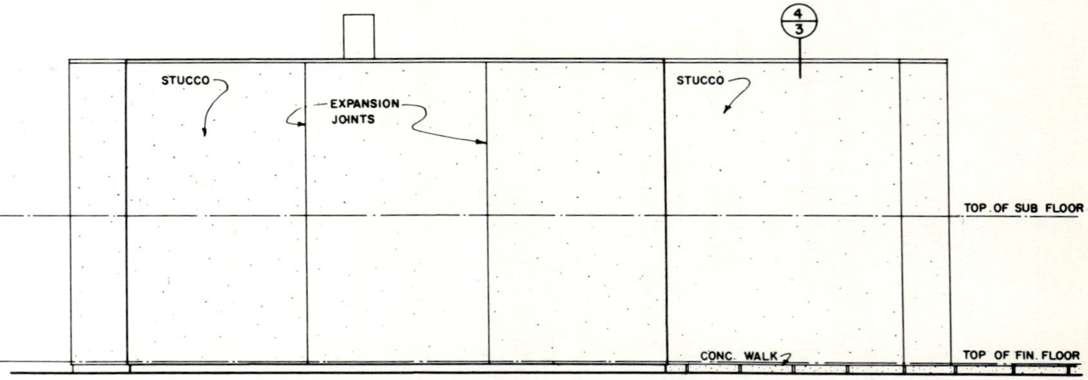

EAST ELEVATION

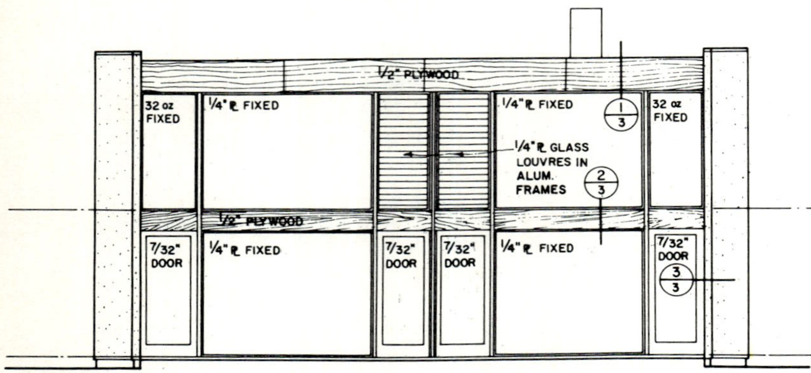

SOUTH ELEVATION

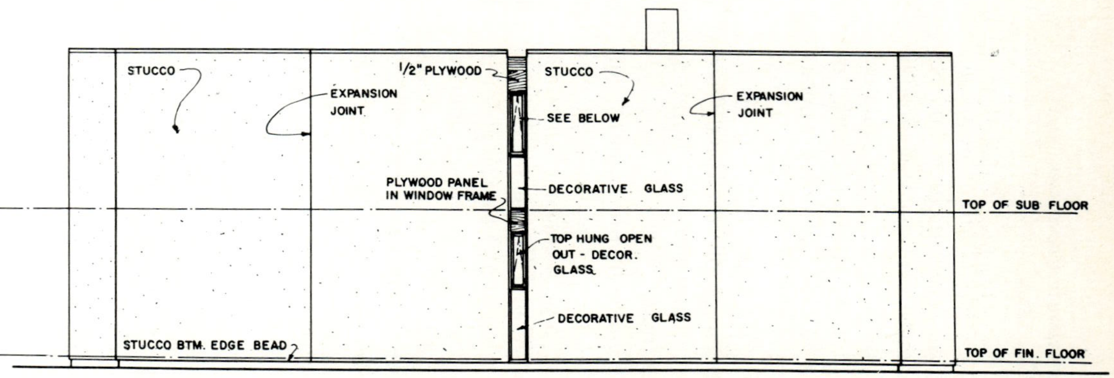

WEST ELEVATION

RESIDENCE 3

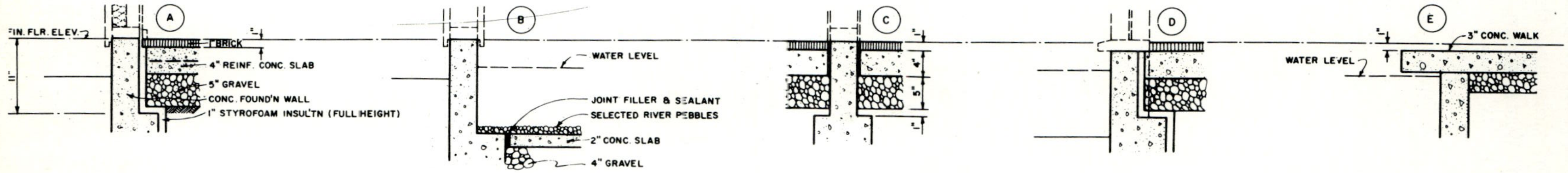

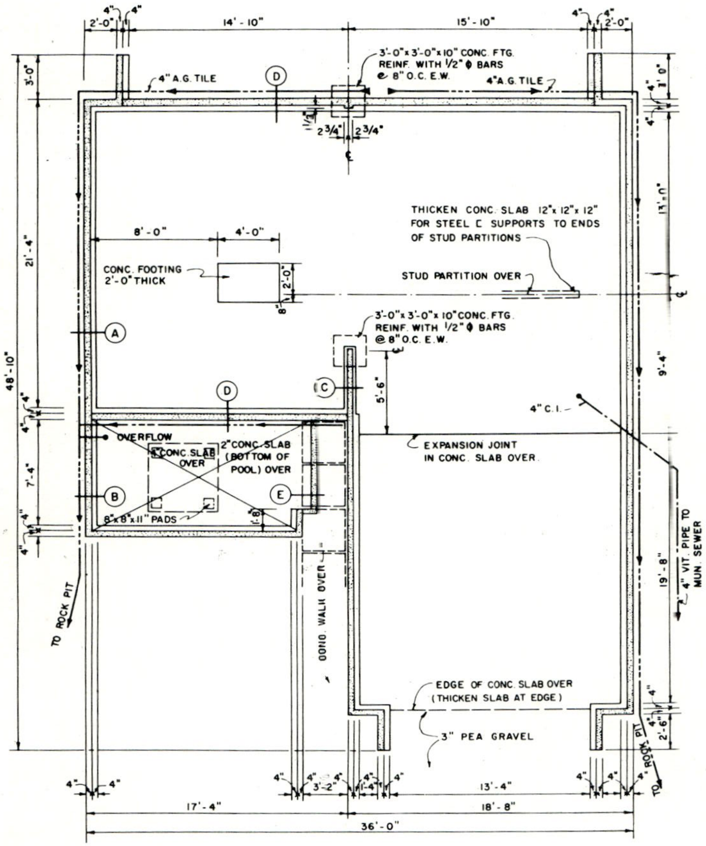

FOUNDATION PLAN

FINISH SCHEDULE

ROOM	BASE	FLOOR	N-WALL	E-WALL	S-WALL	W-WALL	CEILING
ENTRANCE	WOOD	BRICK	GLAZING	1/2" GYPROC	1/2" GYPROC	1/2" GYPROC	1/2" GYPROC
GALLERY	"	"	1/2" GYPROC	O	"	"	"
LIVING	"	"	"	1/2" GYPROC	GLAZING	O	"
DINING	"	"	"	O	"	1/2" GYPROC	"
KITCHEN	"	"	"	O	1/2" GYPROC	"	"
STORAGE	"	CONC.	"	1/2" GYPROC	"	"	"
W.C.	"	BRICK	"	"	"	"	"
GARAGE	O	CONC.	"	"	"	"	2 LAYERS 3/8" GYP.
UPP. GALLERY	WOOD	CARPET	"	"	"	"	1/2" GYPROC
BEDROOM No 1	"	"	"	"	GLAZING	"	"
BEDROOM No 2	"	V.A. TILE	"	"	1/2" GYPROC	"	"
BEDROOM No 3	"	"	"	"	"	"	"
DRESSING	"	"	"	"	"	"	"
BATH	"	"	"	"	"	"	"

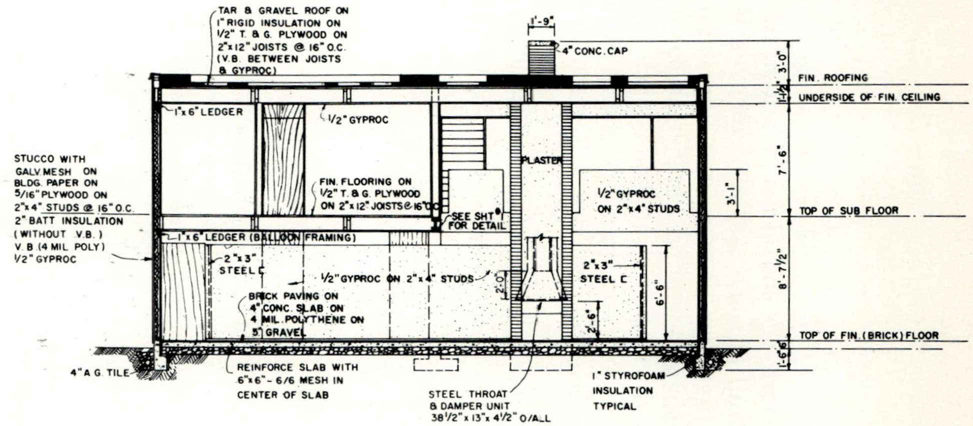

SECTION A - A

RESIDENCE 4

methods of measurement,[4] but with the growing use of microcomputers this need no longer be a constraint. Therefore, one should not be put off by the amount of detail in some of these examples, especially in this first one. All except this first one were real estimates, made with pen and pencil on paper and bought and paid for by satisfied clients. If they were then valid and economic, how much more so now if more efficiently produced with the help of a computer?

As with all the case studies and for all the estimates illustrated and discussed, we shall proceed sheet by sheet through the estimate, making comments on many of the items of work and with general comments about estimating procedures:

Site Works 1/3

Remove topsoil. The site is practically level, and the topsoil more than 6 inches deep. It was possible to minimize the handling of earth—always important on most projects—but such economies are not always clear (especially with bigger projects) until a preliminary estimate with fairly accurate quantities has been made: a good reason for such an estimate, because foresight and planning make savings possible, as shown shortly. This kind of site work can be priced only by estimating the costs of equipment and operator; needed only for one day for this item. Meaningful unit prices for excavation work are not common as there are so many variables; in small projects the quantities are often too small and costs are determined by other things.

Replace topsoil. Some advocate measuring in strict order of the work, something not always easily determined in an estimate. It is better to measure in whichever way facilitates better estimating. Sometimes this means writing down an item's description and then leaving actual measurement until another item later has been measured and can provide the dimensions or quantities needed for the previous item. (See the next excavation item.) Here we measure topsoil by calculating the approximate area (a rough triangle) and apply one-third of the maximum depth (height) to obtain a pyramid's volume. The unit price is based on that of the first item. Actually this item may be superfluous if the topsoil were stockpiled in this location, but we cannot be sure, and we cannot make too fine a point of it in a preliminary estimate.

Sale of surplus topsoil. No credit is allowed because it is a preliminary estimate.

[4]Such as those in the British *Standard Method of Measurement of Building Works* listed in the Bibliography.

Excav(ate) surface trenches. The main dimension, the length of 189 feet, comes from *concrete*, measured later. The unit price is from similar jobs; it could be estimated from equipment time, often the better way. Trenches for services are often overlooked. Often they are done at a different time and this complicates their pricing and cost. Costs of larger projects are often easier to estimate because their items have larger quantities and there is more latitude for an estimator. How do you estimate costs of excavating an irregular-shaped pool about 12 inches deep? You do so not too thoroughly and not too cheaply. If all excavation can be done at the same time, it is cheaper, but how can one be sure that it can be?

Site Work 2/3

Excav(ate) rock pits. You cannot dig a hole 4 feet by 3 feet by 6 feet deep except with some unusual equipment not usually found on a small project; and even if dug by hand it may have to be larger, so measure and price it according to a practical size. If it is eventually dug by machine (and it probably will be, at least in part), then some savings may result but here we have to assume the worst case. (Actually the worst case is to find solid rock already in the pit.)

Spread over site. The object of spreading spoil is to minimize the amount of material to be trucked away (always an expensive item) as well as to restore the original level, more or less. See *remove*, following.

Pea gravel. In preliminary estimates we combine items wherever possible; the ultimate combinations occur in an elemental estimate. The 20 percent shrinkage is allowed for in the unit price ($10 + $2 = $12) for the pea gravel material, not in the measured quantities. (This is discussed in principle at *remove*, following). The polythene membrane (which is unnecessary and ultimately a source of problems) is priced separately.

Rounded beach pebbles. Such items as this are a problem, and probably the only way to estimate their costs is to estimate the costs of a truck and labor to go to the nearest creek bed to collect them.

Site Works 3/3

Remove surplus. Here is the classic procedure for all excavation estimates: Total the quantities of all excavation items; total the backfill and other disposal items, such as spreading on site; deduct one

PRACTICAL "STANDARDIZED" FORMS FOR CONTRACTORS
Form 514 MFD. IN U. S. A.

GENERAL ESTIMATE

BUILDING: RESIDENCE
LOCATION: Vancouver, B.C.
ARCHITECTS: E. Kuckein, AIBC.
SUBJECT: SITE WORK 1/3.

PRELIM.
ESTIMATE NO.
SHEET NO. 1/20
ESTIMATOR: KC
CHECKER: FC
DATE: 1 April 83

FRANK R. WALKER CO., PUBLISHERS, CHICAGO

DESCRIPTION OF WORK	NO. PIECES	DIMENSIONS	EXTENSIONS	EXTENSIONS	TOTAL ESTIMATED QUANTITY	UNIT PRICE M'T'L	~~TOTAL ESTIMATED MATERIAL COST~~	UNIT PRICE LABOR	TOTAL ESTIMATE ~~LABOR~~ COST
Remove topsoil avg 6" deep over site & stockpile for reuse (or remove) & rough grade site									
117 / 125 / 2) 242 = 121 AVG		121 × 50	6050 SF (×6") =		112 CY	@ $3.00 =	336		
					L/SUM	1 day e	180		
						Labr.	150	(say)	350 ✓
Replace topsoil (only @ South end) × 12" AVG depth	½	50 × 30	750 SF (×12") =		27 CY	@ $7.00			190 ✓
						{mach. $3.00 / Labr. +4.00}			
Sale surplus topsoil (112 − 27)					85 CY	NO COST			—
Excav surface trenches ft founds									
(conc fnds)		189 × 2 × 1½	567 ✓						
	1/	4 × 2 × 2½	20 ✓						
	2/	4 × 4 × 1	32	= 619 ✓	23 CY	@ 7.00			160 ✓
Ditto, for services, incl backfilling later									
water − 80' / drains − 80' / elec & gas − 70'		230 × 1½ × 2	690 ✓		excav. } 26 CY	@ 7.00			180 ✓
					b'fill }	@ 3.00			80 ✓
Excav pools × 12" dp									
− rectangular)	1/	14 × 7	+ 98						
− irregular)	1/	36 × 18	+648						
deduct wants)	3½/	9 × 9	− 120	= 626 ✓	23 CY	@ 7.00			160 ✓
							TOTAL	$	1120 ✓

PRACTICAL "STANDARDIZED" FORMS FOR CONTRACTORS
Form 514 MFD. IN U. S. A.

GENERAL ESTIMATE

BUILDING RESIDENCE
LOCATION Vancouver BC
ARCHITECTS E. Kuckein
SUBJECT SITE WORK 2/3

ESTIMATE NO. PRELIM
SHEET NO. 2/20
ESTIMATOR KC
CHECKER FC
DATE 1 April '83

FRANK R. WALKER CO., PUBLISHERS, CHICAGO

DESCRIPTION OF WORK	NO. PIECES	DIMENSIONS	EXTENSIONS	EXTENSIONS	TOTAL ESTIMATED QUANTITY	UNIT PRICE M'T'L	~~TOTAL ESTIMATED MATERIAL COST~~	UNIT PRICE LABOR	TOTAL ESTIMATE ~~LABOR~~ COST
Excav rockpits & connecting trenches by Hand and fill with gravel/rock					2/Nº	labr. / mat.	500 / 100	(4/5)	600 ✓
		2/4 x 5 x 6	240 =	9cy					
Spread over site selected earth and roll to consolidate to x 3" dp (under pea gravel) (AND) Pea gravel x 3" dp spread on 4mil. polythene barrier over site (allow 20% shrinkage)									
		1/121 x 50	6,050 ✓	ddt					
Deduct wants)		(house area)		1165 ✓					
		(exter. walls) (190 x 0-8)		127 ✓		spread & roll	8.00		
		(pools)		626 ✓					
		(¢ bank)		750 ✓		spread gravel	8.00		
		(@ pebbles)		220 ✓		mater	12.00		
			−2888 ✓						
			3,162 x 3" =	790 ✓	30 ✓/cy	@	28.00 ✓		840 ✓
			(4mil poly)	4000 sf		@	5¢		200 ✓
Rounded beach pebbles (⌀ 3" dia) x 5" dp on 4mil. poly									
		30 x 5	150 ✓						
		10 x 7	70 ✓						
			220 ✓ x 5"		4 ✓/cy	@	25.00		100 ✓
					300	@	5¢		20 ✓
Ditto (except 4mil. poly) in pools placed by hand									
			626 x 5"		10/cy	@	30.00		300 ✓
							TOTAL →	$	2,060 ✓

"STANDARDIZED" FORMS FOR CONTRACTORS
PRACTICAL
Form 514 MFD. IN U.S.A.

GENERAL ESTIMATE

BUILDING: RESIDENCE
LOCATION: Vancouver BC
ARCHITECTS: E. Kuckein
SUBJECT: SITE WORK 3/3

ESTIMATE NO.: PRELIM
SHEET NO.: 3/20
ESTIMATOR: KC
CHECKER: FC
DATE: 1 April '83

FRANK R. WALKER CO., PUBLISHERS, CHICAGO

DESCRIPTION OF WORK	NO. PIECES	DIMENSIONS	EXTENSIONS	EXTENSIONS	TOTAL ESTIMATED QUANTITY	UNIT PRICE M'T'L	TOTAL ESTIMATED MATERIAL COST	UNIT PRICE LABOR	TOTAL ESTIMATE LABOR COST
Remove surplus excav earth (not topsoil) to dump									
		TOTAL EXCAV							
		s.trenches)	23 cy						
		" ")	26						
		pools)	23						
		R.pits)	9						
		trenches ")	1 =	82					
Deduct: (wants)		spread)	30						
		backfill)	14						
		")	26 =	−70	12 cy	@ 20.00			240
Backfill @ founds (by machine & hand)									
		189 x 1-4 x 1-6	378		14 cy	@ 7.00			100
Install only 3'x3'x3" precast conc paving slabs (in pathway & terrace) on peagravel bed (see Conc.)									
		17/1	17						
		6/10	60						
			77						
Deduct: want)			−6		71 No	@ 6.00 (labor only)			430
4" dia agric. tile perimeter drains, covered by peagravel (12"x12" o/a), w/ covered open joints									
@ fnd perimeter)	148			1.60					
to rockpit)	12			.50	160 LF	@ 3.00			480
				.75					
						TOTAL	$		1,250

total from the other. The result tells you how much surplus you have to truck away or how much imported fill material you need. Generally for this purpose quantity surveyors measure all excavation and fill work in net, *bank measure*[G] quantities, and allow for *shrinkage and swell*[G] in the only alternative place, in the unit prices. The other way leads to chaos and confusion.

Concrete 1/2

Perim(eter) foundations. Measurement is by an easily calculated, mean-perimeter length. Often it is as easy to use precise dimensions as it is to use dimensions rounded off to the nearest foot. The unit price is synthesized in the first money column: the prime cost for material ($54 per yard), plus a waste allowance, plus labor and equipment costs.

Isolated found(ations). These are kept separate because they are much more expensive to install, although for small quantities the difference is not great. (In making examples, one does tend to observe more of the niceties.)

Forms. Notice how expensive these are. Yet some estimators consider forms to be less important than the concrete, presumably because concrete is concrete—and forms are more or less formless and not really part of the building, so they deserve less attention, it is wrongly believed.

Concrete 2/2

4-inch slab. Here it is treated almost like a subelement; everything possible is included; and the unit price is built up on the estimate sheet nearby.

Site precast concrete. This is made on site with forms held ready to receive any leftover concrete each time there is a concrete pour on the site.

Masonry and Metal Work 1/1

Chimney stacks are awful things to measure in detail, and having done it, one wonders, Was it worth it? They are complicated and also hard to price. Even bricklayers have more problems estimating them than they do building them. Three hundred years ago an English measurer complained about the troubles caused by measuring brick chimneys: He wondered what to do about the flues. He decided, contrary to some modern counterparts, to deduct for the flues. We have taken the easier way (often taken by masonry contractors) and roughly estimated the number of bricks. (Actually according to detailed measurement the stack requires about 1,750, plus some for waste, but considering the costs involved, even a thousand or so bricks in a preliminary estimate is not something to be too concerned about.) Chimney stacks are the kind of item for which an estimator should obtain historical costs to avoid detailed measurement.

Steel beam. This is another item difficult to price, but like the stack, there is only one, and it is not a common item, so we need not make a great effort to analyze the costs. We must keep a proper perspective and remember that we are making a preliminary estimate (not a bid estimate) and that there are limitations to the degree of accuracy achievable by an estimator's analysis of quantities and costs. To price fabricated metal items: First price the metal by weight (at more than the bulk price), add for labor based on a number of hours for fabrication and installing, and allow for transportation.

Rough Carpentry 1/3

Sheathing to roof. With framing materials, waste[G] is always important, and often it is greater than many realize. Waste allowances are generally better made, as in this case, in the unit price. Although the measurements here are rounded off, the final quantity is accurate to within about 10 or 11 percent. (Accurate measurement shows 1,340 square feet total.)

Sheathing to floor. A similar degree of accuracy is obtained with this item, and an additional amount is acceptable in a preliminary estimate, providing that it is not excessive.

Sheathing to walls. Detailed measurement of this item gave a total of 2,940 square feet compared to the 3,000 square feet in this estimate. Measurement is by a simple perimeter-length times the height; the absence of openings in the sheathed walls simplifies it.

Rough Carpentry 2/3

Framing. This work is often problematic; it is difficult to measure accurately, and accurate pricing can be elusive. Another way to price framing is to measure the lumber (as done here) and to price the labor at so much a square foot of building area. Often that method is accurate enough because framing contractors, especially those building houses, often price framing that way themselves.

Wall framing. Here it is fairly simple: the exterior walls are measured by plate-length (130 feet times 3 for three plates; two at the top) and the studs are

PRACTICAL "STANDARDIZED" FORMS FOR CONTRACTORS
Form 514 MFD. IN U.S.A.

GENERAL ESTIMATE

BUILDING: RESIDENCE
LOCATION: Vancouver BC
ARCHITECTS: E Kuckein
SUBJECT: CONCRETE 1/2

PRELIM
ESTIMATE NO.
SHEET NO. 4/20
ESTIMATOR: KC
CHECKER: FC
DATE: 1 APR '83

FRANK R. WALKER CO., PUBLISHERS, CHICAGO

DESCRIPTION OF WORK	NO. PIECES	DIMENSIONS	EXTENSIONS	EXTENSIONS	TOTAL ESTIMATED QUANTITY	UNIT PRICE M'T'L	TOTAL ESTIMATED MATERIAL COST	UNIT PRICE LABOR	TOTAL ESTIMATE LABOR COST
Concrete (3000 psi) ready mix P.C. $54.00 per CY:—									
Perim foundations 8" x 24" o/all (types A thru E)									
4 / 4 11 / 24 13 / 8 2#5	1/142-8		143				Mat 54		
	2/3-0		6				waste +2		
	2/2-6		5				=56		
43-4 158-8	1/4-6		5				Labr 10		
+36-0 4/=-2-8	2/7-4		14				Equp +2		
79-4 156-0 ×2	1/15-4		16						
158-8 -13-4			189 (×0-8×2-0)		10/CY	@ 70.00			700
142-8									
Isolated founds (chmny, piers)									
@ Chmny)	1/4 ×2×2		16						
@ piers)	2/3×3-0-10		15						
			31		31/CF	@ 3.00			100
Forms (use & waste) contact areas:—									
Perim founds	2/189-0 × 2-0		756		760/SF	@ 2.00			1,520
Ditto inserted size 4" × 11"	1/189				190/LF	@ 2.00			380
Isolated found, 4' × 2' × 2' (24 SFCA)					1/No	@ 120.00			120
Ditto, 3' × 3' × 10" (10 SFCA)					2/No	@ 110.00			220
12/-189-0 -24-0									
#5 Rebar in fnds	213 LF × 1.043 lb				250/lb	@ 50¢			125
									+5
						TOTAL	$		3,170

PRACTICAL "STANDARDIZED" FORMS FOR CONTRACTORS
Form 514 MFD. IN U. S. A.

GENERAL ESTIMATE

BUILDING: RESIDENCE
LOCATION: Vancouver BC
ARCHITECTS: E Kuckein
SUBJECT: CONCRETE 2/2.

ESTIMATE NO.: PRELIM
SHEET NO.: 5/20
ESTIMATOR: KC
CHECKER: FC
DATE: 1 APRIL '85

FRANK R. WALKER CO., PUBLISHERS, CHICAGO

DESCRIPTION OF WORK	NO. PIECES	DIMENSIONS	EXTENSIONS	EXTENSIONS	TOTAL ESTIMATED QUANTITY	UNIT PRICE M'T'L	TOTAL ESTIMATED MATERIAL COST	UNIT PRICE LABOR	TOTAL ESTIMATE LABOR COST
Concrete (3000 psi) as before:—									
4" Slab on ground (GF) reinf wi 6"x6" x 6/6 WWF laid on 6 mil poly membrane & steel float finish									
						M conc 67¢ / WWF +15 / 6 poly 3 / = 85¢			
house)	1/	35-4 x 22-8	801			L place conc. +34			
garage)	1/	20-0 x 18-0	360			WWF 10			
"	1/	14-0 x 2-10	41			poly 4			
Deduct:			1202			Finish 30			
want @ entrance)		17-0 x 1-4	−23		1,180 SF	Equip 15			
			1179			@ 1.80			
									2,130 ✓
2" Ditto, in small areas of pool (@ front)									
	1/	14 x 7			100 SF	@ 2.00			200 ✓
Sundry concrete in post-holes, etc.					(Allow)	—			100 ✓
Slab screeds x 4"					150 LF	@ 50¢			80 ✓
Site precast concrete (3000 psi):—									
4" pad slab 5' x 5' (assume 3 5/8" actual)					1/No	@			100 ✓
8"x8"x11" slab pads					4/No	@			40 ✓
Chimney Cap, 3" x 21" x 21"					1/No	@			40 ✓
Paving Slabs, 3'x3'x3" thick (installed in Div. 2)					71	@ 10.00			710 ✓
Perimeter Insulation, 1" Styrofoam on conc fnds					No				
	1/	143 x 2-4	335		340 ✓ SF	@ 50¢			170 ✓
						TOTAL		$	3,570

PRACTICAL "STANDARDIZED" FORMS FOR CONTRACTORS
Form 514 MFD. IN U. S. A.

GENERAL ESTIMATE

ESTIMATE NO. PRELIM.

SHEET NO. 6/20

BUILDING RESIDENCE

LOCATION Vancouver BC

ESTIMATOR KC

ARCHITECTS E Kuckein

CHECKER FC

SUBJECT MASONRY & METAL WORK-1/1

DATE 1 APRIL '83

FRANK R. WALKER CO., PUBLISHERS, CHICAGO

DESCRIPTION OF WORK	NO. PIECES	DIMENSIONS	EXTENSIONS	EXTENSIONS	TOTAL ESTIMATED QUANTITY	UNIT PRICE M'T'L	TOTAL ESTIMATED MATERIAL COST	UNIT PRICE LABOR	TOTAL ESTIMATE LABOR COST
BRICK CHIMNEY STACK & FIREPLACE 20' high o/all (3' above roof) including:—									
Materials:							(Materials)		
Facing bricks approx. 2,000					2000	@ 300	=600 ✓		
Firebricks " 200					200	@ 600.00	=120 ✓		
Mortar 2cy					2/cy	@ 60.00	=120 ✓		
Steel Damper & Throat Unit (as specs) & Lintol bar					1/No	@	540 ✓		
Flue liner units 13"x13"x12"					16/No	@ 5.00	80 ✓		
Scaffolding					(Allow)		50 ✓		
						=	1,510 ✓		
Labor:									
Bricklayer & helper @ $32.00/hr x 40 hrs. =							1,280		
							2,790		
Mark-up: (job overhead + operating o'head + profit)							+710		
							$ 3500		
CHIMNEY: TOTAL								= $	3,500 ✓
10WF45 Steel Beam x 17½ long complete w/ 2-drift pins, 8-angle brackets (1/4"x4"x6") and 16 - 3/8" ϕ x 6" Bolts (approx 850lb.) (@ 1.00 + handling)					1/No	@			900 ✓
3/8"x3½"x42" Flat steel bar Anchors (detail 'F')					10 (=170/lb)	@ 7.00 EA			70 ✓
2"x3"x7'0" long steel ⊏ Stiffeners (to partition framing)					2 (=85/lb)	@ 50.00 EA			100 ✓
							TOTAL	$	4,570 ✓

PRACTICAL "STANDARDIZED" FORMS FOR CONTRACTORS
Form 514 MFD. IN U.S.A.

GENERAL ESTIMATE

BUILDING: RESIDENCE
LOCATION: Vancouver BC
ARCHITECTS: E. Kuckeun
SUBJECT: ROUGH CARPENTRY - 1/3.

ESTIMATE NO.: PRELIM.
SHEET NO.: 7/20
ESTIMATOR: KC
CHECKER: FC
DATE: 1 APRIL '83

FRANK R. WALKER CO., PUBLISHERS, CHICAGO

DESCRIPTION OF WORK	NO. PIECES	DIMENSIONS			EXTENSIONS	EXTENSIONS	TOTAL ESTIMATED QUANTITY	UNIT PRICE M'T'L	TOTAL ESTIMATED MATERIAL COST	UNIT PRICE LABOR	TOTAL ESTIMATE LABOR COST
Douglas Fir (Exterior) Plywood Sheathing: —											
1/2" 'Standard' Sheathing to ROOF deck (10% waste)											
13-6 5-8 / 8-4 8-4 / 21-6 2/= 4-8		43	x	18	774 ✓			M 300/MSF, 10% +30, nails 40			
43-4 18-8		25	x	17	425 ✓			=370			
2/= -0-8 2/= -8		14	x	3	42 ✓			L +200			
42-8 ✓ 18-0 ✓		32	x	4	128 ✓						
2 (43+18) =		122	x	1-0	122 ✓						
					1491		1500 ✓ SF	@ 57¢ ✓			860 ✓
1/2" 'Select' Sheathing to FLOOR deck (10% waste)											
42-8		43	x	18	774 ✓			M 500, +50, 40			
(say) 43-0		17 1/2	x	8 1/2	149 ✓			=590			
					923 ✓		1,000 ✓ SF	@ 80¢	L +200		800 ✓
5/16" 'Standard' Sheathing to WALLS (7 1/2% waste)											
30 134 / 14 / 2 4/= 8 / 23 4/= 12 / 22 / 22 4/= 10 / 21 / 134 ↑ 164		164	x	18	2952		3,000 ✓ SF	@ 33¢	M 240, +18, 40, =298; L +27, =325		990 ✓
Black Bldg. Paper to Extr. Walls (laps measd.)											
Area-Wall shtng					3000						
" ply siding (Fin. Carp)					+264						
					3,264						
Add: 25% (for laps & waste)					+816		4,000 ✓ SF	@ 6¢			240 ✓
								TOTAL		$	2,890 ✓

"STANDARDIZED" FORMS FOR CONTRACTORS
PRACTICAL
Form 514 MFD. IN U. S. A.

GENERAL ESTIMATE

BUILDING: RESIDENCE
LOCATION: Vancouver BC
ARCHITECTS: E. Kucken
SUBJECT: ROUGH CARPENTRY 2/3.

ESTIMATE NO.: PRELIM
SHEET NO.: 8/20
ESTIMATOR: KC
CHECKER: FC
DATE: 1 APRIL '83

FRANK R. WALKER CO., PUBLISHERS, CHICAGO

DESCRIPTION OF WORK	NO. PIECES	DIMENSIONS	EXTENSIONS	EXTENSIONS	TOTAL ESTIMATED QUANTITY	UNIT PRICE M'T'L	TOTAL ESTIMATED MATERIAL COST	UNIT PRICE LABOR	TOTAL ESTIMATE LABOR COST
Douglas Fir "Construction" grade framing lumber – unless o/wise indicated:—									
2x4s in balloon framing to exterior walls					(allow 15% waste)				
130									
5/= 15		3/130	390✓						
6/3 = 18		164/18	2950✓	LF	M (MBF)		300		
.163✓				= 3,340	waste		+45		
							345		
					Labour		365		
2x4s in interior walls:—							710		
57									
25	200								
106	6. 15/2 = 42	3/200	600✓						
12	3	245/8	1,960✓						
200↑	245	4/36	144✓						
				+ 2,704✓					
				6,044✓					
				= 6,000 =	4,000✓ BF @	710.00/M			2,840✓
Sundry 2x4 framing @ stairs, cabinets					1,000 BF @	1100.00			1,100✓
2x12 Joists framed in FLOOR & ROOF (waste incl.)					M		350		
					L		+175		
@ floor)	50/20		1,000✓				= 525		
@ roof)	75/20		1,500✓						
			2,500✓ × 2BF =		5,000✓ BF @	530.00/M			2,650✓
Cross-bridgings to 2x12s (measd. over joists)									
		2:2/50	200✓		(260✓ LF @)				
		2/28	56✓		= 194 SETS @	2.50 (SET)			490✓
							TOTAL: $		7,080✓

enumerated, one stud per foot of length—an old rule-of-thumb that in many instances can let you down—plus some extra studs because of the short lengths of the projecting walls and the extra corners. The one stud per foot rule would have come up short here, and it only seems to work on standard types of houses with simple rectangular or L-shaped plans.

Interior walls are always more trouble: total length times 3 for plates, plus the studs at 1 per foot, plus extra studs at doorways, and 4 times 36 feet for backings at edges of gypsum drywall sheets where top plates of partitions are parallel to the floor joists above—often overlooked. With three simple sets of dimensions we can measure all the interior partitions; with two sets we can measure the exterior walls; and the resultant quantities are within less than 50 board-feet of the total quantity measured elsewhere in much greater detail. The trick to measuring framing accurately is to avoid *errors of omission.*

Joists. These are easily measured in this project because the designer obviously designed for simplicity and economy in his own house; a lesson for all designers. Is not simplicity often a mark of excellence, especially in design? Consequently no allowance has been made in the unit price for waste as none should occur—unless somebody steals lumber from the site.

Cross-bridging. There is a certain amount of minor framing that practically cannot be measured, even in detailed estimates, and one usually makes a lump-sum allowance to cover it, as in this estimate. Cross-bridging, however, should be measured and priced. It is too expensive to ignore and it is easily measured.

Rough Carpentry 3/3

There are several hundred dollars worth of framing on this sheet that might easily be overlooked by an estimator. Because most are not ordinary framing members, they are better priced by the linear foot, and at a higher rate.

For commonplace, typical buildings, framing can often be measured sufficiently accurately in a preliminary estimate using the building area at so many board-feet of material and so many dollars' worth of labor per square foot, for example, in the preliminary estimates for speculative housing and rental apartments for which such cost data exist. (Some framing contractors make bid estimates that way.) This approach cannot be used with a house such as this, with its—in the West—atypical balloon framing and its window walls, north and south. It is not the typical and average Pacific Northwest house. Why are not more houses designed by architects? Most houses are so poorly designed and poorly orientated on their sites that an architect-designed house usually stands out like a rosebush in a dump.

Finish Carpentry 1/4

Douglas fir exterior plywood siding. This is on the north and south walls in fairly small panels that are expensive to cut and install because of the vertical V-joints indicated to be placed symmetrically. Consequently the waste for this minor item is quite high. Items such as this are a problem, even in detailed bid estimates, because an estimator cannot ignore the details that increase the costs and because the measurement and pricing techniques are often inadequate——even the best of them. Elsewhere, detailed measurements show a calculated waste factor of 25 percent and include several ancillary *labor items*[G] measured in an attempt to quantify the abnormal amounts of labor involved in this fairly unusual item. The alternative approach is perhaps the more common one—at least in the United States, where no official standard method of measurement exists—and that is to deal with such items as this by measuring the material gross and by allowing an estimated crew time for the job (in man-hours). Frequently, this may be the better way. This example uses an intermediate method: net measurement of the work and a labor estimate based on crew time. This is a similar approach but with a different way of dealing with waste.

Finish Carpentry 2/4

Wall panelling. The same comments made about plywood siding apply equally to interior paneling, and in this example the latter's costs are more critical because of the higher material cost and the location and shape of the paneling, with its *raking cutting and waste.*[G] Here, the plywood is measured gross; elsewhere, detailed measurement included raking, cutting and waste as a separate item of work.

It is quite usual in an estimate to price an unusual item (for which no historical unit prices are available) or an item of small quantity such as the handrails on this sheet, by treating it as a single, enumerated item (e.g., a 3 by 8 Douglas fir, C grade, vertical grain, 4-foot length of handrail) to estimate a total price (of \$20), and *then* to calculate the unit price by dividing the total cost by the total quantity (\$20 divided by 4 feet = \$5 per foot). It is the reverse of the normal and expected order of pricing when a unit price is shown. The purpose is to enable the

PRACTICAL "Standardized" Forms for Contractors
Form 514 MFD. IN U.S.A.

GENERAL ESTIMATE

BUILDING: RESIDENCE
LOCATION: Vancouver BC
ARCHITECTS: E. Kucken
SUBJECT: ROUGH CARP. 3/3

ESTIMATE NO.: PRELIM.
SHEET NO.: 9/20
ESTIMATOR: KC
CHECKER: FC
DATE: 1 APRIL '83

FRANK R. WALKER CO., PUBLISHERS, CHICAGO

DESCRIPTION OF WORK	NO. PIECES	DIMENSIONS	EXTENSIONS	EXTENSIONS	TOTAL ESTIMATED QUANTITY	UNIT PRICE M'T'L	TOTAL ESTIMATED MATERIAL COST	UNIT PRICE LABOR	TOTAL ESTIMATE LABOR COST
1x6 Ledgers set into 2x4 studs @ 16" o/c (5%)							M 15 LF / L +100		
		2/2/43	172 ✓						
		2/2/3	12 ✓						
		2/2/2½	10 ✓		190 ✓ LF @		1.15/LF		220 ✓
2x12 Headers, Blockings, etc									
(@ fasias)		2/34	68				M 350 (MBF)		
(@ mullions)		2/32	64				L 350		
")		2/18	36						
			168 ✓		170 ✓ LF @		1.40/LF		240 ✓
2x10 Ditto									
(@ valances)		2/2/34	136		140 ✓ LF @		1.30/LF		180 ✓
2x4s, attached to 2x10s & 2x12s							M 20 / 3		
		1/50	50				L 42		
		2/2/34	136		190 ✓ LF @		65¢/LF		120 ✓
2x6, Ditto									
(@ valances)		2/36	72 ✓				30 / 4		
		2/18	36 ✓		110 ✓ LF @		80¢/LF / 46		90 ✓
Sundry framing (Cedar planter, etc.)					(Allow)		—		400 ✓
Rough hardware (nails, connectors, etc.)					(Allow)		—		400 ✓
							TOTAL: $		1,650 ✓

PRACTICAL "STANDARDIZED" FORMS FOR CONTRACTORS
Form 514 MFD. IN U. S. A.

GENERAL ESTIMATE

BUILDING: RESIDENCE
LOCATION: Vancouver BC
ARCHITECTS: E. Kuckein
SUBJECT: FINISH CARP. 1/4.

ESTIMATE NO.: PRELIM.
SHEET NO.: 10/20
ESTIMATOR: KC
CHECKER: FC
DATE: 1 APRIL '83

FRANK R. WALKER CO., PUBLISHERS, CHICAGO

DESCRIPTION OF WORK	NO. PIECES	DIMENSIONS		EXTENSIONS	EXTENSIONS	TOTAL ESTIMATED QUANTITY	UNIT PRICE M'T'L	TOTAL ESTIMATED MATERIAL COST	UNIT PRICE LABOR	TOTAL ESTIMATE LABOR COST
1/2" D.Fir Exterior Plywood (AB) Siding (20% WASTE)					(4x8)	(4x10)	SHEETS			
		31-4 x	2-3	71	4	–				
		31-4 x	1-1	34			M	600		
9/4x8 = 288		14-0 x	2-3	32			waste	+120		
1/4x10 = 40		14-0 x	5-0	70	4	–	L	500		
328		17-0 x	2-3	38						
-264		17-0 x	1-1	19	1	1				
64 = 20%				264 SF	9 NO.	1 NO.				
						270 SF	@ 1.22			330 ✓
1/2" Ditto @ eaves Soffites (35% waste)					(4x8)		M	600		
		31-4 x	2-9	86	4		W	200		
8/ = 256		17-0 x	2-9	47			L	800		
165										
91 = 35%		14-0 x	2-3	32	4					
				165	8	170 SF	@ 1.60			270 ✓
1/2" Ditto Interior Panelling (detail #2/3) (12½%)							M	600		
								80		
		31-4 x	1-2	37			L	800		
		17-0 x	1-2	20 ✓		60 SF	@ 1.50			90 ✓
1/2" Ditto x 10"–11" wide in Bases & Valances (15%)							M	600		
		2/2/10-3		41			W	100		
		1/35-4		36			L	800		
		1/18-0		18		100 LF	@ 1.50			150 ✓
							TOTAL: $			840 ✓

PRACTICAL "STANDARDIZED" FORMS FOR CONTRACTORS
Form 514 MFD. IN U. S. A.

GENERAL ESTIMATE

BUILDING RESIDENCE
LOCATION Vancouver BC
ARCHITECTS E. Kuckein
SUBJECT FIN. CARP. 2/4

PRELIM. ESTIMATE NO.
SHEET NO. 11/20
ESTIMATOR KC
CHECKER FC
DATE 1 APRIL '83

FRANK R. WALKER CO., PUBLISHERS, CHICAGO

DESCRIPTION OF WORK	NO. PIECES	DIMENSIONS	EXTENSIONS	EXTENSIONS	TOTAL ESTIMATED QUANTITY	UNIT PRICE M'T'L	TOTAL ESTIMATED MATERIAL COST	UNIT PRICE LABOR	TOTAL ESTIMATE LABOR COST
1/2" 'CUSTOM' Grade AB Walnut plywood									
Wall Panelling w/ Vertical V-joints (meas'd. GROSS)									
	4/4/	9 x 12	432	(4'x8') 4 No.	(4'x12') — @	$96.00/SHEET			380 ✓
Deduct wants	2/2/1/2/	7 1/2 x 7 1/2	−113		4 No. @	168.00 "			670 ✓
			319 sf	Labor	20 @ hr	30.00			600 ✓
									1,650
Panelling @ K. Cabinets (meas'd. GROSS)									
4 4: 5/13	4/6 = 24 4: 8 3	13 x 6	78		(4x8)				
		35 x 0-4	12		4 No. @	$96/SHT			400
			90 sf	(20 hr) Labor	@	30.00			600
2x8 D Fir ("C" VG) Handrails (on brackets)									
		2/12			24 LF @	3.50/LF			80 ✓
3x8 Ditto		1/4			4 LF @	5.00			20 ✓
1/2" x 1 1/2" P Mahog (stained) Base Trim (5% waste ~~GROSS~~)									
(@ extr. walls)		2/130	260						
(@ intr. walls)		2/57	114						
(" ")		2/106	212						
31 14 (@ wdws)		2/59	118						
14 59			704	Ddt					
DEDUCT wants)		2/24		48					
		2/38		76					
		2/13	−150	26	550 LF @	34¢			190 ✓
						TOTAL	$		2,940 ✓

estimator to see the item's costs finally expressed as a unit price and to permit a mental judgment as to the reasonableness of the price; also, to produce a unit price for possible future use in this or another estimate.

Base trim. Such items always give trouble, and their measurement can be tedious. Yet how else can their costs be estimated? For common building types one may have design data as to the average number of square feet of floor area for each linear foot of base trim. Here, simple overall lengths produce a perfectly accurate result (the same as that produced by much more detailed and piecemeal measurement): twice times the length of exterior walls for two floors; twice times the total interior wall length on each floor for two sides of the walls; plus the lengths of the window walls; less the *wants*[G] at exterior walls around the front pool, at the garage, and at the living room (2 times 13 feet). Minor deductions, such as for doorways, are ignored so as to allow for minor additions, and the final result is accurate enough. Allow 5 percent for waste in cutting, as there is bound to be some.

Finish Carpentry 4/4

Windows and doors. The only good source of prices for these is a manufacturer or distributor, and given sufficient information usually they can readily quote prices. Types, qualities, and prices vary so much from place to place that cost books often are not able to offer prices for the specific type for which you may be looking. It is usually quicker to call a local manufacturer. For speculative housing, actual prices for items such as these are often surprisingly low compared to prices published in cost books. But beware; specifications are critical and highly variable. Nonstandard sizes (often favored by designers) may increase prices threefold or more.

Sundry fixtures. For 1 house, make an allowance; for 50 houses, find out what the fixtures are to be and exactly how much they will cost per house.

Protection 1/1

Roofing. Simple dimensions are rounded off for such a relatively cheap item, but do not try to include the cost of metal flashings unless they are minimal. Here they are dealt with separately as they are extensive and amount to more than half of the cost of the roofing itself.

Insulation. Other such relatively cheap items (like roofing) can be easily measured from dimensions derived from other items.

Finishes 1/3

Exterior stucco. Simply measured by total perimeter times height; plus reveals and ends of walls (that amount in total to over 6 square yards in area, please note). Metal beads are measured separately because in good quality work (such as required here; otherwise beads would not appear in such abundance) they are not inconsiderable in cost, especially if they are of zinc alloy.

Ceramic tiling. Tiling around standard bath tubs is such a common item there is no need to measure it in more detail.

Finishes 2/3

Drywall ceilings. Trade contractors usually measure this work gross according to the numbers of boards of different lengths needed for each room or area. For detailed estimates I prefer to measure it both ways, gross and net, and to compare the results and to see the calculated amount of waste.[5] For common types of buildings one usually has design data from past projects whereby the approximate area of drywall to walls and ceilings can be calculated from a factor (of 3.0 to 4.0, depending on building density[G]) applied to the gross floor area. Here we have measured by approximate quantities because of the building's unique design.

No other worthwhile method is possible—except fully detailed measurement of both gross and net quantities, as done elsewhere, where the results were: for ceilings, gross area = 1,568 square feet and net area = 1,400 square feet—a difference (of waste) of exactly 12 percent of the net area. This approximate estimate shows a net figure of 1,500 square feet and a waste allowance of 10 percent, giving a gross total of 1,650 square feet, which is about 5 percent higher than the equivalent figure in the detailed estimate. That is quite accurate enough. Some would say, unnecessarily accurate for a preliminary estimate. But here we are working from completed drawings not usually available for a preliminary estimate. Nevertheless, comparable accuracies can be obtained from incomplete drawings because, as in this example, we generally use overall dimensions in the measurement of such major finishing items as drywall.

The ceiling over the living room is higher than normal and therefore costs more for scaffolding and labor. This could be dealt with by an *extra-over*[G] item.

---

[5]As in the detailed example for this work in the earlier book.

PRACTICAL "STANDARDIZED" FORMS FOR CONTRACTORS
Form 514 MFD. IN U. S. A.

GENERAL ESTIMATE

BUILDING: RESIDENCE
LOCATION: Vancouver BC
ARCHITECTS: E. Kuckein
SUBJECT: FIN. CARP. 3/4

ESTIMATE NO.: PRELIM.
SHEET NO.: 12/20
ESTIMATOR: KC
CHECKER: FC
DATE: 1 APRIL '83

FRANK R. WALKER CO., PUBLISHERS, CHICAGO

DESCRIPTION OF WORK	NO. PIECES	DIMENSIONS	EXTENSIONS	EXTENSIONS	TOTAL ESTIMATED QUANTITY	UNIT PRICE M'T'L	TOTAL ESTIMATED MATERIAL COST	UNIT PRICE LABOR	TOTAL ESTIMATE LABOR COST
WINDOW WALLS, of 'structural' grade D. Fir Frames, Posts, Mullions in long lengths (16') w/ "C" VG. stops planted on (as specs)									
							MATERIAL		
6x8 Posts (x8')		2/8			16 LF	@ 4.50	72 ✓		
2x6 Mullions (x16')		4/16	64						
		2/16	32		96 LF	@ 1.25	120 ✓		
2x4 Frames (16')		2/2/16			64 LF	@ 1.00	64 ✓		
2x2 Trim (16')		1.6/16			112 LF	@ 50¢	56 ✓		
							= 312	=	310 ✓
1x2 (Nominal) Stops & Trim, planted on							LABOR (complete frames)		750 ✓
		5/2/8	80						
	10/	2/2/8	320						
		3/2/32	192						
	1	3/2/18	144						
		1/36	36						
			772		800 LF	@ 20¢			160 ✓
1/4" Float Glass & glazing to wood frames									
[① louvres 3'x6'8]		2/31 × 7½	465						
		2/17 × 7½	255						
deduct. want)	3·	1/3 × 7½	(720 − 90) =		630 SF	@ 3.50			2,200 ✓
2x6 (NOMINAL) 'C' VG Sills			(32+18)	MATERIAL	50 LF	@ 2.00			100 ✓
2x8 (DITTO) DITTO			(32+18)	MATERIAL	50 LF	@ 2.50			130 ✓
							TOTAL	$	3,650 ✓

PRACTICAL "STANDARDIZED" FORMS FOR CONTRACTORS
Form 514 MFD. IN U. S. A.

GENERAL ESTIMATE

ESTIMATE NO. PRELIM

BUILDING RESIDENCE
SHEET NO. 13/20
LOCATION Vancouver BC
ESTIMATOR KC
ARCHITECTS E. Kuckein
CHECKER FC
SUBJECT FIN. CARP. 4/4 (WDWS & DOORS)
DATE 1 APRIL '83

FRANK R. WALKER CO., PUBLISHERS, CHICAGO

DESCRIPTION OF WORK	NO. PIECES	DIMENSIONS	EXTENSIONS	EXTENSIONS	TOTAL ESTIMATED QUANTITY	UNIT PRICE M'T'L	TOTAL ESTIMATED MATERIAL COST	UNIT PRICE LABOR	TOTAL ESTIMATE LABOR COST
Supply & Install Wood Windows (factory made) & Doors (frames meas'd. separately)									
Wood Windows in 5-lights 14' x 3'10" o/all (55 SF) @ 350.00									350 ✓
Ditto vertical in 2-lights 13" x 15'6" o/all (17 SF) @ 150.00									150 ✓
2x8 Sill x 14" long (to last wdws) 1/No @ 10.00									10 ✓
1x4 Stool x 14'0" long (to 5-light wdws) 1/No @ 20.00									20 ✓
S & Install Glass Louvres in Alum. frames (in Wdw Walls) ea. 3' x 6'8" o/all 3/No @ 80.00									240 ✓
									770 ✓
Supply & Install Wood Doors w/ matching Transom panels over (to 7'6" clg. ht):							DOORS & FRAMES (MATERIAL)		
(D) Hollow 4/ 2'6 x 6-8+ (T. panel) (paint finish)					4	@	20		80 ✓
(E) " 3/ 2'0 x 6-8+ (") " "					3	@	20		60 ✓
(F) " 2/ 3'0 x 6-8+ (") " "					2	@	25		50 ✓
(G) Solid 2/ 2'10" x 6-8+ (") " "					2	@	70		140 ✓
(H & T) Hollow 1/ 1'6" x 6-8+ (") (paint finish)					1	@	20		20 ✓
(B) " 5/ 2'8 x 6-8+ (") " "					5	@	25		130 ✓
(A) Solid entrance 1/ 3'-0" x 7-4 (stain finish)					1	@	120		120 ✓
(C) Glazed (solid) 4/ 3'-0" x 7-4 " "					4	@	115		460 ✓
					22/No	@	(LABOR) 18		400 ✓
Overhead Garage Door 14' x 6'6" x 1¾", complete					1 No	@	—		600 ✓
CASH ALLOWANCE for Finish Hardware & Small Fixtures									1,100
Sundry fixtures installed (bathrooms etc)									200
							TOTAL	$	4,130 ✓

PRACTICAL "STANDARDIZED" FORMS FOR CONTRACTORS
Form 514 MFD. IN U. S. A.

GENERAL ESTIMATE

PRELIM. ESTIMATE NO. ____

SHEET NO. 14/20

BUILDING RESIDENCE

LOCATION Vancouver BC.

ARCHITECTS E Kuckein

SUBJECT Cabinets & Fixtures - 1/1.

ESTIMATOR KC

CHECKER FC

DATE 1 APRIL '83

FRANK R. WALKER CO., PUBLISHERS, CHICAGO

DESCRIPTION OF WORK	NO. PIECES	DIMENSIONS	EXTENSIONS	EXTENSIONS	TOTAL ESTIMATED QUANTITY	UNIT PRICE M'T'L	TOTAL ESTIMATED MATERIAL COST	UNIT PRICE LABOR	TOTAL ESTIMATE LABOR COST
CASH ALLOWANCE for KC Cabinets supplied and installed, including:—							(CA)		2,200 ✓
sink counter		x 8' incl. top							
wall cupbds		x 8'							
ditto		x 8'							
work counter		x 2'9" incl. top							
CASH ALLOWANCE for Vanity Cabinets supplied and installed including:—							(CA)		700 ✓
Counter		x 4'x1'8"x 2'4"							
"		x 4'8x1'8"x 2'4"							
18" Closet Shelves & coat rails, w/ bearers									
Unit x4'6					1 No @	18			
" x6'6					1 No @	26			
" x7'6					1 No @	30			
" x3'0"					1 No @	12	}		90 ✓
					3 hr Labor @	30.00			90 ✓
Linen closet shelving (on bearers) of ½x1 W R. Cedar Slats and Frame, size 12"x30"					4 No @	10.00			40 ✓
						TOTAL	$		3,120 ✓

"STANDARDIZED" FORMS FOR CONTRACTORS PRACTICAL Form 514 MFD. IN U.S.A.

GENERAL ESTIMATE

PRELIM.
ESTIMATE NO. ______

SHEET NO. 15/20

BUILDING RESIDENCE

LOCATION Vancouver BC

ARCHITECTS E. Kuckein

SUBJECT PROTECTION - 1/1

ESTIMATOR KC

CHECKER FC

DATE 1 APRIL '83

FRANK R. WALKER CO., PUBLISHERS, CHICAGO

DESCRIPTION OF WORK	NO. PIECES	DIMENSIONS		EXTENSIONS	EXTENSIONS	TOTAL ESTIMATED QUANTITY	UNIT PRICE M'T'L	TOTAL ESTIMATED MATERIAL COST	UNIT PRICE LABOR	TOTAL ESTIMATE LABOR COST
ROOFING, 4-ply AG Built-up on 1" thick Rigid Insulation + VB.										
	1/36	x	43	1,548						
	1/32	x	3½	112						
	1/14	x	3	42						
				1,702						
Deduct: want)	1/17	x	18	-306	=	14 SQ	@ 120.00			1,684 +6
DITTO, on 8" parapet incl. 0.020" Alum Cap Flashing x 12" girth & Counter Flashing x 9" girth										
2(42-8)+2(3-4)+3(3-4)+2(2-10)+4(2-0)+4(1-9)						123 LF	@ 5.00			620
4x4 △ Cants on roof deck										
(112+62+10+)						190 LF	@ 1.10			210
0.020" Aluminum Cant Flashings x 9" girth										
2(17½+14+ =)						63 LF	@ 2.00			130
0.020" Alum Wdw Head Flashings x 6" girth										
(32+18+2+ =)						52 LF	@ 1.10			60
3lb. Lead Pipe Flashings, 3" & 4" dia.						2 NO	@ 8.00			20
Batt Insulation & poly. VB :-						SF				
6" (R19) to Roof (btwn joists)						1,500	@ 45¢			680
3" (R11) to Walls (btwn studs)						3,000	@ 25¢			750
3" (R11) to Garage ceiling (btwn joists)						360	@ 25¢			90
								TOTAL	$	4,250

PRACTICAL "STANDARDIZED" FORMS FOR CONTRACTORS
Form 514 MFD. IN U. S. A.

GENERAL ESTIMATE

PRELIM
ESTIMATE NO. ______
SHEET NO. 16/20

BUILDING RESIDENCE
LOCATION Vancouver BC
ARCHITECTS E. Kuckein
SUBJECT FINISHES 1/3. STUCCO

ESTIMATOR KC
CHECKER FC
DATE 14 APRIL '83

FRANK R. WALKER CO., PUBLISHERS, CHICAGO

DESCRIPTION OF WORK	NO. PIECES	DIMENSIONS	EXTENSIONS	EXTENSIONS	TOTAL ESTIMATED QUANTITY	UNIT PRICE M'T'L	TOTAL ESTIMATED MATERIAL COST	UNIT PRICE LABOR	TOTAL ESTIMATE LABOR COST
Exterior Stucco to walls, w/ white finish									
30-0 / 14-0 / 2-0 / 22-8 / 21-4 / 20-8 / 21-8 = 132-4 ✓; 132-0 ✓, 4/=8-0 ✓, 4/=12-0 ✓, 4/=10-0 ✓ = 162-0 ✓		1/ 162-0x18-0	2,916 ✓						
		2/ 0-4x6-6	4 ✓						
		3/ 0-3x18-0	9 ✓						
		5/ 0-6x18-0	45 ✓						
			2,974 ✓ (÷9)		330 ✓ /SY	@ 16.00			
									5,280 ✓
Interior Stucco on bk. chmny, tinted									
3x13					4 ✓ /SY	@ 20.00			80 ✓
Zinc-alloy Stucco beads, in stucco									
(@ exp. joints)		2/4/18	144 ✓						
(@ corners)		5/2/18	180 ✓						
"		2·6/18	144 ✓						
(@ edges)		1/162	162 ✓						
		2/22½	45 ✓						
			675 ✓		680 ✓ /LF	@ $1.00			680 ✓
								$	6,040 ✓
CERAMIC TILING									
Tiling around bathtub x 5½' high (P.C. $3.00/SF)					1 /No.	@ $240.00			240 ✓
Extra for Soap Dish (ceramic)					1 /No.	@ $20.00			20 ✓
						TOTAL		$	6,300 ✓

PRACTICAL "STANDARDIZED" FORMS FOR CONTRACTORS
Form 514 MFD. IN U. S. A.

GENERAL ESTIMATE

BUILDING: RESIDENCE
LOCATION: Vancouver B.C.
ARCHITECTS: E. Kuckein
SUBJECT: FINISHES 2/3. DRYWALL

ESTIMATE NO.: PRELIM.
SHEET NO.: 17/20
ESTIMATOR: KC
CHECKER: FC
DATE: 1 APRIL '83

FRANK R. WALKER CO., PUBLISHERS, CHICAGO

DESCRIPTION OF WORK	NO. PIECES	DIMENSIONS	EXTENSIONS	EXTENSIONS	TOTAL ESTIMATED QUANTITY	UNIT PRICE M'T'L	TOTAL ESTIMATED MATERIAL COST	UNIT PRICE LABOR	TOTAL ESTIMATE LABOR COST
½" Gypsum Drywall to CEILINGS (10% waste)									
@ u. floor)	1/	43 x 18	774 ✓						
@ G. floor)	1/	36 x 23	828 ✓						
@ u floor gallery)	1/	17½ x 8	140 ✓						
Deduct wants)			1,742 ✓	Ddt					
@ DR ceiling)	1/	17 x 13	–	221 ✓					
@ stairwell)		9 x 4	–	36 ✓					
			−257 ✓						
			1,485 ✓		1,500 ✓ SF	@ 55¢/SF			830 ✓
½" Ditto to CEILING 16' above floor (25% waste)									
		17 x 13-0			220 ✓ SF	@ 75¢/SF			165 ✓
Double ⅜" Drywall to GARAGE CEILING (10%)									
		20 x 18			360 ✓ SF	@ 1.20/SF			435 ✓
½" Gypsum Drywall to WALLS (allow 10% waste)									
extr. wall shtg.)		(area)	2,952 SF ✓						
intr. walls)	2/	200 x 7½	3,000 ✓						
			5,952 ✓	Ddt					
Deduct wants)									
@ N. screen wall)	2/	24 x 7½	–	360 ✓					
	3/2/	3 x 7½	–	90 ✓					
	2/2/	2-8 x 7½	–	80 ✓					
@ panelling)		(542 + 57 SF)	–	600 ✓					
@ int. doors)	11/2/	2-8 x 7½	–	440					
			−1,570 ✓						
			4,382 ✓		4,400 ✓ SF	@ 45¢/SF			1,980 ✓
Galv. Beads for drywall									
4400/10 (general areas)			440		1,000 ✓ LF	@ 50¢			500 ✓
extra @ doors)	17/2/2/	7½	510						
							TOTAL	$	3,910 ✓

Drywall to walls. Again, this includes overall dimensions and simple measurement with deductions for the major wants, including interior doors, in this case because the doors and their transom panels extend right up to the ceiling. The results are 4,840 square feet gross area here, compared with 4,704 square feet in a detailed estimate—an insignificant difference. Again, beads are measured separately, but this time by using a factor applied to the total drywall area, plus extra beads required on either side of interior doorways in an unusual detail.

Finishes 3/3

Brick paviors to ground floor. This is another unusual item that in a more detailed estimate for a bid would justify detailed measurement of all the cutting (and resultant waste) at perimeters. Here we try to allow for that in the unit price, not an easy thing to do well. (The detailed estimate elsewhere shows that up to a maximum of 230 feet of straight cutting might be needed at perimeters, depending on job dimensions, and it is not a cheap item.)

V.A.tiling. Even if the displacement (want) caused by interior walls is ignored the total area measured would be about 900 square feet, only 5 percent higher than a more accurate measurement.

Painting. Even in bid estimates for houses and apartments few estimators actually measure areas to be painted; they usually estimate on the basis of overall area, number of rooms, coats, complexity, etc., in a rather subjective manner. Only for a large project can we justify measurement of painting for a preliminary estimate, and even then the areas would be *reflected quantities*[G] obtained from other work.

Services 1/1

All of these costs are estimated as lump-sums based on the number of units (fixtures, etc.) where applicable. Plumbing cannot be estimated from the building area; sometimes electric lighting can be, especially in larger buildings where the fixed costs (of service connection, meters, and such) are less significant than in a small building. But lump-sum does not mean that we can ignore the scope and quality of the work. Plumbing fixtures are priced at above average, and there are three WCs (water closets) in the house. Light fixtures can cost anything from $5 to $105 each, or more, so a cash allowance[G] is the only way to deal with them, sometimes even in bidding documents.[G]

Summary

The total costs (for 1982) are estimated to be about $85,000, and if we add about 8 percent to increase the costs to those of 1983 (almost a guess, because 1983 costs are as yet not fully known), we get a total cost of about $92,000, which is about $4,600 less than the estimated costs based on building area. This is an acceptable difference, and it certainly substantiates the original classification of average quality rather than that of custom which surely most would choose if they looked at the drawings of the residence and were told that it was built for an architect from his own design drawings. Based on the custom-quality base costs in the manual, the estimate by building area would have been far too high.

Exercises

1. Using a copy of the manual, *Means Square Foot Costs*, estimate the costs of this same residence by classifying it as custom quality and by making appropriate deductions and additions for differences between the manual's model building and the subject. Make any necessary adjustment to your estimated costs for differences in time and location and compare your result with the total costs in the estimate of the example. Explain their similarity or difference, as the case may be.

2. Compare and discuss the results of the two estimates shown in this chapter; explain the difference between them and offer detailed proposals for estimating procedures and information that might reduce the difference. Alternatively explain in detail why you believe that the difference cannot be eliminated.

3. Using the information provided, make an elemental estimate of the costs of the residence to demonstrate the efficacy and economy of the elemental method.

4. Describe what you have learned about the use of cost manuals containing illustrations of model buildings for estimating, with particular reference to the initial process of classification of buildings.

5. (a) Examine and compare details of both the estimates in this chapter, list specific differences in the costs of the same items in the two estimates, and explain why and how these differences have occurred.

(b) Explain the significance of time, location, and market conditions to the results of these two estimates.

PRACTICAL "STANDARDIZED" FORMS FOR CONTRACTORS
Form 514 MFD. IN U. S. A.

GENERAL ESTIMATE

BUILDING: RESIDENCE
LOCATION: Vancouver BC
ARCHITECTS: E. Kuckein
SUBJECT: FINISHES 3/3. FLOORINGS

ESTIMATE NO.
SHEET NO. 18/20
ESTIMATOR: KC
CHECKER: FC
DATE: 1 APRIL '83

FRANK R. WALKER CO., PUBLISHERS, CHICAGO

DESCRIPTION OF WORK	NO. PIECES	DIMENSIONS	EXTENSIONS	EXTENSIONS	TOTAL ESTIMATED QUANTITY	UNIT PRICE M'T'L	TOTAL ESTIMATED MATERIAL COST	UNIT PRICE LABOR	TOTAL ESTIMATE LABOR COST
1"x 9"x 4½" Brick Paviors in ct. mtr. to GFLOOR									
(16-2+15-2+2/2)		35-4 x 22-10	809						
(13-8+3-10+5-4)				ddt					
Deduct: wants)		17 x 1	–	17					
@ LR. pvg.)		7 x 9	–	63					
@ storage)		10-8 x 5	–	53					
			-133						
			676		680 SF	@ 4.00			2,720
1" Fine port. cement topping on GF slab t/smooth									
		9 x 7			60 SF	@ 2.00			120
0.080" Vinyl-Asbestos Tiling on Upper Floor deck									
		43 x 18	774						
		17½ x 8½	149						
Deduct: wants)			923	ddt					
@ stairs)		8 x 4	–	32					
92-18 @ walls)		74 x 0-4	–	25					
			-57						
			866		870 SF	@ 1.10/SF			960
							SUB-TOTAL	$	3,800
PAINTING									
							CASH ALLOWANCE	$	3,800
Interior painting (Basic)			$ 1,800						
extra: for panelling, trim, etc.			300						
Exterior (not stucco)			1,500						
extra: for priming			200						
							TOTAL	$	7,600

PRACTICAL "STANDARDIZED" FORMS FOR CONTRACTORS
Form 514 MFD. IN U. S. A.

GENERAL ESTIMATE

BUILDING: RESIDENCE
LOCATION: Vancouver BC
ARCHITECTS: E. Kuckein
SUBJECT: SERVICES – 1/1.

ESTIMATE NO.: PRELIM.
SHEET NO.: 19/20
ESTIMATOR: KC
CHECKER: FC
DATE: 1 APRIL '83

FRANK R. WALKER CO., PUBLISHERS, CHICAGO

DESCRIPTION OF WORK	NO. PIECES	DIMENSIONS	EXTENSIONS	EXTENSIONS	TOTAL ESTIMATED QUANTITY	UNIT PRICE M'T'L	TOTAL ESTIMATED MATERIAL COST	UNIT PRICE LABOR	TOTAL ESTIMATE LABOR COST
PLUMBING									
Water main supply & connection					(allow)				350 ✓
Hose bibs & connections					4/No @ 25.00				100 ✓
Fixtures: –									
WCs	(2 No + 1 No)		= 3						
Lav. basins	(2 (twin) + 1)		= 3						
Kit. sink	(1 (double))		= 1						
Bathtub	(1)		= 1						
Shower stall	(1 (complete))		= 1		(CASH ALLOWANCE)				1,500
Piping (incl. pools drains)						(allow)			1,800
Roof drains & RWLs						"			200
Gas HW Tank complete						"			400
						(sub-total:)			4,350 ✓
GAS HEATING (incl. stack) Mains connection; Connection to gas range & HW Tank, complete						(allow)			1,300 ✓
ELECTRIC LIGHTING									
Mains connection, panel, etc.					(underground service)	(allow)			200 ✓
Service recept.	(10 + 12 + 2 + 2)		= 26		40' underground duct @ 2.50				100 ✓
ceiling lights	(10 + 5)		= 15						
switches (int.)	(10 + 8)		= 18						
" (ext.)	(1)		= 1		Wiring to fixtures 64/No @ $8				500 ✓
exterior lights	(2 + 2)		= 4						
			64						
LIGHTING FIXTURES					(CASH ALLOWANCE)				400 ✓
						TOTAL $			6,850 ✓

PRACTICAL "STANDARDIZED" FORMS FOR CONTRACTORS
Form 514 MFD. IN U. S. A.

GENERAL ESTIMATE

BUILDING: RESIDENCE
LOCATION: Vancouver BC.
ARCHITECTS: E. Kuckein
SUBJECT: SUMMARY

PRELIM.
ESTIMATE NO.
SHEET NO. 20/20
ESTIMATOR: KC
CHECKER: FC
DATE: 1 APRIL '83

FRANK R. WALKER CO., PUBLISHERS, CHICAGO

DESCRIPTION OF WORK	~~NO. PIECES~~ PAGE	DIMENSIONS (SHEET No)		EXTENSIONS	EXTENSIONS	TOTAL ESTIMATED ~~QUANTITY~~	~~UNIT PRICE M'T'L~~	TOTAL ESTIMATED ~~MATERIAL~~ COST	~~UNIT PRICE LABOR~~	TOTAL ESTIMATE ~~LABOR~~ COST
SITEWORK	1	1	(Excavation)				$	1,120		
"	2	2	(Gravel, etc)					2,060		
"	3	3	(Removal, backfill, paving)					1,250		
CONCRETE	1	4	(Foundations)					3,170		
"	2	5	(Slabs, Insulation)					3,570		
MASONRY &	1	6	(Chimney)					3,500		
METAL WORK	1	6	(Beam, etc.)					1,070		
R. CARP	1	7	(Sheathing)					2,890		
"	2	8	(Framing walls, decks)					7,080		
"	3	9	(Framing decks)					1,650		
FIN. CARP.	1	10	(Plywood siding/panels)					840		
"	2	11	(Hwd. Plywood)					2,940		
"	3	12	(Window walls)					3,650		
"	4	13	(Openings, Hardware)					4,130		
CABINETS ETC	1	14	(Kitchen, Closets, Linen)					3,120		
PROTECTION	1	15	(Roofing, flashings, Insul.)					4,250		
FINISHES	1	16	(Stucco, Ceramic tiling)					6,300		
"	2	17	(Drywall)					3,910		
"	3	18	(Flooring, Painting)					7,600		
SERVICES	1	19	(Plumbing Gas Heat Electric)					6,850		
							$	70,950		
ADD: Contractor's General Requirements, & Indirect Costs (Overhead & Profit) (20%)								14,250		
							$	85,200		
TOTAL COSTS: (1982)										

6. From the point of view of an owner, discuss the relative costs of making these two estimates (in this chapter) and their relative worth to an owner. (Assume that it required about 6 man-hours to peruse the drawings and the manual and to make the shorter estimate, and that it required about 24 man-hours to make the longer estimate.)

7. Answer the following questions about the residence and the two estimates in this chapter:

(a) Which physical feature of the residence do you consider the most uncommon?

(b) What is the relative effect on total costs of the item that you consider most uncommon?

(c) Make your own evaluation of the difference between a flat roof and a pitch roof (4 in 12). How much difference is there between your evaluation and the amount stated in the example shown?

(d) From the shorter of the two estimates shown has any significant modification been omitted? Explain your answer.

(e) In the designed details of the residence, point out those that that you believe have effected savings in costs. Do you think that these features have any significant effect on the difference between the amounts of the two estimates shown? Explain why and how.

2

Store and Warehouse Building

This reproduction of an actual preliminary estimate illustrates how a fairly detailed estimate can be made from rough preliminary sketches (reproductions are included), and it provides good material for preliminary estimating exercises using approximate quantities of commercial work conceptualized from preliminary design drawings.

The purpose of the proposed building was to integrate the storage of wholesale products (such as appliances) with a display area for wholesale and retail sales. The sketches show, in section, a parking floor at the lowest level, a main floor with a warehouse area in the rear, a mezzanine floor with warehouse area, and a proposed third floor for warehousing to be added in a second phase. The construction is shown as typical, heavy warehouse construction. (A proposed third phase would have added a wood-frame penthouse apartment.) Let us review the estimate sheets.

MAIN FLOOR

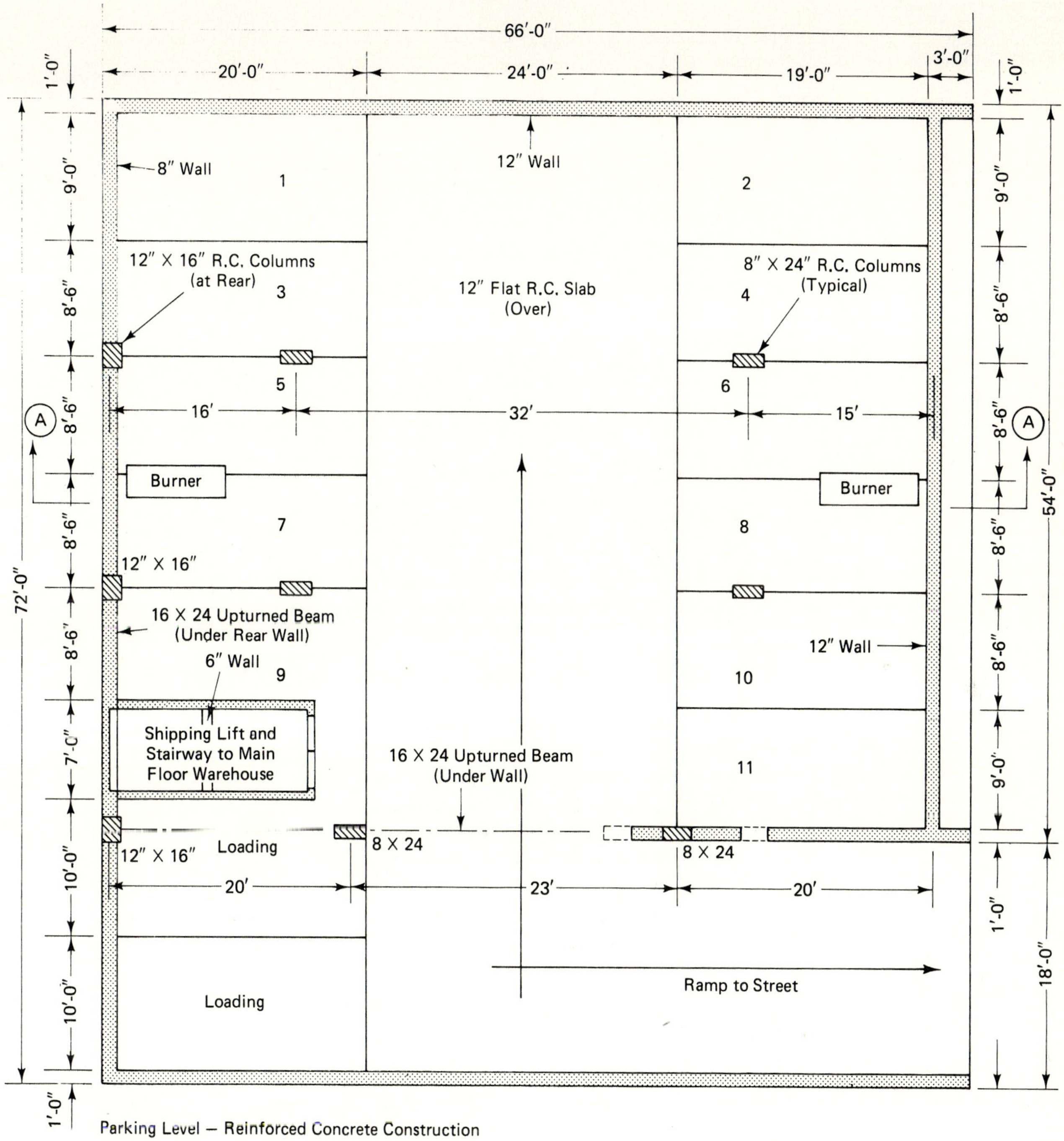

Parking Level — Reinforced Concrete Construction
Capacity 11 Cars, 2 Trucks (One Large and One Small)

PRELIMINARY DRAWING — NOT FOR CONSTRUCTION

SECOND FLOOR

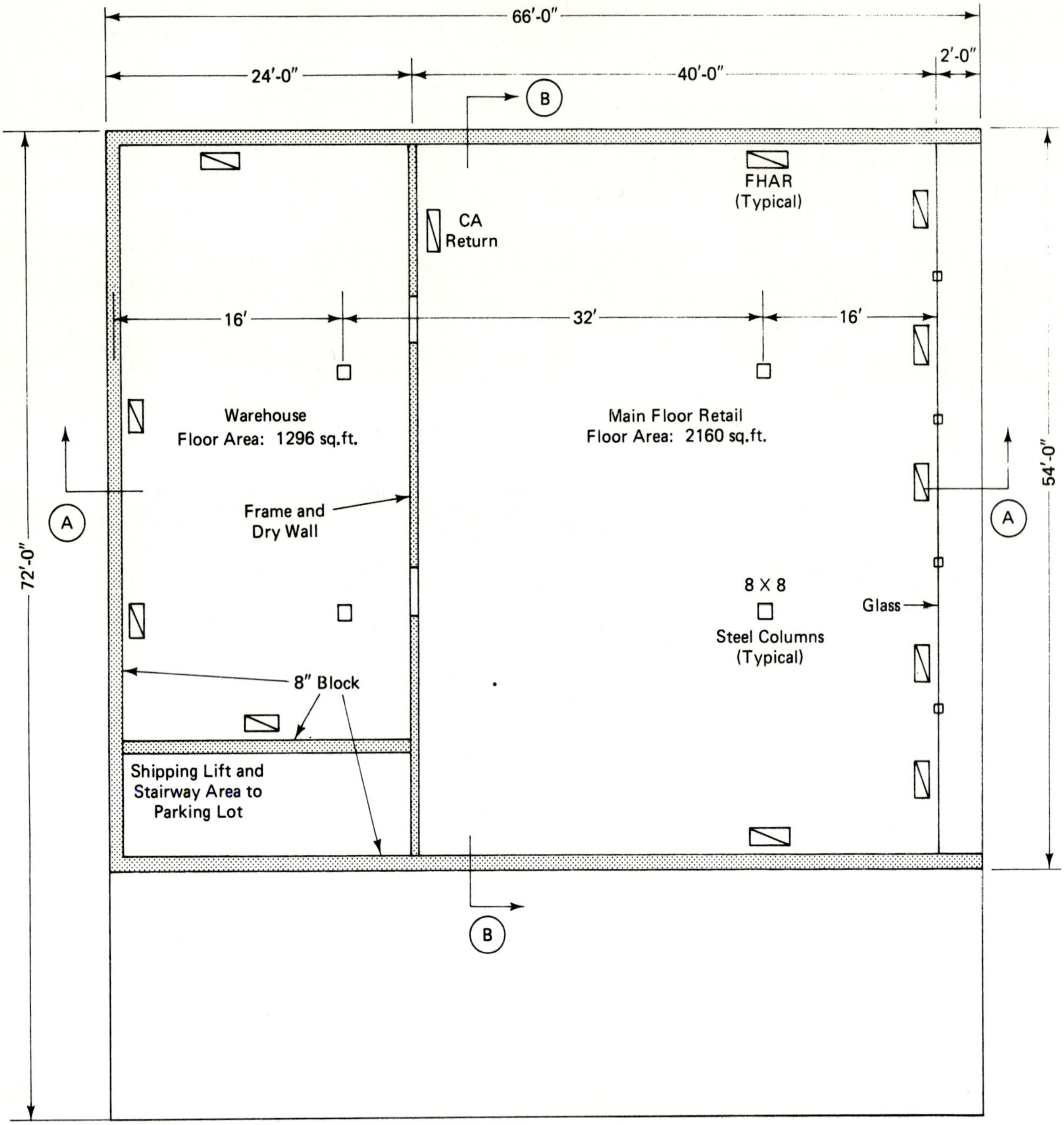

PRELIMINARY DRAWING – NOT FOR CONSTRUCTION

THIRD FLOOR

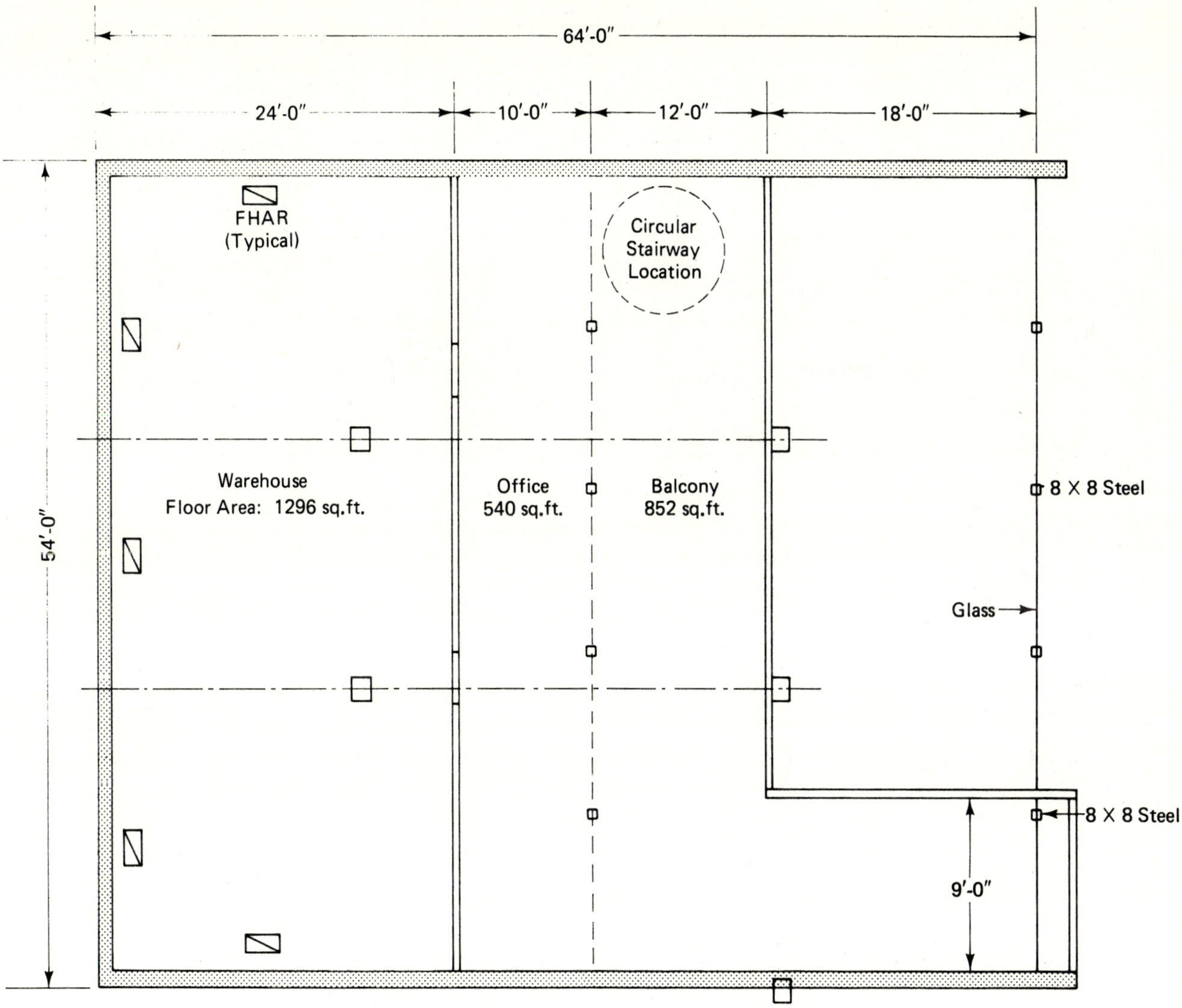

PRELIMINARY DRAWING – NOT FOR CONSTRUCTION

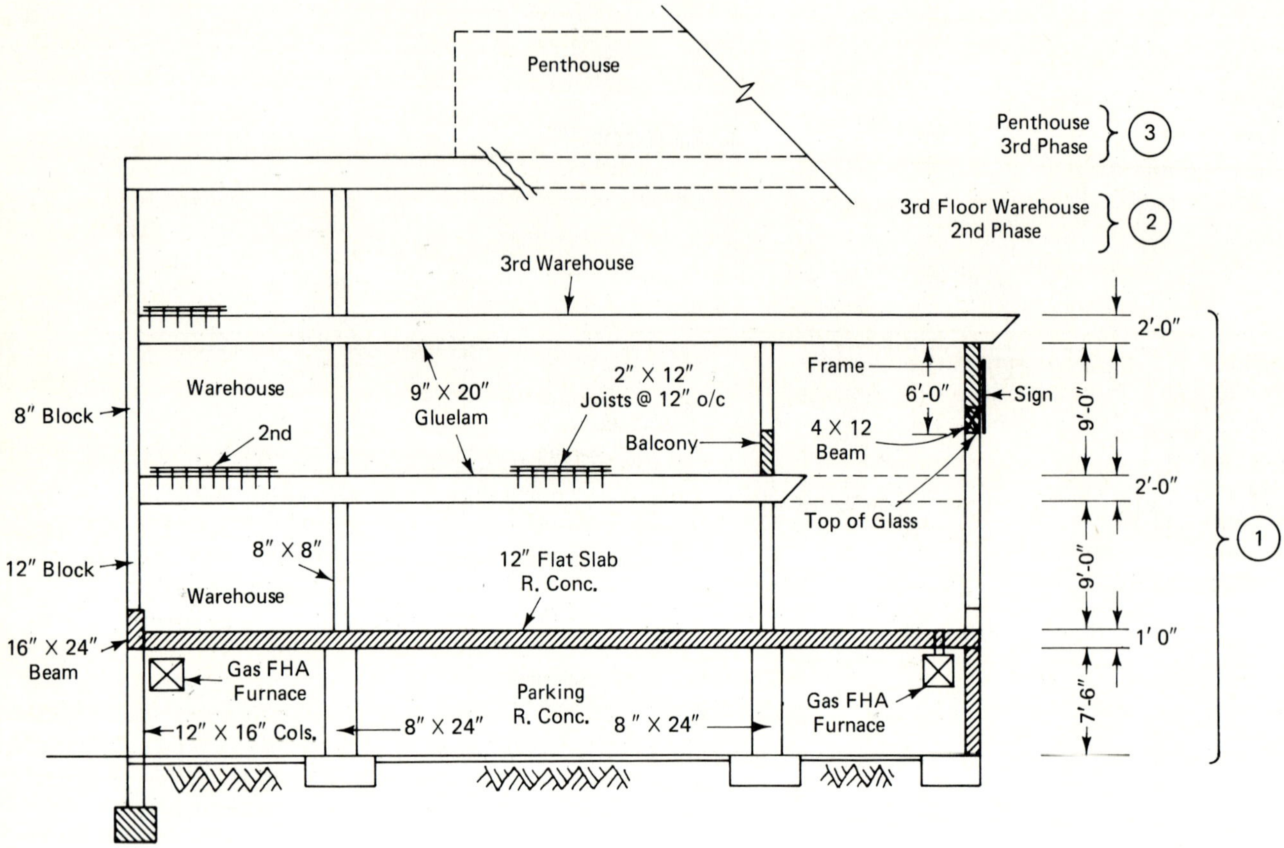

Section A-A

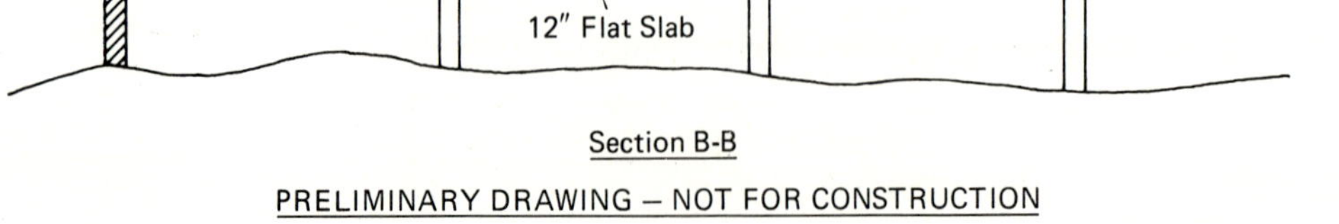

Section B-B

PRELIMINARY DRAWING – NOT FOR CONSTRUCTION

Summary sheet. No markup figures appear here because in this estimate trade markup[G] (for overhead and profit) is included with each division of trade work[G] (15 to 20 percent is added; assuming that project was to be done through Construction Management (CM),[G] the fee for the construction manager's services is also excluded from this summary. Typically it might be 5 to 10 percent of total construction costs.)

Review of the percentages of total costs. The proportion for job overhead costs (general requirements) is typical. Site works are minimal because the building occupies all the urban site area. The proportion for structural work is higher than usual as this is a strong, simple building designed for 100-pound live loads on the floor decks, with few finishes and fixtures. The electrical costs are typical; the mechanical work is minimal.

General requirements. Supervision costs are low because on such a simple building project it was deemed excessive to have a fulltime superintendent. Connections to sewers and to other utilities are typically overlooked in preliminary estimates, and they can be expensive, especially when the local authority requires an owner to be partly responsible for some off-site work, as in this instance. It is prudent to note such items as soon as they come to mind, even if they are better placed elsewhere in an estimate, to ensure that they are not missed.

Site works. The measurement is kept simple and overall. The little niceties of measurement in which an estimator may take pride of accuracy in making a bid estimate are not indulged in here. Yet accuracy is little impaired because the small omissions and additions that are neglected tend to offset each other, providing that the estimator stays aware of them. However, certain items that one may never find in some bid estimates are included in this preliminary estimate, including hand trimming and leveling bottoms of excavations to receive concrete, for example. In this estimate there is over $1,000 worth of trimming, and it is all necessary. Concrete cannot be laid on surfaces left by an excavator's shovel or by a bulldozer's blade.

Reinforced concrete. This is the largest division of work in this estimate, and it deserves comment because concrete is important so often. The measurement is simple and needs no explanation. (Try doing it yourself, by eye, from the drawings, as we proceed. I have omitted some of the side-calculations from the estimate—the aggregation of dimensions, for example—both for the sake of clarity and also to give the serious student something more to think about.) Foundation work (concrete, forms, and rebar) is grouped together to facilitate checking or revision.

Concrete work is generally first measured by length or by area and the total length or total area is used to calculate the total volume so that the total length or total area is available to measure forms and finishes. Small quantities of expensive items such as concrete stairs are measured in cubic feet. Items that are unique and difficult to price, such as stair forms, are not given much time because detailed analysis is impractical, and instead the relatively high but minor costs are estimated intuitively.

Costs of sundry items are allowed for by lump-sums instead of trying to measure and price them individually. This is valid even in bid estimates, providing good judgment is used. The omission of such items leads to short estimates.

Summarizing the costs of concrete (on sheet 6) provides an overall check on the estimate and knowledge and information for future estimates. The estimator must always be creating and supplying himself with this kind of information. Generally we may expect to see the costs of a concrete structure more or less equally divided among the three main ingredients: concrete, formwork, and rebar. In this case these costs are about equal, but we may expect the steel rebar to be higher than usual because of the building's design for heavy live loads. Actually the finishing costs appear a bit too high, although some concrete hardeners can be expensive. But our concern here is primarily with method, not with actual costs.

Masonry. Have you noticed how often the name of this ancient trade is wrongly spelled, with two *a*'s? Few words (I cannot think of another) end with a consonant plus *ry* rather than with *ary*.

Extra over items[G] simplify measurement, and that is their purpose. Because the masonry walls are so important, some costs for additional, special supervision are included.

Metal work. The structural steelwork is priced by weight but at unit prices higher than for framed buildings, as this work is unframed and small in amount. Items such as steel fire escapes are a bother; nobody ever seems to know what they cost. If there is only one in a project, give it a good price and go on. Always allow for the sundry minor items; they always exist, and usually they cannot be found and priced individually (certainly not in a preliminary estimate). I remember that I was amazed the first time in North America that I saw a detailed estimate for a major building project contained several thousands of dollars for miscellaneous metals and that the items were not individually identified and

PRACTICAL "STANDARDIZED" FORMS FOR CONTRACTORS
Form 514 MFD. IN U. S. A.

GENERAL ESTIMATE

BUILDING: STORE & WAREHOUSE
LOCATION:
ARCHITECTS:
SUBJECT: SUMMARY

ESTIMATE NO.: PRELIM. #1
SHEET NO.: Sum/1
ESTIMATOR: KC
CHECKER:
DATE: 04/01/83

FRANK R. WALKER CO., PUBLISHERS, CHICAGO

DESCRIPTION OF WORK	NO. PIECES	DIMENSIONS	EXTENSIONS	EXTENSIONS	TOTAL ESTIMATED QUANTITY	~~UNIT PRICE M'T'L~~	TOTAL ESTIMATED ~~MATERIAL~~ COST	~~UNIT PRICE LABOR~~	TOTAL ESTIMATE ~~LABOR~~ COST
1. General Requirements						#1	15,600	6.0 %	
2. Site Works (Excav; Drains						2	7,380	2.8	
3. Concrete (Forms; Rebar; Finish						3-6	75,480	28.6	
4. Masonry (C Blockwork						7	16,050	6.0	
5. Metal Work (Steel columns; F.E Escape						8	8,355	3.2	
6. Finish Carpentry (Doors; Stairs; Fixtures						8	5,115	1.9	
Rough Carpentry (Decks;						9-10	27,780	10.5	
7. Protection (Roofing & Flashings;						11	7,530	2.9	
8. Metal Windows & Doors (AS QUOTATION)						–	13,800	5.2	
9. Finishes (Drywall; Partitions; Painting)						12	24,210	9.2	
10. Specialties (see Cash Allowances)						–	–	–	
11. Equipment (" " ")						–	–	–	
14. Elevator (AS QUOTATION)						–	20,700	7.8	
15/ Mechanical }							–	–	
16 Electrical } Services: Plumbing						13	7,200	2.7	
Heating						13	10,950	4.2	
Electrical						13	19,320	7.3	
						$	259,470 ✓		
CASH ALLOWANCES									
Work to Sewer & Connection					$ 3,000 ✓				
Store Sign (as quoted)					1,530 ✓		4,530 ✓	1.7	
TOTAL ESTIMATED COSTS (PHASE ONE)						$	264,000 ✓	100.0	

PRACTICAL "STANDARDIZED" FORMS FOR CONTRACTORS
Form 514 MFD. IN U.S.A.

GENERAL ESTIMATE

ESTIMATE NO. PRELIM #1
SHEET NO. 1/13
ESTIMATOR KC
CHECKER
DATE 04/01/83

BUILDING STORE/W'HOUSE
LOCATION
ARCHITECTS
SUBJECT GEN. REQU'S

FRANK R. WALKER CO., PUBLISHERS, CHICAGO

DESCRIPTION OF WORK	NO. PIECES	DIMENSIONS	EXTENSIONS	EXTENSIONS	TOTAL ESTIMATED QUANTITY	UNIT PRICE M'T'L	TOTAL ESTIMATED MATERIAL COST	UNIT PRICE LABOR	TOTAL ESTIMATE LABOR COST
Bonds (bid and 50% performance)							(Allow)		1,200 ✓
Permit (excl plumbing)							"		300 ✓
Layout (by Surveyor)							"		900 ✓
Hoardings									1,500 ✓
Light & power (temporary)									600 ✓
Telephone									600 ✓
Water (temporary supply)									150 ✓
Office supplies									300 ✓
Insurance (Builder's risk)									300 ✓
Clean-up									600 ✓
Supervision (partial) [6 months x 50% x $3,000]									9,000 ✓
Connections to City Sewers ('Storm' in "Site Works")									–
('Soil' as C. Allowance)									–
Documentation									150 ✓
							To Summary $		15,600 ✓

PRACTICAL "STANDARDIZED" FORMS FOR CONTRACTORS
Form 514 MFD. IN U. S. A.

GENERAL ESTIMATE

BUILDING: STORE & W'HOUSE

LOCATION:

ARCHITECTS:

SUBJECT: Site Works

ESTIMATE NO.: PRELIM. #1

SHEET NO.: 2/13

ESTIMATOR: KC

CHECKER:

DATE: 04/01/83

FRANK R. WALKER CO., PUBLISHERS, CHICAGO

DESCRIPTION OF WORK	NO. PIECES	DIMENSIONS	EXTENSIONS	EXTENSIONS	TOTAL ESTIMATED QUANTITY	UNIT PRICE M'T'L	TOTAL ESTIMATED MATERIAL COST	UNIT PRICE LABOR	TOTAL ESTIMATE LABOR COST
Bulk Machine Excavation & Removal					CY				
		70 x 76 x 4	21,280		800	-	-	3.75	3,000
Imported Gravel Fill at founds.					CY				
		300 x 3 x 4	3600		134	9.00	1200	4.50	600
Handtrim & level for Slab					SY				
		70 x 76	5320		600	-	-	1.50	900
Trench Excavtn & Removal									
		120 x 4 x 1	480		CY				
		160 x 3 x 1	480		35		-	7.50	270
Handtrim Bottoms trenches									
		120 x 4	480		SY				
		160 x 3	480		107		-	1.50	150
							$ 1200		$ 4,920
4" Tile Drain @ perimeter (gravel in. sep.)									
		2/66	132						
		2/72	144						
		1/20	20		LF 300	.75	240	1.00	300
Storm Drain Connection (to City Sewer)							(ALLOW)		600
4" Vertical Connections x 36" incl Tee					No. 6	3.75	30		30
4" Bends					8	4.00	30		30
							1,500		5,880
									1,500
							To Summary		7,380

PRACTICAL "STANDARDIZED" FORMS FOR CONTRACTORS
Form 514 MFD. IN U. S. A.

GENERAL ESTIMATE

BUILDING: STORE & W'HOUSE
LOCATION:
ARCHITECTS:
SUBJECT: R CONCRETE - 1/4

ESTIMATE NO.: PRELIM #1
SHEET NO.: 3/13
ESTIMATOR: KC
CHECKER:
DATE: 04/01/83

FRANK R. WALKER CO., PUBLISHERS, CHICAGO

DESCRIPTION OF WORK	NO. PIECES	DIMENSIONS	EXTENSIONS	EXTENSIONS	TOTAL ESTIMATED QUANTITY	UNIT PRICE M'T'L	TOTAL ESTIMATED MATERIAL COST	UNIT PRICE LABOR	TOTAL ESTIMATE LABOR COST
READY-MIXED CONCRETE (3500)									
Continuous Footings									
	1/	120 x 2 x 1-0	240		CY				
	1/	160 x 1½ x 0-9	180		16	46.50	750	7.50	120
Isolated Footings (cols)	4/	5 x 5 x 2½	250		10	51.00	510	15.00	150
Forms to Ftgs < 12" high									
	2/	120	240		LF				
	2/	160	320		560	.30	180	.75	420
Ditto, > 12" high	4/	20 x 2½	200		200	.30	60	.75	150
Rebar in Ftgs [26 cy x 100 lb]					lb 2600	.30	780	.15	390
							2,280		1,230
4" Slab on Grd (tamped) to Parking		64 x 70	4,480 x 0-4		CY 56	46.50	2,610	7.50	420
12" Susp Slab (flat) to Main Floor		64 x 52	3,328 x 1-0		123	51.00	6,270	10.50	1,290
8"x24" Columns	6/	7½		45 x 1-4	CF 60	2.00	120	1.00	60
12"x16" Columns	3/	7½	23						
extra depth)	3/	4½	13 x 1-4		60	2.00	120	1.00	60
12" Walls 66/54		120 x 7½	900 x 1-0		CY 34	51.00	3,270	13.50	860
8" Walls		120 x 7½	900		30				
		40 x 7½	300 x 0-8		(carry fwd.)		$ 14,670		$ 3,920

PRACTICAL "STANDARDIZED" FORMS FOR CONTRACTORS
Form 514 MFD. IN U. S. A.

GENERAL ESTIMATE

BUILDING STORE & W'HOUSE — ESTIMATE NO. PRELIM #1
SHEET NO. 4/13
LOCATION
ESTIMATOR KC
ARCHITECTS
CHECKER
SUBJECT R.CONCRETE 2/4 — DATE 04/01/83

FRANK R. WALKER CO., PUBLISHERS, CHICAGO

DESCRIPTION OF WORK	NO. PIECES	DIMENSIONS	EXTENSIONS	EXTENSIONS	TOTAL ESTIMATED QUANTITY	UNIT PRICE M'T'L	TOTAL ESTIMATED MATERIAL COST	UNIT PRICE LABOR	TOTAL ESTIMATE LABOR COST
R/MX CONCRETE (3500)									
Upstand Beam 16"x12" wide			120(x1-4)		6 CY	51.00	300 ✓	13.50	75 ✓
6" Walls (@ lift)	2/	15 x 7½	225						
	1/	7½ x 7½	56	x 0-6	5 CY	51.00	255 ✓	15.00	75 ✓
Stairs & Landings (8½"R x 12" GOING)									
7 ◺ 11½ 9	1/	4 x 4 x 1	16						
	15/	4 x 1 x 1¼	17		90 CF	3.00	270 ✓	1.50	135 ✓
Sundry Concrete for Trades (Allow)					8 CY	46.50	375 ✓	9.00	75 ✓
							1,200 ✓		360 ✓
FORMWORK									
To 12" Slab (flat)			3330		3330 SF ✓	.51	1,700 ✓	2.00	6,660 ✓
4" Slab Screeds	5/	70	350						
(material only)	6/	64	384		800 LF	.22	180 ✓		–
12" Ditto (ditto)	5/	52	260						
	4/	64	256		600	.30	180 ✓		–
12" Wall Forms	2/	900 SF			1800 SF	.18	330 ✓	1.10	1,980 ✓
8" Ditto	2/	1200 SF			2400	.18	435 ✓	1.08	2,600 ✓
6" Ditto	2/	225 + 2/55			560	.18	100 ✓	1.05	560 ✓
Column Forms					540	.75	405 ✓	2.20	1,200 ✓
Stair Forms					(Allow)		100 ✓		450 ✓
Sundries (Holes, etc)					(")		100 ✓		200 ✓
					To Summary $		4,730 ✓	$	14,010 ✓

PRACTICAL "STANDARDIZED" FORMS FOR CONTRACTORS
Form 514 MFD. IN U.S.A.

GENERAL ESTIMATE

BUILDING STORE & W'HOUSE

LOCATION

ARCHITECTS

SUBJECT R. CONCRETE - 3/4

ESTIMATE NO. PRELIM. #1

SHEET NO. 5/13

ESTIMATOR KC

CHECKER

DATE 04/01/83

FRANK R. WALKER CO., PUBLISHERS, CHICAGO

DESCRIPTION OF WORK	NO. PIECES	DIMENSIONS	EXTENSIONS	EXTENSIONS	TOTAL ESTIMATED QUANTITY	UNIT PRICE M'T'L	TOTAL ESTIMATED MATERIAL COST	UNIT PRICE LABOR	TOTAL ESTIMATE LABOR COST
CONCRETE FINISHING / SUNDRIES									
Spade Finish and Hardner (to ramp)					SF ✓ 4,500	.15	675 ✓	.45	2,025 ✓
Steel Trowel Slab					3,300	–	– ✓	.30	990 ✓
Curing Slabs					7,800	.06	480 ✓	.03	240 ✓
" Walls (and sack Finish) (1800 + 2400 + 450 + 360)					5,000	–	–	.06	300 ✓
6" Gravel Beds		4500 SF / 2250	) x 0.6		CY 84	10.50	900 ✓		–
Consolidate and level Beds					SY 500	–	–	1.50	750 ✓
Control Joints ½" x 3½" & sealant 2(64 + 70 + [2 x 15] + [6 x 4½]) = 382					LF 406	.60	240 ✓	.60	240 ✓
Concrete Testing & Inspection					(Allow)		65 ✓		1,015 ✓
Patching Exposed conc. surfaces					(Allow)		200 ✓		670 ✓
							2560 ✓		6,230 ✓
REBAR IN CONC									
6" x 6" x 10/10 WWF (mesh)		4500 SF x 21# + 10%			16 / 1000 / SQ 45	12.00	540 ✓	.07½	340 ✓
Total CONCRETE Volume 207 CY x 200#					TON 21	600.00	12,600 ✓	240.00	5,040 ✓
Equipment Rental for Concrete (incl. pump)					(Allow)				1,000 ✓
							$ 15,700 ✓		$ 12,610 ✓

PRACTICAL "STANDARDIZED" FORMS FOR CONTRACTORS
Form 514 MFD. IN U. S. A.

GENERAL ESTIMATE

BUILDING STORE & W'HOUSE

LOCATION

ARCHITECTS

SUBJECT R. CONCRETE - 4/4

ESTIMATE NO. PRELIM. #1

SHEET NO. 6/13

ESTIMATOR KC

CHECKER

DATE 04/01/83

FRANK R. WALKER CO., PUBLISHERS, CHICAGO

DESCRIPTION OF WORK	NO. PIECES	DIMENSIONS			EXTENSIONS	EXTENSIONS	TOTAL ESTIMATED QUANTITY	UNIT PRICE M'T'L	TOTAL ESTIMATED MATERIAL COST	UNIT PRICE LABOR	TOTAL ESTIMATE LABOR COST
SUMMARY											
		SHEET	1/4						14,670 ✓		3,920 ✓
		"	2/4						4,730 ✓		14,010 ✓
		"	3/4						15,700 ✓		12,610 ✓
TOTALS		"	4/4						$ 35,100		$ 30,540
COSTANALYSIS					CONC	FORMS	REBAR				
SHEET 1/4	M				13,650	240	780		14,670 ✓		–
	L				2,960	570	390		–		3,920 ✓
" 2/4	M				1,200	3,530	–		4,730 ✓		–
	L				360	13,650	–		–		14,010 ✓
" 3/4	M				2560	–	13,140		15,700 ✓		–
	L				6,230	–	5,380				11,610 ✓
	E				–	–	1,000				1,000 ✓
TOTALS					$ 26,960	17,990	20,690		35,100 ✓		30,540 ✓

CONCRETE	18,170	
FORMS	17,990	
REBAR	19,690	
	$ 55,850	= $270.00/CY
FINISH'G	8,790	
EQUIPM'T	1,000	
	$ 65,640	= $317.00/CY
ADD: CONTRACTOR'S MARK UP (15%)	9,840	
To Summary	$ 75,480	

PRACTICAL "STANDARDIZED" FORMS FOR CONTRACTORS
Form 514 MFD. IN U. S. A.

GENERAL ESTIMATE

ESTIMATE NO. PRELIM #1
SHEET NO. 7/13
BUILDING STORE & W'HOUSE
LOCATION
ESTIMATOR KC
ARCHITECTS
CHECKER
SUBJECT MASONRY
DATE 04/01/83

FRANK R. WALKER CO., PUBLISHERS, CHICAGO

DESCRIPTION OF WORK	NO. PIECES	DIMENSIONS			EXTENSIONS	EXTENSIONS	TOTAL ESTIMATED QUANTITY	UNIT PRICE M'T'L	TOTAL ESTIMATED MATERIAL COST	UNIT PRICE LABOR	TOTAL ESTIMATE LABOR COST
Exterior Grade Hollow Concrete Blockwork											
(2/66-0 + 52)							SF				
12" Walls	1/	184	x	8-0	1472		1470	1.56	2,295 ✓	2.46	3,615 ✓
8" Walls	1/	184	x	11-0	2024						
(up to 3rd.)	1/	23	x	9-0	207		2230	1.11	2,475 ✓	1.50	3,345 ✓
							LF				
EXTRA OVER, 12" Wall for 12"x8" Bond Beam							54	1.50	80 ✓	1.50	80 ✓
DITTO, 8" wall for ditto							54	1.00	55 ✓	1.00	55 ✓
							LF				
Damp course 2/184							370	.30	120 ✓	1.50	555 ✓
									5,025 ✓		7,650 ✓
							SF				
Scaffolding rental							3700	@.15	555 ✓		–
Supervision							(Allow)		–		720 ✓
Sub-totals											
									5,580 ✓		8,370 ✓
											5,580 ✓
											13,950 ✓
ADD: Contractor's MARK-up								(15%)			2,100 ✓
									To Summary		$ 16,050 ✓

PRACTICAL "STANDARDIZED" FORMS FOR CONTRACTORS
Form 514 MFD. IN U. S. A.

GENERAL ESTIMATE

BUILDING: STORE & W'HOUSE
LOCATION:
ARCHITECTS:
SUBJECT: METAL WORK

ESTIMATE NO.: PRELIM. #1
SHEET NO.: 8/13
ESTIMATOR: KC
CHECKER:
DATE: 04/01/83

FRANK R. WALKER CO., PUBLISHERS, CHICAGO

DESCRIPTION OF WORK	NO. PIECES	DIMENSIONS	EXTENSIONS	EXTENSIONS	TOTAL ESTIMATED QUANTITY	UNIT PRICE M'T'L	TOTAL ESTIMATED MATERIAL COST	UNIT PRICE LABOR	TOTAL ESTIMATE LABOR COST
8x8 I Steel Columns					lb				
	4·4/20		160 x 31 lb		5000	.60	3,660 ✓	.20	1220 ✓
Base plate x50# x 8/1					400				
Top plate x50# x 8/1					400				
Glulam Beam Connections				6 No	300				
					6100				
Steel Fire Escape					(ALLOW)		900 ✓		300 ✓
Sundry minor metal items					(ALLOW)		550 ✓		130 ✓
							5,310		1,650 ✓
									5,310 ✓
									6,960 ✓
ADD: Subcontractor's MARK-UP					20%				1,395 ✓
							To Summary	$	8,355 ✓
FINISH CARPENTRY									
1¾" x 3' x 6'8" Intr. flush doors & frames (inc. h'ware)					6 No @ 75.00				450 ✓
Wood Staircase; 4' wide x 11' rise x 18' going, complete						(Allow)			500
Washroom Vanity Table (for lav. basin) 8' x 2' (no cpds.)					1 No @ 200.00				200
3/8" Sanded D Fir ply flooring underlay									
		2547 SF 3456			6000 SF	[.18 .32 .50]		@ .50	3,000 ✓
Sunday trim and finish / h'ware									110 ✓
									4,260 ✓
ADD: SUBCONTRACTOR'S MARK-UP (20%)									855
							To Summary	$	5,115 ✓

priced. Now, I sometimes wonder about the merits of the detailed measurement that goes into contracts with quantities.[G]

Carpentry (second floor). The two floors were measured separately in case changes in design were needed. One must expect revisions while making preliminary estimates. Glued-laminated (glulam) timbers are a mystery to some because of the way that they are priced. Even the cost data books appear to avoid details about them. It seems that they are more common in the West than in the East. One way to price glulam timbers is to measure the laminations in board-feet as though they were made of laminations of nominal size (either 1 1/2 or 1 5/8 inches thick, but measured as 2 nominal inches); add about 20 to 25 percent for laps and waste, and price the gross board measure, as done in this case. A 9-inch by 21-inch beam contains fourteen 1 1/2-inch thick laminations, nominally 2 inches (times 10 inches). (The laps in shiplap sheathing are best allowed for as a constant in the unit price along with an allowance for waste.)

Protection. Dampproofing and roofing are easily measured using the areas of other work. The roofing is only three-ply because another floor was to be added to the building within a few years. Sheet-metal flashings are best measured separately, otherwise they are usually underpriced. Always allow for sundry (and unmeasured) flashings. Insulation also is easily measured from reflected dimensions (areas) of other work.

Finishes. Here the drywall is measured as installed, but no openings are deducted. The unit prices allow for waste. High costs of lath and plaster to columns (for fireproofing) are more dependent on the isolation of the small quantities of work than on the actual amount of area covered; in a preliminary estimate it is as better to make a judgment as to the total costs than it is to try to measure accurately and analyze them. However, there are limits to this approach beyond which estimating becomes a guessing game. The effective estimator knows when to "jump at a price" and when to make the effort to analyze. Painting is measured entirely from either reflected areas or by jumped-at prices based on intuition and experience, especially for painting on complex metal fabrications that are practically impossible to measure.

M & E services. Plumbing costs are estimated by fixtures, not by building area; heating costs, by the heating plant and the distribution systems, not by building volume. Electrical costs are estimated by the number of outlets and switches, plus lump-sum allowances for lighting fixtures, mains, and hot-water tank, not by building area. Building area (or volume) is only useful as a basis for estimating M & E costs when data are available for similar buildings with similar systems. This building is atypical.

Generally. This estimating example is more true to life than the first one because it was a real preliminary estimate; the previous example is artificial in that it was derived from and compared with an earlier quantity survey. The measurement in this example is simpler and less detailed, and more reliance is placed on the unit pricing for the estimate's accuracy and effectiveness. Such an estimate would take about 15 hours of work for an experienced estimator, plus the initial time spent receiving instructions and information from the client. At $50 per hour it would cost a client about $800, that is, about 0.30 percent of the estimated cost of the work.

Although the drawings for this project are rudimentary, they are typical of those with which an estimator often must work, and they are adequate for a preliminary estimate based on fairly accurate but approximate quantities of commonplace work. The serious student should be able to use them for practical exercises.

Exercises

1. Answer the following questions about the project building shown in this chapter:

(a) What is the unit cost per square foot of building area based on the estimated costs provided? Explain your answer in detail.

(b) In the estimate shown, why were not both floor decks measured together, since they are of the same design and construction?

(c) Identify those common job overhead items *not* included in this estimate. Describe them. Can you offer practical reasons why these items were probably omitted?

(d) What other common items were omitted from the estimate? Give valid reasons or reasonable suppositions for these omissions.

(e) How do these omissions affect your result in (a) and how might you adjust that result in order to make the cost information more useful in making future estimates?

2. Suggest another method of estimating job overhead costs and explain why it is not applicable to this project.

3. Suppose that you were asked to make a preliminary estimate for the same building (the store and warehouse) within a period of seven hours; how would you go about it in order to produce the best

PRACTICAL "STANDARDIZED" FORMS FOR CONTRACTORS
Form 514 MFD. IN U. S. A.

GENERAL ESTIMATE

ESTIMATE NO. PRELIM. #1

BUILDING STORE & W'HOUSE

SHEET NO. 9/13

LOCATION

ESTIMATOR KC

ARCHITECTS

CHECKER

SUBJECT CARPENTRY (2ND FLOOR)

DATE 04/01/83

FRANK R. WALKER CO., PUBLISHERS, CHICAGO

DESCRIPTION OF WORK	NO. PIECES	DIMENSIONS	EXTENSIONS	EXTENSIONS	TOTAL ESTIMATED QUANTITY	UNIT PRICE M'T'L	TOTAL ESTIMATED MATERIAL COST	UNIT PRICE LABOR	TOTAL ESTIMATE LABOR COST
9"x20" GLULAM BEAMS					BF				
	2/49 (='4/2x10x98' (+20%)) =				2,750	@ 1260	3,465 ✓	@ 180	495 ✓
D.FIR "CONSTR" GRADE									
2x12 Joists	3/50/20		3000						
	20/10		200	x 2BF =	6,500	@ 400	2,600 ✓	@ 200	1,300 ✓
3/2x12 Laminated Beam (bolted)									
	3·2/20		100	x 4BF	500	@ 400	200 ✓	@ 400	200 ✓
2x4 Ledgers (bolted to beams)									
	2/2/49		196						
	18		18		150	@ 300	50 ✓	@ 300	50 ✓
2x4 Wall Plates (62+46+)			108		75	@ 300	25 ✓	@ 300	25 ✓
1" Shiplap Sheath'g									
(63 - 18 =)		45 x 53	2385						
		18 x 9	162		3000	@ 300	900 ✓	@ 210	630 ✓
2x12 Solid Bridging									
	2/3/45		270		LF				
	1/18		20		300	@ .50	150 ✓	@ 1.00	300 ✓
Bolts & Hangers (material only)					(Allow)		50 ✓		—
				To Sub-Summary (next sheet)		$	7,440 ✓	$	3,000 ✓

PRACTICAL "STANDARDIZED" FORMS FOR CONTRACTORS
Form 514 MFD. IN U.S.A.

GENERAL ESTIMATE

BUILDING STORE & W'HOUSE
LOCATION
ARCHITECTS
SUBJECT CARPENTRY (3RD FLOOR) (ROOF OR)

ESTIMATE NO. PRELIM #1
SHEET NO. 10/13
ESTIMATOR KC
CHECKER
DATE 04/01/83

FRANK R. WALKER CO., PUBLISHERS, CHICAGO

DESCRIPTION OF WORK	NO. PIECES	DIMENSIONS	EXTENSIONS	EXTENSIONS	TOTAL ESTIMATED QUANTITY	UNIT PRICE M'T'L	TOTAL ESTIMATED MATERIAL COST	UNIT PRICE LABOR	TOTAL ESTIMATE LABOR COST
9"x20" GLULAMS a.b. 2/64 (=14/2x10x128'+20%) BF					= 3,600	@1260	4,540	@210	755 ✓
2x12 Joists, ab.	3/65/20		2600	x 2BF =	5,200	@400	2,080	@200	1,040 ✓
2x4 Ledgers ab	2/2/64		256		170	@300	50	@300	50 ✓
2x4 W. Plates, ab.	2/64		128		85	@300	25	@300	25 ✓
1" Shiplap, a.b	64 x 54		3,456		4200	@300	1,260	@210	880 ✓
2x12 S. Bridg'g	2/3/64		384		380 LF	@.50	190	@1.00	380 ✓
Bolts & Hangers					(Allow)		50		—
2x12 + 2x6 Fascia	2/64 2/54		128 108	x 3BF =	800 BF	@500	400	@250	200 ✓
SUB-TOTALS	ROOF/3RD FLOOR (SHEET /)						8,595 ✓		3,330 ✓
	2ND " (" /)						7,440 ✓		3,000 ✓
							16,035 ✓		6,330 ✓
2x4 Framing in Partitions, Balcony-balustrad'g									
partitions' plates)	2/3/54-0		324						
studs)	2/8.42/10-0		1,000						
	3/22-0		66		2,500 BF	@360	750 ✓	@360	900 ✓
	18/10-0		180						
balc. balustrade)	3/62-0		186	R. H'ware (Allow)			135		—
	50/4-0		200				16,920		7,230 ✓
	3/22-0		66						16,920 ✓
	18/4-0		72						
@ toilets)	2/3/10-0		60						24,150 ✓
	2/10/10-0		200						
ADD: CONTRACTOR'S MARK-UP					(15%)				3,630 ✓
							To Summary $		27,780 ✓

PRACTICAL "STANDARDIZED" FORMS FOR CONTRACTORS
Form 514 MFD. IN U. S. A.

GENERAL ESTIMATE

BUILDING STORE & W'HOUSE
LOCATION
ARCHITECTS
SUBJECT PROTECTION

ESTIMATE NO. PRELIM. #1
SHEET NO. 11/13
ESTIMATOR KC
CHECKER
DATE 04/01/83

FRANK R. WALKER CO., PUBLISHERS, CHICAGO

DESCRIPTION OF WORK	NO. PIECES	DIMENSIONS	EXTENSIONS	EXTENSIONS	TOTAL ESTIMATED QUANTITY	UNIT PRICE M'T'L	TOTAL ESTIMATED MATERIAL COST	UNIT PRICE LABOR	TOTAL ESTIMATE LABOR COST
Damp proof conc. fnds			(900+100+)		SF 1,000 @	.45			450 ✓
Wpf. membrane under Slab					4,500 @	~~.25~~ .20			900 ✓
3 Ply HG b/up Roofing (only) (TEMPORARY)									
~~4 Ply HG b/up ROOFING & 1" Insultn~~					SQ				
		64 x 54			35 @	60.–			2,100 ✓
Galv. gravel stop and blocking @ Eaves					LF				
(2/64 + 2/54)					240 @	2.00			480 ✓
Roof Drains					NO 4 @	34			135 ✓
S. Metal Roof Vents (revolving)					NO 2 @	300			600 ✓
Sundry minor flashings (@ stacks)						(ALLOW)			110 ✓
									4,775 ✓
2" Batt Insulation in Roof joists & framing incl. VB									
@ roof)	1/	64 x 54	3456						
@ walls)	2/	53 x 10	1060		SF				
	2/	10 x 10	200		5,000 @	.30			1,500 ✓
									6,275 ✓
ADD: SUBCONTRACTOR'S MARK-UP					(20%)				1,255 ✓
					To Summary				7,530 ✓

PRACTICAL "STANDARDIZED" FORMS FOR CONTRACTORS
Form 514 MFD. IN U. S. A.

GENERAL ESTIMATE

BUILDING STORE & W'HOUSE

LOCATION

ARCHITECTS

SUBJECT FINISHES

ESTIMATE NO. PRELIM. # 1

SHEET NO. 12/13

ESTIMATOR KC

CHECKER

DATE 04/01/83

FRANK R. WALKER CO., PUBLISHERS, CHICAGO

DESCRIPTION OF WORK	NO. PIECES	DIMENSIONS	EXTENSIONS	EXTENSIONS	TOTAL ESTIMATED QUANTITY	UNIT PRICE M'T'L	TOTAL ESTIMATED MATERIAL COST	UNIT PRICE LABOR	TOTAL ESTIMATE LABOR COST
1/2" Gypsum Drywall to Ceilings & Walls taped & filled									
@ ceilings)		(floor areas)	2547						
			3456	6003					
@ walls)	2/2/	53 x 10	2120						
	2/	62 x 4	496						
	2/2/	10 x 10	400	3016	SF 9,000 @	.60			5,400 ✓
Galv. metal Beads		9000/10			(say) LF 1000 @	.45			450 ✓
5/8" G. D'wall (@ Stairs)	2/	22 x 10			SF 440 @	.75			330 ✓
Metal Lath & Plaster Fire Protectn on Cols.						(Allow)			1,400 ✓
Demountable Gypsum / Aluminum Partitions x10'0" high (@ offices)									
	4/	22	88						
	2/	12	24		LF 112 @	60.00			6,720 ✓
Extra for Doorways 3'x7'					No 6 @	90.00			540 ✓
Painting: 2-cts Latex: – on extr. concrete					SF 10,000 @	.25			2,500 ✓
– on intr. conc & block					4,000 @	.20			1,000 ✓
– on drywall					9,000 @	.20			1,800 ✓
– on intr. doors trim						(allow)			500 ✓
– on extr. wood trim						(allow)			300 ✓
– on intr. metal work						(allow)			110 ✓
									21,050 ✓
[FLOORING – by Owner!]					ADD: MARKUP 15%			=	3,160 ✓
							To Summary $		24,210 ✓

Painting quantity take-off:

(on conc.)	(on block)	(on d'wall)	(on wood)
240	1500	9,000	300
120	2200		390
240	3700		65
160			335
200			1090
900			
1800			
2400			
600			
500			
600 = 5800			
+3700			
6800 – 1000 / 9,500			
10,000			

PRACTICAL "STANDARDIZED" FORMS FOR CONTRACTORS
Form 514 MFD. IN U.S.A.

GENERAL ESTIMATE

ESTIMATE NO. PRELIM. #1
BUILDING STORE & W'HOUSE
SHEET NO. 13/13
LOCATION
ESTIMATOR KC
ARCHITECTS
CHECKER
SUBJECT M&E. SERVICES
DATE 04/01/83

FRANK R. WALKER CO., PUBLISHERS, CHICAGO

DESCRIPTION OF WORK	TOTAL ESTIMATED QUANTITY	UNIT PRICE M'T'L	TOTAL ESTIMATE LABOR COST
PLUMBING: W.C.S, complete incl piping	No 2	@ 1,000	2000 ✓
Lav. basins & sinks, ditto	4	@ 750	3000 ✓
Fire hose standpipe (22')	(ALLOW)		300 ✓
3" C I RWLs (4/24) and connections	4	@ 50	200 ✓
Hose bibs and connections	2	@ 100	200 ✓
Testing & permit	(ALLOW)		300 ✓
			6,000 ✓
ADD: SUBCONTRACTOR'S MARK-UP	(20%)		1,200 ✓
	To Summary		$ 7,200 ✓
HEATING:			
Horiz. domestic-type FHA oil HEATERS, (150-200,000 BTU) susp below conc floor slab; incl. metal flue stack x 27'0" high, complete	No 2	@ 3,000.00 EA	6,000 ✓
Add: zone controls	2	@ 300.00	600 ✓
S. METAL DUCTWORK:			
HA outlets & ducts (Main Floor) (4+7)	11	@ 120	1,320 ✓
Ditto (2nd ")	5	@ 120	600 ✓
Vertical Riser x 12'0"	1	@ 60	60 ✓
CA inlets & ducts (Main Floor)	3	@ 60	180 ✓
Ditto (2nd ")	3	@ 30	90 ✓
Vertical Riser x 12'0"	1	@ 60	60 ✓
Oil Storage Tank & feed line	1	@ 600	600 ✓
			9,510 ✓
ADD: SUBCONTRACTOR'S MARK-UP	(15%)		1,440 ✓
	To Summary		$ 10,950 ✓
ELECTRICAL:			
Commercial Wiring complete {100 points 10,000 SF}		@ .60 = $	6,000 ✓
Fixtures, Mains Connect, HW tank; (Allow)			10,800 ✓
			16,800 ✓
ADD: SUBCONTRACTOR'S	(15%)		2,520 ✓
	To Summary		$ 19,320 ✓

outlets switches
3/= 60 2/= 8
3/= 10 4 6 6/2/= 12
80 20
L/sums
Fixtures $ 9,000
Connectn. = 750
HWT = 1050

estimate possible within the available time? Give specific details of your estimating method. Explain its shortfall (if any) when compared with the estimate shown here.

4. Discuss critically and in detail the method and result of estimating the costs of reinforcing steel found in the example.

5. Discuss the estimate's cost analysis of reinforced concrete work, its purpose, and its results. Can a similar approach be applied to any other major part of the work? If you believe not, explain why. If you believe so, give a detailed example using such hypotheses as you need.

6. Explain in detail the reasoning behind the method of applying indirect costs[G] in the estimate. Is the reasoning valid? Explain your answer. Give valid reasons for the variation in markup added to the direct costs of work of the several trades.

7. Criticize the estimating methods applied to the following trade work in the estimate and propose a more effective procedure for each: (a) plumbing; (b) heating; and (c) electrical work.

3

A Masonry Apartment Building

This kind of apartment building is common on the Pacific Coast and in Hawaii. It is built of reinforced concrete slabs and columns with concrete block walls and with such reinforced concrete walls as needed for structural stability; it usually contains up to 20 or 30 suites and is simple and not architecturally prepossessing. The building shown in this case study was the idea of a developer who brought along a thumbnail sketch of the building's plan, and little more—and I believe that he had used his thumbnail to make the drawing. Because of the choice site he proposed a building of better than average quality and wanted to know what it would cost. We discussed specifications. I made some notes and then produced this first preliminary estimate.

Because masonry apartments there were not then common and because I had no directly applicable data with which to work, I decided to use approximate quantities. I copied the rough plan and gave it to a structural engineer to estimate the dimensions and costs of the

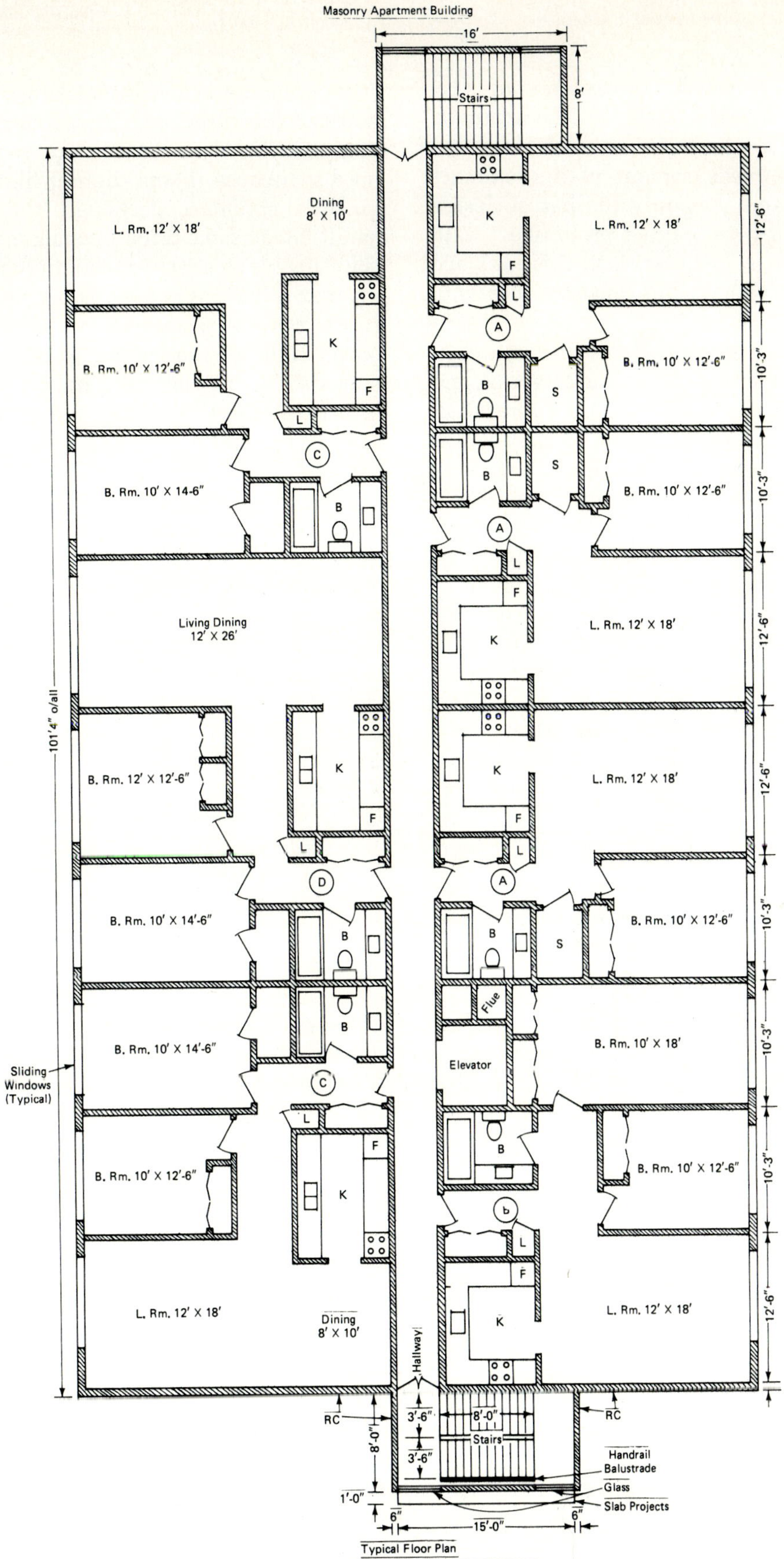

Typical Floor Plan

PRELIMINARY – NOT FOR CONSTRUCTION

Collier – Masonry Apartment Building

structural concrete required. The sketch plan is reproduced here. Following is a copy of the estimate, with comments. This example is included because it shows how simple measurement of approximate quantities can be and that only the barest design information is needed to make a preliminary estimate of an ordinary building or, as in this case, one that was not quite ordinary but consisted mostly of ordinary items of work.

The building contained four floors and 22 apartments. The lowest floor contained a utility area and some apartments. Because the site sloped, the building's utility area was located on the lowest floor, at the back, and below ground, while the apartments at the front were practically at ground level. It was this natural slope that gave the developer the idea of sneaking a four-story building into a site zoned for a maximum of three stories—by calling the lowest floor a basement. When I made the first cost estimate, the developer had not yet established with the local authorities that the lowest floor was, under their interpretation of the law, a basement, but our primary concern is to examine estimating techniques, not building law. To get the general picture, we begin with the estimate's summary sheet.

The costs of *structural concrete work* and *mechanical and electrical services* were obtained from engineering consultants who got them from projects of a similar nature, and the costs were gross amounts that already included all the subcontractors' markup amounts. Consequently it was necessary to add for job overhead costs only to the other remaining trade work. (In retrospect, and especially for a preliminary estimate, I think that was too fine a point and that I should have added a markup to all the costs for *general requirements*.) At the time of making the estimate I had recently negotiated several maximum-cost-plus-fee contracts and I knew the local market and current profit margins fairly well; it was a buyers' market, and I knew that 5 percent was an adequate fee.

Site works 1/2. The actual measurements of the work are not reproduced here as they were done separately from the pricing. The rock excavation averaged about 30 inches in depth with a maximum of about 8 feet in one corner to 0 at the other end. Obviously the excavation work would be expensive. Rock blasting is commonplace on the rugged Pacific Northwest coastline.

Site works 2/2. The parking shelter was designed as a big open carport to be built as cheaply as possible.

Reinforced concrete 1/1. I had asked for an estimate of the costs of this structural work from an engineer, but I also made my own estimate—and used it because it was slightly higher than the engineer's estimate. Although the difference was small, I took some pride in using my own figure and counting his as a second opinion that only confirmed my own. I had based my costs on approximate quantities (totaling 830 cubic yards) while his were based on a single unit price, but admittedly a price supported by plenty of cost data. I had to rely on the structural engineer to tell me which walls needed to be of reinforced concrete and which could be masonry.

Walls and finishes 1/3. Although working with approximate quantities, the items were measured as quasi-elements, that is, as combinations of items of work of more than one trade (the first item—small masonry panels—was an exception because it was expedient: Know what you are doing and do what you will). For example, 8-inch ground floor walls (concrete block) with 1-inch rigid insulation and plaster and paint (two coats) inside; paint (two coats) outside, is an element. The first measurement (370 x 8 feet) is the building's perimeter times the story-height; the end walls, which are of concrete, are deducted later. This as opposed to piecemeal measurement, which can lead to confusion and errors.

Walls and finishes 3/3. For the interior framed partitions there is no detailed measurement of framing members in this case (unlike the residence example). One does whatever appears expedient at the time in making preliminary estimates, and this flexibility is important. There are no absolute rules in estimating, only the precepts proposed in Part 3, and they are not sacrosanct and may be contravened with good reason.

Finishes 1/1. The rest of the finishes are included here. The basic flooring is not described; only the prime cost[G] (P.C.) is stated (actually based on the costs of resilient tiling). A cheaper flooring is assumed for the bedrooms. Proposed carpeted areas are priced at the additional cost extra-over the basic flooring's prime cost.

With finishes on exterior walls introduced at the outset, it was logical to conclude all finishes in the estimate before passing on to other trade work; consequently the standard trade order has been superseded. Some may think this heresy.

Finish carpentry and cabinets 1/1. On the basis of length, cabinets can only be roughly priced. It is better to enumerate the different kinds or to get a lump-sum price per kitchen.

Roofing 1/1. The sheet numbering system (1/1) used in these estimates tells if all sheets are present

PRACTICAL "STANDARDIZED" FORMS FOR CONTRACTORS
Form 514 MFD. IN U.S.A.

GENERAL ESTIMATE

ESTIMATE NO. PRELIM. # 1

BUILDING MASONRY APARTM'TS

SHEET NO. 14/14

LOCATION

ESTIMATOR KC

ARCHITECTS

CHECKER

SUBJECT SUMMARY - 1/1

DATE 1 APRIL '83

FRANK R. WALKER CO., PUBLISHERS, CHICAGO

DESCRIPTION OF WORK	NO. PIECES / DIMENSIONS # #		TOTAL ESTIMATED MATERIAL COST	UNIT PRICE LABOR	TOTAL ESTIMATE COST
Site Works	1/2 (1)	(Excavation / Paving / L'cape)	$ 23,380 ✓		
" "	2/2 (2)	(Parking Shelter)	10,640 ✓		
				=	34,020 ✓
R. Concrete	1/1 (3)	(Structure)			174,300 ✓
Walls & Finishes	1/3 (4)	(Exterior walls)	20,295 ✓		
" "	2/3 (5)	(" " sills, finishes)	19,215 ✓		
" "	3/3 (6)	(Interior walls, block & frame)	66,975 ✓		
				=	106,485 ✓
Finishes	1/1 (7)	(Floors, Painting)			80,300 ✓
Fin. Carp & Cabinets	1/1 (8)				50,345 ✓
Roofing	1/1 (9)				11,190 ✓
Openings (Wdws)	1/2 (10)		18,610 ✓		
" (Doors)	2/2 (11)		24,885 ✓		
				=	43,495 ✓
Specialties	1/1 (12)	(Metalwork, Appliances, Elev.)			68,010 ✓
Services (M&E)	1/1 (13)	(H&V, Plumb'g, Electric'l)			136,400 ✓
					704,545 ✓
General Requirements: 4% of $394,000✓* (approx)			(say)	=	15,755 ✓
*excl. Structure & M&E Services ($704,545 - 310,700)					720,300 ✓
TOTAL DIRECT COSTS OF WORK					720,000 ✓
ADD. INDIRECT COSTS (Contractor's Fee - 5%)					+36,000 ✓
TOTAL ESTIMATED COSTS OF WORK				$	756,000 ✓

PRACTICAL "STANDARDIZED" FORMS FOR CONTRACTORS
Form 514 MFD. IN U. S. A.

GENERAL ESTIMATE

BUILDING MASONRY APARTM'TS
LOCATION
ARCHITECTS
SUBJECT SITE WORKS 1/2

PRELIM ESTIMATE NO.
SHEET NO. 1/14
ESTIMATOR KC
CHECKER
DATE 1 APRIL '83

FRANK R. WALKER CO., PUBLISHERS, CHICAGO

DESCRIPTION OF WORK	NO. PIECES	DIMENSIONS	EXTENSIONS	EXTENSIONS	TOTAL ESTIMATED QUANTITY	UNIT PRICE M'T'L	~~TOTAL ESTIMATED MATERIAL COST~~	UNIT PRICE LABOR	TOTAL ESTIMATE LABOR COST
Excavation in ROCK by blasting and machine, incl. removal (or backfill)									
Bulk (up to 8' dp)					344/CY @		24.00		8260 ✓
Trenches (shallow)					50/CY @		36.00		1800 ✓
Trenches excav by machine and hand in HARDPAN, incl removal					25/CY @		18.00		450 ✓
									10,510 ✓
Concrete (3500 psi) incl. forms & rebar: —									
4" Paving slabs, tinted, incl. dividers					240/SF @		2.00		480 ✓
Steps, in small quantities					45/CF @		7.00		315 ✓
8" Retaining Wall (low ht)					250/SF @		3.00		750 ✓
6" Ditto (low ht, @ gardens)					480/SF @		3.00		1440 ✓
Pavings, incl. beds, 1½" Asphaltic Concrete									
@ Driveway, Parking Area)					930/SY @		4.50		4,185 ✓
Painted Parking Lines							(Allow)		300 ✓
4" dia. Perimeter Drains, incl. gravel					500/LF @		1.80		900 ✓
Landscaping,							(CASH ALLOWANCE)		4,500 ✓
							To Summary $		23,380 ✓

PRACTICAL "STANDARDIZED" FORMS FOR CONTRACTORS
Form 514 MFD. IN U. S. A.

GENERAL ESTIMATE

BUILDING: MASONRY APARTM'TS
LOCATION:
ARCHITECTS:
SUBJECT: SITE WORKS 2/2

PRELIM ESTIMATE NO.:
SHEET NO.: 2/14
ESTIMATOR: KC
CHECKER:
DATE: 1 APRIL '83

FRANK R. WALKER CO., PUBLISHERS, CHICAGO

DESCRIPTION OF WORK	NO. PIECES	DIMENSIONS	EXTENSIONS	EXTENSIONS	TOTAL ESTIMATED QUANTITY	UNIT PRICE M'T'L	TOTAL ESTIMATED MATERIAL COST	UNIT PRICE LABOR	TOTAL ESTIMATE LABOR COST
The following work in PARKING SHELTER:—									
2" T&G. D fir decking, & A G builtup Roofing					1120 SF @	3.90			4370 ✓
5"x10" Glued-laminated Beams					160 LF @	9.00			1,440 ✓
3" dia Tubular Steel Columns, painted					(8 No.=) 60 LF @	1.50			90 ✓
Sets of Steel Plate Caps & Bases					8 No. @	60.00			480 ✓
W. Red Cedar Siding and 2x2 framing in Parapet, w/ 1/2" D. fir ply sheathing to backside and stain finish					500 SF @	3.00			1500 ✓
Parapet Metal Flashings at Cap and Counter flashing (9"+9" girth) & mineral coated Roofing (90lb) to parapet face, incl 3x3 cant & 2x3 cap					200 LF @	7.50			1500 ✓
2" dia x 8'0" long RWLs, incl parapet outlet					2 No. @	30.00			60 ✓
1x1 WR Cedar Battens in vertical screening on 2x2 WRC framing top, bottom, & @ 24" o/c					400 SF @	3.00			1,200 ✓
To Summary $									10,640 ✓

Analysis

$$\frac{\$10{,}640}{1120\,sf} = \$9.50/sf$$

(excluding: Pavings)

PRACTICAL "STANDARDIZED" FORMS FOR CONTRACTORS
Form 514 MFD. IN U.S.A.

PRELIM.
ESTIMATE NO.

GENERAL ESTIMATE

BUILDING MASONRY APARTM'TS
SHEET NO. 3/14
LOCATION
ESTIMATOR KC
ARCHITECTS
CHECKER
SUBJECT R. CONCRETE (STRUCTURE) 1/1
DATE 1 APRIL '83

FRANK R. WALKER CO., PUBLISHERS, CHICAGO

DESCRIPTION OF WORK	NO. PIECES	DIMENSIONS	EXTENSIONS	EXTENSIONS	TOTAL ESTIMATED QUANTITY	UNIT PRICE M'T'L	TOTAL ESTIMATED MATERIAL COST	UNIT PRICE LABOR	TOTAL ESTIMATE LABOR COST
			SF		CF				
6" Flat Slabs (floors/roof)			25,300	(x 0.5)	12,650	✓			
4" Slab on Gr. (floor)			6,200	(x 0.34)	2,066	✓			
8" End Walls			4,300	(x 0.67)	2,867	✓			
6" Intr. Shear Walls (elevator partitions)			1,243	(x 0.5)	622	✓			
			LF						
Columns (8"x24")			706	(x 1.34)	941	✓			
Col. Bases (5'x5'x2') (18 No) =					900	✓			
			LF						
Footings, continuous (24"x12")			545	(x 2.0)	1090	✓			
" (elevator)			144	(x 1.0)	144				
			SF						
8" U/standing Beam/Wall (x30")			379	(x 8" x 30")	633	✓	(@ Parapet & 2nd. floor)		
Staircases (2 No)					440	✓			
					22,353 CF	✓			
				(say)	830 CY @ $210.00*		= 174,300 ✓		

[$174,300 / 24,500 SF = $7.11/SF]

STRUCT ENGINEER'S ESTIMATE:—

24,580 SF @ $6.90 = $169,600
extra: for beam 1,500
$ 171,100

* Unit price from previous projects, includes Gen. Requirements' costs.

To Summary $ 174,300 ✓

PRACTICAL "STANDARDIZED" FORMS FOR CONTRACTORS
Form 514 MFD. IN U. S. A.

GENERAL ESTIMATE

BUILDING MASONRY APARTM'TS

LOCATION

ARCHITECTS

SUBJECT WALLS & FINISHES 1/3

PRELIM. ESTIMATE NO.

SHEET NO. 4/14

ESTIMATOR KC

CHECKER

DATE 1 APRIL '83

FRANK R. WALKER CO., PUBLISHERS, CHICAGO

DESCRIPTION OF WORK	NO. PIECES	DIMENSIONS	EXTENSIONS	EXTENSIONS	TOTAL ESTIMATED QUANTITY	UNIT PRICE M'T'L	TOTAL ESTIMATED MATERIAL COST	UNIT PRICE LABOR	TOTAL ESTIMATE LABOR COST
Hollow Conc Block Extr. Walls									
8" Small panels btwn RC. cols, reinforced (finishes m. sep.)									
	2/3/4/	2-0 x 5-6	264		270 SF	@ 4.80			1,295 ✓
8" Ground Flr Walls wi 1" rigid insulation + plaster + 2-paint (inside) & 2-paint (outside)									
	1/	370-0 x 8-0	2960						
	1/	137-0 x 3-6	480						
	1/	38-0 x 11-6	437	DDT.					
			3877						
DEDUCT (wdws)	1/	34-0 x 5-6	–	187					
"	1/	24-0 x 3-0	–	72					
doors)	3/	3-0 x 7-0	–	63					
			–322						
DEDUCT			3555						
@ end walls	2/	58-0 x 8-0	–928		2630 SF	@ 4.35			11,440 ✓
8" Upper Flrs walls, as last, but wi 2" high x 3/4" projecting block courses (btwn 8" courses) (below windows)									
	2/2/	105-4 x 2-6	1053						
	2/2/	6-0 x 2-6	60						
	2/3/	41-0 x 1-4	328		1440 SF	@ 5.25			7,560 ✓

[NB: R. Conc @ 2nd floor. Finishes = from GF.]

To Summary $ 20,295 ✓

PRACTICAL "STANDARDIZED" FORMS FOR CONTRACTORS
Form 514 MFD. IN U. S. A.

GENERAL ESTIMATE

ESTIMATE NO. PRELIM

BUILDING MASONRY APARTMTS

SHEET NO. 5/14

LOCATION

ESTIMATOR KC

ARCHITECTS

CHECKER

SUBJECT WALLS & FINISHES 2/3

DATE 1 APRIL '83

FRANK R. WALKER CO., PUBLISHERS, CHICAGO

DESCRIPTION OF WORK	NO. PIECES	DIMENSIONS	EXTENSIONS	EXTENSIONS	TOTAL ESTIMATED QUANTITY	UNIT PRICE M'T'L	TOTAL ESTIMATED MATERIAL COST	UNIT PRICE LABOR	TOTAL ESTIMATE LABOR COST
Conc Block Walls, a.b.									
8" Plain Sidewalls @ Entr.									
		31-0 x 9-0	279		280 SF	@ 5.25			1,470 ✓
8" Decorative Screen Walls (of open flue and lintol blocks) painted 2-crs b/sides, in small panels (@ Entr & Parking)									
(@ stairs)	3/	13-6 x 8-0	324						
(@ parking)	1/	35-0 x 9-0	315		640 SF	@ 10.50			6,720 ✓
Two-Flue Chimney Stack of conc. brick w/ fire brick lining, incl plaster finish & 2-painting.									
					40 LF	@ 33.00			1,320 ✓
									9,510 ✓
4"x12" Extr. precast Conc. window sills, w/ 6" wide Intr. quarry tile sills									
	2/	3/100	600						
	3/	4/2	24						
	3/	2/6	36		660 LF	@ 10.50			6,930 ✓
Extr. Stucco finish & 2-paint in small panels btwn windows									
	2/	3/8/2-6 x 5-6			660 SF	@ 3.00			1,980 ✓
Hardwall Plaster & 2-paint in ditto (inside)									
	2/	3/8/2-0 x 5-6			530 SF	@ 1.50			795 ✓
							To Summary $		19,215 ✓

or if any are missing; for example, a sheet numbered 1/1 definitely stands alone, but a sheet numbered (1) has how many sheets following?

Many roofers quote builtup roofing costs including the metal flashings, but because I am so aware of the fallacy of relating perimeter lengths to areas, I invariably measure and price flashings separately. Roofing costs are notoriously variable and while roofing costs usually are a small part of the total costs, it is as well to price roofing high enough. Flashings and their dimensions and specifications vary greatly. Some have standard profiles; others have to be custom made.

Openings 1/1. A quotation for windows was obtained from a window manufacturer, but I also measured and priced them to see how the costs were made up.

Openings 1/2. The doors were required by the developer to be of much better than average quality. The prices include for finish hardware and finishes to doors and frames. Nevertheless, I know that the unit prices could be lower.

Mechanical and electrical services 1/1. These costs were obtained from engineering consultants.

Cost analysis 1/3. On this sheet is an estimate based on a unit cost per square foot of building area derived from similar buildings in another city (about 100 miles away) with an allowance for increased costs due to location. This was to ensure that the detailed estimated costs were within the general range of other known costs. Also shown is a table of the costs calculated as percentages of the total costs. The structure's cost is about 24 percent of total costs, a typical proportion. This is the largest single cost among the divisions of work if we were to separate the trade costs of walls and their finishes.

Cost analysis 2/3. On this sheet all costs are allocated to one of three classes: (1) costs of items *not* related to building area or to suites—the so-called isolated items—(2) costs related to the building's area, and (3) costs directly related to the number of suites. Go through the estimate, sheet by sheet, and you will see how quickly this allocation can be done, providing that one does not attempt to allocate every last dollar to the precisely correct class. For example the parking shelter's size depends on the number of parking stalls and therefore on the number of suites. There is no need to apply the same degree of analysis to the parking shelter building as one is applying to the apartment building. In the concrete structure the end walls and stairs are put into the isolated class because they are not affected by the number of suites or by the building's area and so on. Most of the items in the estimate are naturally grouped in the right class and not much further analysis is needed.

Cost analysis 3/3. On this sheet the allocated costs are resummarized to show the total costs as well as the totals for each of the three classes, or groups of items, and to check the previous analysis. (Rounding off unit prices causes minor discrepancies.) Also on this sheet is shown another analysis using what are called frame-feet to measure the building in another way and to calculate another unit cost, per frame-foot of building. This analysis is included here because it was in the original estimate and because it is of some interest and utility.

The *frame-foot method of cost analysis* was developed in England during the 1950s, when quantity surveyors were searching for quick ways to produce preliminary estimates of educational buildings. This method never became a common tool because it soon developed into the more sophisticated elemental method now widely used in most English-speaking countries. Nevertheless, its existence is worth noting, and its use is sometimes more effective than that of the common building-area method. The principle underlying the frame-foot method is a weighting given to major elements according to their importance to the total costs. For example, above the lowest floor, each floor is given an increment to allow for building height. The roof element is weighted by a factor of 1.50, foundations by 2.0, and so on, as shown.[1] This method is effective for estimating using data from similar buildings but of different heights; otherwise this method might not be much more effective than the simpler building-area method. This method can be used in combination with the three groups of basic elements shown earlier.

Exercises

1. Answer the following questions about the case study in this chapter:
   (a) Which was the better way of estimating the costs of the reinforced concrete structure, and why?
   (b) How are job overhead costs allowed for in this estimate? Is the method used satisfactory? What might be a better method?
   (c) Why is the estimate arranged in an unorthodox order of work? Is this acceptable in this estimate? In any estimate? Explain your answers.

[1] Such factors should be established for different kinds of buildings by the analysis of actual costs, even if the results are "smoothed" and regularized to simplify their application.

"STANDARDIZED" FORMS FOR CONTRACTORS
PRACTICAL
Form 514 MFD. IN U. S. A.

GENERAL ESTIMATE

BUILDING: MASONRY APARTM'TS

ESTIMATE NO.: PRELIM #1

SHEET NO.: 6/14

LOCATION:

ESTIMATOR: KC

ARCHITECTS:

CHECKER:

SUBJECT: WALLS & FINISHES 3/3

DATE: 1 APRIL '83

FRANK R. WALKER CO., PUBLISHERS, CHICAGO

DESCRIPTION OF WORK	NO. PIECES	DIMENSIONS	EXTENSIONS	EXTENSIONS	TOTAL ESTIMATED QUANTITY	UNIT PRICE M'T'L	TOTAL ESTIMATED MATERIAL COST	UNIT PRICE LABOR	TOTAL ESTIMATE LABOR COST
4" Conc Blk Intr Partition Walls flush painted & 2-paint b/sides									
@ lowest flr)	1/	217-0 x 8-0	1736						
+ extra ht.)	1/	127-0 x 3-6	445						
@ upper flrs)	3/	890-0 x 8-0	21360	DDT					
			23,541						
DEDUCT: doors)	101/	2-8 x 8-0	–	2155					
		(4" stud walls)	–	5640					
		(6" c block)	–	1920					
			–9,715		13,800 SF	@ 3.30			45,540 ✓
6" DITTO, but plast'r'd & painted b/sides									
@ L. rooms)	3/5/	16-0 x 8-0			1900 SF	@ 4.65			8,835 ✓
2x4 Framed Stud Partitions w/ ½" Gypsum Drywall & 2-paint b/sides									
9/20-6 = 185 13/40-0 = 520	1/	705-0 x 8-0			5600 SF	@ 2.25			12,600 ✓
							To Summary $		66,975 ✓

PRACTICAL "STANDARDIZED" FORMS FOR CONTRACTORS
Form 514 MFD. IN U. S. A.

GENERAL ESTIMATE

BUILDING: MASONRY APARTM'TS

LOCATION:

ARCHITECTS:

SUBJECT: Finishes 1/1

ESTIMATE NO.: PRELIM #1

SHEET NO.: 7/14

ESTIMATOR: KC

CHECKER:

DATE: 1 APRIL '83

FRANK R. WALKER CO., PUBLISHERS, CHICAGO

DESCRIPTION OF WORK	NO. PIECES	DIMENSIONS	EXTENSIONS	EXTENSIONS	TOTAL ESTIMATED QUANTITY	UNIT PRICE M'T'L	TOTAL ESTIMATED MATERIAL COST	UNIT PRICE LABOR	TOTAL ESTIMATE LABOR COST
Hardwall Plaster & 2-Paint conc soffites									
					21,400 SF	@ 1.05			22,470 ✓
Floor Finish (general) @ P.C. $2.25/SF on conc.									
					15,650 SF	@ 2.25			35,215 ✓
Extra Over last item for Carpet & U/lay (corridors, entrances) @ P.C. $5.25/SF (= $3.00 extra)									
					1500 SF	@ 3.00			4,500 ✓
Floor Finish (bedrooms) @ P.C. $1.65/SF on conc.									
					4850 SF	@ 1.65			8,000 ✓
Floor Finish (landings) @ P.C. $3.60/SF on conc.									
					300 SF	@ 3.60			1,080 ✓
									71,265 ✓
1/4" thick Rubber Stair Treads approx. 42" x 10" incl. 1/8" thick Rubber Risers approx. 42" x 7" high installed									
					60 No	@ 75.00			4,500 ✓
Aluminum Non-Slip Tread Nosings x 36" wide set in flush into conc steps									
					36 No	@ 51.00			1,835 ✓
Painting on concrete (2-cts latex outside)							(ALLOW)		2,100 ✓
Calking & Sealing							(ALLOW)		600 ✓
							To Summary $		80,300 ✓

PRACTICAL "STANDARDIZED" FORMS FOR CONTRACTORS
Form 514 MFD. IN U. S. A.

GENERAL ESTIMATE

BUILDING MASONRY APARTM'TS

LOCATION

ARCHITECTS

SUBJECT FIN. CARPENTRY & CABINETS 1/1

ESTIMATE NO. PRELIM #1

SHEET NO. 8/14

ESTIMATOR KC

CHECKER

DATE 1 APRIL '83

FRANK R. WALKER CO., PUBLISHERS, CHICAGO

DESCRIPTION OF WORK	NO. PIECES	DIMENSIONS	EXTENSIONS	EXTENSIONS	TOTAL ESTIMATED QUANTITY	UNIT PRICE M'T'L	TOTAL ESTIMATED MATERIAL COST	UNIT PRICE LABOR	TOTAL ESTIMATE LABOR COST
Storage Lockers, 2x4 Framing (one side & front) wi hinged door (wi. padlock) covered by galv. 2x2 mesh, size 6'x7'x 8' high					22 No	@ 105			2310 ✓
P. Mahog. Closet Doors fully louvred & compl. wi hardware + clothes rail + two shelves + shoe rack, fitting, complete:					No				
sizes:		3' x 8' o/all			31	@ 135			4,185 ✓
		4' x 8' "			31	@ 171			5,300 ✓
		7' x 8' "			26	@ 270			7,020 ✓
Matching P Mahog solid returned end panel 3/4" thick, 2' x 8' o/all on 1½" x 1½" light frame					35	@ 60			2,100 ✓
									20,915 ✓
Hardwood Kit. Cabinets, "CABCO" (or equal) complete wi hardware, installed:—									
Kit. counter cpd wi. Formica top wi square edge & coved upstand at back edge 24" x 36" nominal size					264 LF	@ 75			19,800 ✓
Kit. Wall cupbds, 12" x 24" nom. size					176 LF	@ 45			7,920 ✓
Utility counter cpbd. wi top a.b.					12 LF	@ 60			720 ✓
Serving hatch fixture wi slide-up door & pocket h'ware + shelf, 36" x 32" nom. size					22 No	@ 45			990 ✓
								To Summary $	50,345 ✓

PRACTICAL "STANDARDIZED" FORMS FOR CONTRACTORS
Form 514 MFD. IN U. S. A.

GENERAL ESTIMATE

BUILDING MASONRY APARTM'TS

LOCATION

ARCHITECTS

SUBJECT ROOFING 1/1

ESTIMATE NO. PRELIM. #1

SHEET NO. 9/14

ESTIMATOR KC

CHECKER

DATE 1 APRIL '83

FRANK R. WALKER CO., PUBLISHERS, CHICAGO

DESCRIPTION OF WORK	NO. PIECES	DIMENSIONS	EXTENSIONS	EXTENSIONS	TOTAL ESTIMATED QUANTITY	UNIT PRICE M'T'L	TOTAL ESTIMATED MATERIAL COST	UNIT PRICE LABOR	TOTAL ESTIMATE LABOR COST
4-Ply A&G built-up Roofing on 1" rigid insul'tn on VB									
(105×62) + (20×8) + (20×9)					6850 SF @	1.20	8,220 ✓		
Galv. Steel Flashings (26 Ga.) pre-painted									
Eaves fascia flashing x 24" girth + lock strip x 6" + conc insert strip + 3x3 cant strip									
2(105 + 62 + 8)					350 LF @	3.60	1,260 ✓		
Parapet counter flashing x 12" girth wi lock strip in conc. over 90# mineral roofing x 30" high over 3x3 cant strip					32 LF @	9.00	290 ✓		
Brass Roof Drain fittings, complete					4 NO @	30.00	120 ✓		
2½" dia C.Iron RWLs wrapped in 1" fiberglas insulation (installed in ducts in conc.) complete wi 4-bends x 36' of all length				(4×36')	4 NO @	325.00	1,300 ✓		
						To Summary $	11,190 ✓		

PRACTICAL "STANDARDIZED" FORMS FOR CONTRACTORS
Form 514 MFD. IN U. S. A.

GENERAL ESTIMATE

BUILDING MASONRY APARTM'TS
LOCATION
ARCHITECTS
SUBJECT OPENINGS (WDWS) 1/2

ESTIMATE NO. PRELIM. #1
SHEET NO. 10/14
ESTIMATOR KC
CHECKER
DATE 14 APRIL '85

FRANK R. WALKER CO., PUBLISHERS, CHICAGO

DESCRIPTION OF WORK	NO. PIECES	DIMENSIONS	EXTENSIONS	EXTENSIONS	TOTAL ESTIMATED QUANTITY	UNIT PRICE M'T'L	TOTAL ESTIMATED MATERIAL COST	UNIT PRICE LABOR	TOTAL ESTIMATE LABOR COST
Sliding Aluminium Windows (mill finish) w/ 7/32" heavy sheet glass & glazing									
(@ sides)	3/2/	84-0 x 5-6	2772						
(@ ends)	3/4/	2-0 x 5-6	132						
")	3/2/	6-0 x 5-6	198						
(@ Lower floor)	1/	58-0 x 5-6	319						
" "	1/2/	1-6 x 8-0	24						
DEDUCT			3,445	DDT					
wants)	3/2/	17-0 x 1-0	–	102					
")	3/2/	35-0 x 1-0		210					
(@ less kit)	3/	8-0 x 2-6		60					
			−372		3,070 SF	@ 4.43			13,600 ✓
EXTRA OVER for Opening Vents (sliding) size 2' x 4'6", incl. h'ware									
(@ sides only)	3/2/9				54 No	@ 40.00			2160 ✓
									15,760 ✓
Aluminium Lights (@ Stairs) incl. sheet glass									
	4/	3-4 x 23-0			310 SF	@ 6.87			2,130 ✓
EXTRA OVER (last) for hinged opening Vents size 36" x 18", incl h'ware					4 No	@ 30.00			120 ✓
EXTRA OVER (last) for tinted glass							(ALLOW)		600 ✓
[Prices based on manufacturer's quotation.]									
							To Summary $		18,610 ✓

"STANDARDIZED" FORMS FOR CONTRACTORS
PRACTICAL
Form 514 MFD. IN U. S. A.

GENERAL ESTIMATE

BUILDING MASONRY APARTM'TS

LOCATION

ARCHITECTS

SUBJECT OPENINGS (DRS) 2/2

ESTIMATE NO. PRELIM #1

SHEET NO. 11/14

ESTIMATOR KC

CHECKER

DATE 1 APRIL '83

FRANK R. WALKER CO., PUBLISHERS, CHICAGO

DESCRIPTION OF WORK	NO. PIECES	DIMENSIONS	EXTENSIONS	EXTENSIONS	TOTAL ESTIMATED QUANTITY	UNIT PRICE M'T'L	TOTAL ESTIMATED MATERIAL COST	UNIT PRICE LABOR	TOTAL ESTIMATE LABOR COST
Doorways, complete w/ doors, frames, finish hardware, finish (factory & site), & installation ("H" grade)									
1¾" Solid hardwood (oak) half-glazed Entrance Doorway, w/ decorative plate glass, 3'x 7'0"					1 No	@ 390.00			390 ✓
1¾" Ditto, but fully glazed, 3'x 7'0"					1 No	@ 420.00			420 ✓
2⅛" Galv. steel CLASS "A" Fire Doorway, 3'x 7'0" (labelled)					2 No	@ 1050.00			2100 ✓
1¾" Ditto, CLASS "B" w/ glazed panel 3'x 7'0"					4 No	@ 360.00			1,440 ✓
1¾" Solid hardwood interior doorway, 3'x 6'8", w/ transom light, 3'x 1'4					102 No	@ 180.00			18,360 ✓
EXTRA OVER for security locksets at entrances (master keyed)					25 No	@ 15.00			375 ✓
1¾" Solid interior double doorways fully glazed w/ ¼" wired plate 5'x 6'8", w/ double transom light over, 5'x 1'4", & w/ two closers					3 PAIR	@ 600.00			1,800 ✓
								To Summary $	24,885 ✓

"STANDARDIZED" FORMS FOR CONTRACTORS
PRACTICAL
Form 514 MFD. IN U.S.A.

GENERAL ESTIMATE

BUILDING MASONRY/APARTM'TS
LOCATION
ARCHITECTS
SUBJECT SPECIALTIES 1/1

ESTIMATE NO. PRELIM. #1
SHEET NO. 12/14
ESTIMATOR KC
CHECKER
DATE 1 APRIL '83

FRANK R. WALKER CO., PUBLISHERS, CHICAGO

DESCRIPTION OF WORK	NO. PIECES	DIMENSIONS	EXTENSIONS	EXTENSIONS	TOTAL ESTIMATED QUANTITY	UNIT PRICE M'T'L	TOTAL ESTIMATED MATERIAL COST	UNIT PRICE LABOR	TOTAL ESTIMATE LABOR COST
M. Steel Picket-style Handrail-Balustrade w/ solid plastic capping (@ landings etc.)					170 LF @	12.00			2,040 ✓
M. Steel flat handrail on brackets					60 LF @	5.00			300 ✓
									2,340 ✓
Kitchen Appliances: P.C. $1200.00 per suite (incl. Frig, Stove, Garburetor,)					22 No @	1200			26,400 ✓
Garbage Chute (3-story)							(ALLOW)		3,600 ✓
Elevator (Hydraulic, 1200#, 100 fpm) Complete, incl builder's work (x4 stops)							(ALLOW)	(PC)	34,000 ✓
								(b/w)	1,670 ✓
							To Summary	$	68,010 ✓

PRACTICAL "STANDARDIZED" FORMS FOR CONTRACTORS
Form 514 MFD. IN U. S. A.

GENERAL ESTIMATE

BUILDING MASONRY APARTM'TS

LOCATION

ARCHITECTS

SUBJECT MECH & ELEC SERVICES 1/1

ESTIMATE NO. PRELIM #1

SHEET NO. 13/14

ESTIMATOR KC

CHECKER

DATE 1 APRIL '83

FRANK R. WALKER CO., PUBLISHERS, CHICAGO

DESCRIPTION OF WORK	NO. PIECES	DIMENSIONS	EXTENSIONS	EXTENSIONS	TOTAL ESTIMATED QUANTITY	UNIT PRICE M'T'L	TOTAL ESTIMATED MATERIAL COST	UNIT PRICE LABOR	TOTAL ESTIMATE LABOR COST
Engineering Consultants' Estimates:—									
Warm Air Heating & Ventilation									
23145 SF @ $2.10/SF						=	48,600 ✓		
EXTRA for individual thermostats 22 @ 300						=	+ 6,600 ✓	=	55,200 ✓
Plumbing									
22 suites @ $1350.00						=	29,700 ✓		
EXTRA for additional sprinkler heads							+ 900 ✓	=	30,600 ✓
Electrical									
Light & power	22 suites		@ $2,000.00				44,000 ✓		
Fixtures	22 "		@ $300.00				+ 6,600 ✓	=	50,600 ✓
							To Summary	$	136,400 ✓

PRACTICAL "STANDARDIZED" FORMS FOR CONTRACTORS
Form 514 MFD. IN U. S. A.

GENERAL ESTIMATE

ESTIMATE NO. PRELIM. #1

BUILDING MASONRY APARTMENTS

SHEET NO. 1/3

LOCATION

ESTIMATOR KC

ARCHITECTS

CHECKER

SUBJECT COST ANALYSIS – 1/3

DATE 1 APRIL 83

FRANK R. WALKER CO., PUBLISHERS, CHICAGO

DESCRIPTION OF WORK	NO. PIECES	DIMENSIONS	EXTENSIONS	EXTENSIONS	TOTAL ESTIMATED QUANTITY	UNIT PRICE M'T'L	TOTAL ESTIMATED MATERIAL COST	UNIT PRICE LABOR	TOTAL ESTIMATE LABOR COST

Gross Building Area = 6145 SF x 4 = 24580 SF

Basic Cost of Building Work 24,580 @ 27.75/SF (*) = 682100

ADD: non-basic items:

* Site Works (except excavtn)	22,500
* Garburetors	4,900
* Individual Thermostats	6,600
	716,100
ADD: Allowance for extra costs due to location (+5%)	716,000
	+36,000 (approx)
TOTAL ESTIMATED COSTS	$ 752,000 ✓

(*) Basic costs per SF in Vancouver BC.

COST BREAKDOWN / PROPORTIONS

	(net)	(+4%)	(+5%)	(%)	(%)
Site Works	34,020	35,381	37,150	4.91	5
Concrete	174,300	—	183,015	24.20	24
Walls & Finishes	186,785	194,256	203,970	26.97	27
Carpentry & Cabs	50,345	52,359	54,975	7.27	7
Roofing	11,190	11,638	12,220	1.62	2
Openings	43,495	45,235	47,500	6.28	6
Specialties	68,010	70,730	74,270	9.80	10
M & E Services	136,400	—	143,220	18.93	19
Totals:	$ 704,545		$ 756,320	99.98%	100 ✓

PRACTICAL "STANDARDIZED" FORMS FOR CONTRACTORS
Form 514 MFD. IN U.S.A.

GENERAL ESTIMATE

PRELIM. ESTIMATE NO. #1

BUILDING MASONRY APARTMENTS

SHEET NO. 2/3

LOCATION

ESTIMATOR KC

ARCHITECTS

CHECKER

SUBJECT COSTANALYSIS - 2/3

DATE 1 APRIL '83

FRANK R. WALKER CO., PUBLISHERS, CHICAGO

COSTS ALLOCATED TO ELEMENTAL GROUPS:

DESCRIPTION OF WORK	#		(1) ISOLATED	(2) AREA	(3) SUITE
Excavation	1		–	10,510 ✓	–
Pavings, Landscap'g	1		12,870 ✓	–	–
Parking Shelter	2		–	–	10,640 ✓
Conc. End Walls Shear walls, stairs (3,749 cf)	3		29,190 ✓	–	–
Conc. Structure (remain.)	3		–	145,110 ✓	–
Walls & Finishes	4		–	20,295 ✓	–
" " (stairs, stack)	5	(remainder)	9,510 ✓	9,705 ✓	–
" " (partitions)	6		–	66,975 ✓	–
Finishes (floors) (stairs)	7		6,335	71,265 ✓	–
" (end walls)	7		2100 ✓	600 ✓	–
Fin. Carpentry & Cabs.	8		–	–	50,345 ✓
Roofing	9		–	11,190 ✓	–
Openings (windows)	10		2,850 ✓	15,760 ✓	–
" (Doors)	11		–	–	24,885 ✓
Specialties	12		68,010 ✓	–	–
Services: (H&V) (Plumbing)	13		–	48,600 ✓	30,600 ✓
(fixtures) (controls) (Elec)			–	44,000 ✓	13,200 ✓
			130,865	444,010	129,670
ADD: Gen. Requirem'ts (4%)*			4,067	8,252	3,435
			134,932	452,262	133,105
ADD: Contractor's Fee (5%)			6,747	22,613	6,655
			$141,679	$474,875	$139,760
				(÷ 24580 SF)	(÷ 22 suite)
				= $19.32/SF	= $6,353/suite

SUMMATION:

$141,679
474,875
139,760
756,314 ✓

*except on: Structure, Services

PRACTICAL "STANDARDIZED" FORMS FOR CONTRACTORS
Form 514 MFD. IN U.S.A.

GENERAL ESTIMATE

BUILDING MASONRY APARTMENTS — ESTIMATE NO. PRELIM #1
SHEET NO. 3/3
LOCATION — ESTIMATOR KC
ARCHITECTS — CHECKER
SUBJECT COST ANALYSIS – 3/3 — DATE 1 APRIL '83

FRANK R. WALKER CO., PUBLISHERS, CHICAGO

DESCRIPTION OF WORK	NO. PIECES	DIMENSIONS	EXTENSIONS	EXTENSIONS	TOTAL ESTIMATED QUANTITY	UNIT PRICE M'T'L	TOTAL ESTIMATED MATERIAL COST	UNIT PRICE LABOR	TOTAL ESTIMATE LABOR COST

CHECK OF COST ALLOCATIONS

(1) Isolated & Exterior Items (Lump Sum) $ 141,679

(2) Area-related Items 24,580 sf @ $19.32 474,886

(3) Suite-related Items 22 suites @ $6,353 139,766

Total Costs (minor difference due to approximations) $ 756,331

STORY-FRAME CONTENT & ANALYSIS OF 22-SUITE BLOCK

						EXTR. WALLS
FOUNDATIONS	5,870	x 2.00	11,740			
BASEMENT FLR	5,870	x 1.00	5,870		Bmt	2949
FIRST FLOOR	5,870	x 1.15	6,751		1-3	8448
SECOND "	5,870	x 1.30	7,631		part. parapet	145
THIRD "	5,870	x 1.45	8,512			11,542
ROOF-DECK	6,510	x 1.50	9,765			
EXTR. WALLS	11,542	x 1.00	11,542			

TOTAL: STORY-FRAME AREA 61,811 frame-feet

UNIT COST per FF { $756,000 / 61,811 ff = $12.23/ff }

2. Prepare a detailed but hypothetical analysis and estimate of the costs of job overheads for such a project as the one in the example with brief explanations and justifications for each item.

3. Using the estimate in the example, make the best estimate that you can for a similar apartment building containing 30 apartment units. Before doing your estimate, write a descriptive outline of the method that you intend to use, explaining why you believe it the best method available, and a brief outline specification of the work.

4. Using the information in the estimate, reconstruct a truly elemental estimate for the same apartment building according to the standard list of elements provided in this textbook. Write a brief comparison of your elemental estimate with the estimate in this example. What does this comparison tell you about preliminary estimates based on elements and about approximate quantities generally?

5. Looking at the estimate's cost analysis, sheet 1/3, critically discuss the validity and effectiveness of the check estimate made with a unit price and the building's total area. Suggest and demonstrate an alternative way of checking the original estimate.

6. Describe in detail a method of checking the economic feasibility of developing the apartment building in this example by working through a hypothetical example of a feasibility study, using data and information that is valid for your area. In your final results show (a) a hypothetical but realistic cost for the site; (b) total "hard" and "soft" development costs for the project; (c) a projected income flow from the property rented over the next 50 years; (d) the hypothetical financial position of the developer at completion of construction at 1 year later, 2 years later, 10 years later, 25 years later, and 50 years later.

4

A Frame Apartment Building

The three-story frame apartment building standing on a reinforced concrete parking structure (more or less underground) is common on the Pacific Coast. In the East it is more likely to have exterior masonry walls, but that difference affects the estimating of costs only slightly. The apartment block illustrated by the working drawings related to this chapter is typical—except that between the parking level and the apartments there is a commercial floor of retail stores. For our purposes we can include or exclude this commercial floor. We shall ignore it in this chapter and deal with it in the next.

Because this kind of building provides so much of the general rental accommodation and because it is economical to build, economics play a large part in the design, construction, and costs of these apartment buildings. Consequently their costs and the costs of their elements and items of work are common currency, with a few exceptions. Almost always, at least in the Pacific Northwest, they are built by nonunion labor, often by part-time developers who act as their own construction managers and who contract for and coordinate the various trade works. Economic constraints limit variation in building design; consequently costs of the relatively standard elements and items of work are commonly estimated and quoted per suite. Likewise total costs are often expressed per suite or per square foot of building area. There are, however, risks in this practice if certain features are not considered and accounted for, for example the proportions of bachelor, one-bedroom, and two-bedroom suites in a building; the number of corners in a building's plan; the relationship between a building and its site; the amount and kinds of site work required; and the type of balconies, if any.

Because of such variable features it is better to estimate apartment building costs—even for the briefest and earliest of estimates—by *elements*. Here is a typical list with some illustrative costs:

Earthworks	$ 7.50 per CY
Landscaping	$ 3.05 per SF
Site services	cash allowance
Demolition	cash allowance
Piling	cash alowance
Parking (covered) complete	$10.20 per SF
Apartment floors	$20.10 per SF
Apartment exterior walls	$ 5.55 per SF
Apartment roof	$ 2.14 per SF
Balconies	cash allowance
Furnishings, fixtures, appliances	cash allowance
Elevator (four stops)	$45,000. each
General requirements (division 1)	cash allowance

Usually the costs of those elements priced by a cash allowance[G] are too variable to be fixed because they are dependent on the possible choices of the owner-designer or of the buyer or on unpredictable conditions such as those of site or market. Earthwork costs are valid only for a known locality (which is, of course, true for all costs but especially those of site works). The landscaping costs are based on the usual minimum requirements of government-backed mortgagees. The parking costs are for a utilitarian single-deck underground reinforced concrete structure with no finishes other than painted parking lines. Other unit prices are for complete elements *including all finishes* but excluding those items indicated by other elements listed, such as furnishings, fixtures, and appliances.

There is still room for variations in quality and cost among all of these elements and an estimator can use this elemental approach successfully only if he is dealing with a number of such buildings in a known market. Under those conditions such brief elemental estimates can be surprisingly accurate. For a plan with more than four corners, unit prices related to the exterior walls and parking should be increased. Similarly costs of unusual site works must be added, as well as the costs of any of the following:

1. Unusually large footings for poor subsoil conditions
2. Special retaining walls at building or on site
3. Unusually long or complicated access ramps to parking
4. Unusually long spans of beams and slabs
5. Masonry work, other than internal fire-separation walls
6. More than minimal quality handrails and railings
7. Decorative screens, grilles, and such
8. Other than a flat wood-framed roof deck
9. Any finish carpentry work of wood, such as moldings, trim, liners and other millwork
10. Roof membranes other than average felt, asphalt, and gravel
11. Roof gardens, roof decks, and the like
12. Other than standard residential double-glazed metal windows
13. Other than average factory-made doors and pressed steel frames
14. Finishes other than of average quality (painted drywall, cheap carpet)
15. Other than average quality plumbing and lighting fixtures
16. Other than average quality appliances (three per dwelling).

The list is not complete and is intended only to give an idea of the items that an estimator must consider when making such an estimate.

After an initial elemental estimate, which can usually be made in a few hours, the next estimate should be from approximate quantities of trade work, possibly using elemental prices for M & E work; in fact, quite a lot of prices in the intial estimate may be valid for the second estimate. Ways of measuring trade work are described in Part 3, and you may want to read from there before tackling a preliminary estimate using approximate quantities.

Turning to the frame apartment building and its drawings in this book, the careful reader will find some discrepancies among the drawings; yet the building was built from them, and as drawings for

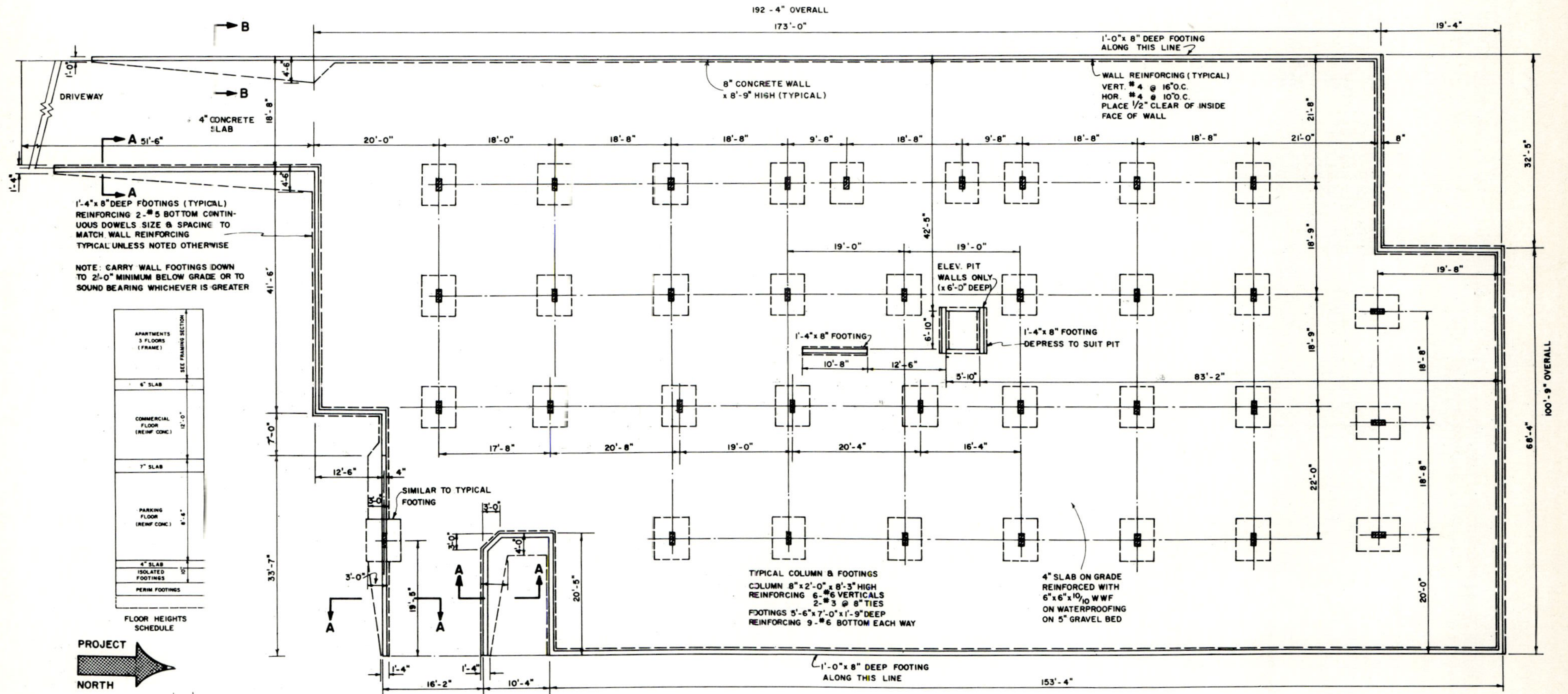

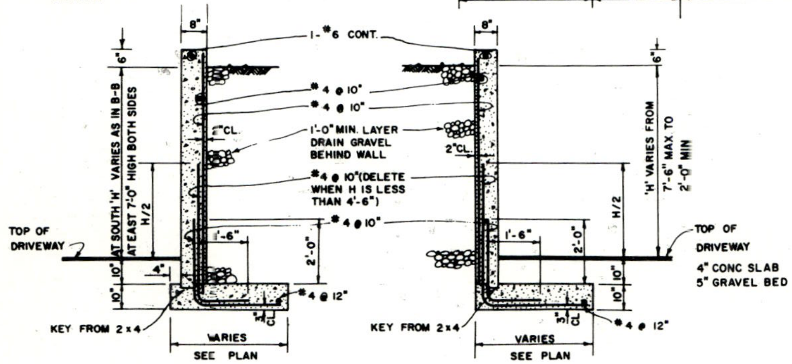

SECTION A-A

SECTION B-B

FOUNDATION PLAN

NOTE: ADDITIONAL CONCRETE WORK NOT SHOWN ON DRAWINGS:

(1) DRAINAGE PITS-SIZE 3'-0" x 2'-0" x 3'-6" DEEP WITH 8" WALLS (TO BE LOCATED AS DIRECTED AT SITE)

(2) ALLOW FOR THE FOLLOWING CONCRETE WORK NOT SHOWN ON DRAWINGS TO BE PLACED AS DIRECTED AT SITE AS SPECIFIED: –

4" THICK EXTERIOR PAVING....................27 CY
FRONT ENTRANCE STEPS..........................3 CY
CONCRETE IN CONJUNCTION WITH MECHANICAL & ELECTRICAL SERVICES UNDER GROUND (SEE MECHANICAL & ELECTRICAL DRAWINGS).........................35 CY

GENERAL NOTES

CONCRETE EXCEPT WHERE NOTED, TO HAVE MINIMUM COMPRESSIVE STRENGTH OF 3000 P.S.I. AT 28 DAYS, MAXIMUM AGGREGATE TO BE 1 1/2"

REINFORCING STEEL TO BE SUPPORTED ON SUITABLE CONCRETE SUPPORTS SPACED NOT MORE THAN 4'-0" IN EITHER DIRECTION AND TO BE FIRMLY FASTENED IN PLACE BEFORE POURING OF CONCRETE IS COMMENCED CLEARANCES AS SHOWN ON THE DRAWINGS MUST BE MAINTAINED.

CONCRETE SHALL HAVE A SLUMP OF NOT MORE THAN 3" AND MECHANICAL VIBRATORS MUST BE USED THROUGHOUT.

REINFORCING STEEL TO BE INTERMEDIATE GRADE DEFORMED BARS IN ACCORDANCE WITH THE LATEST C.S.A. SPECIFICATION G-30 , DEFORMATIONS TO A.S.T.M. SPEC. A-305-507

REINFORCING PROTECTION TO BE: -

- 3" FOR FOOTINGS WHERE CONCRETE IS DEPOSITED AGAINST GROUND;
- 2" FOR FORMED SURFACES EXPOSED TO WEATHER OR IN CONTACT WITH GROUND;
- 1 1/2" FOR BEAMS & COLUMNS
- 3/4" FOR SURFACES OF SLABS AND WALLS NOT DIRECTLY EXPOSED TO GROUND OR WEATHER.

UNLESS OTHERWISE SHOWN OR NOTED SPLICES IN REINFORCING STEEL ARE TO BE: -

- SLABS ON GRADE AND HORIZONTAL WALL STEEL – 36 ϕ;
- VERTICAL COLUMN AND WALL STEEL – 20 ϕ;
- ALL TENSION SPLICES – 36 ϕ
- ALL COMPRESSION SPLICES – 20 ϕ

UNLESS OTHERWISE NOTED, 2-#5 BARS ARE TO BE PLACED AROUND OPENINGS IN SLABS AND WALLS AND SHALL EXTEND 2'-0" BEYOND OPENING

CORNER BARS ARE TO BE PROVIDED TO MATCH HORIZONTAL WALL REINFORCING.

STAGGER ALL LAPS IN HORIZONTAL WALL REINFORCING – 36 ϕ

APARTMENT BLOCK ①

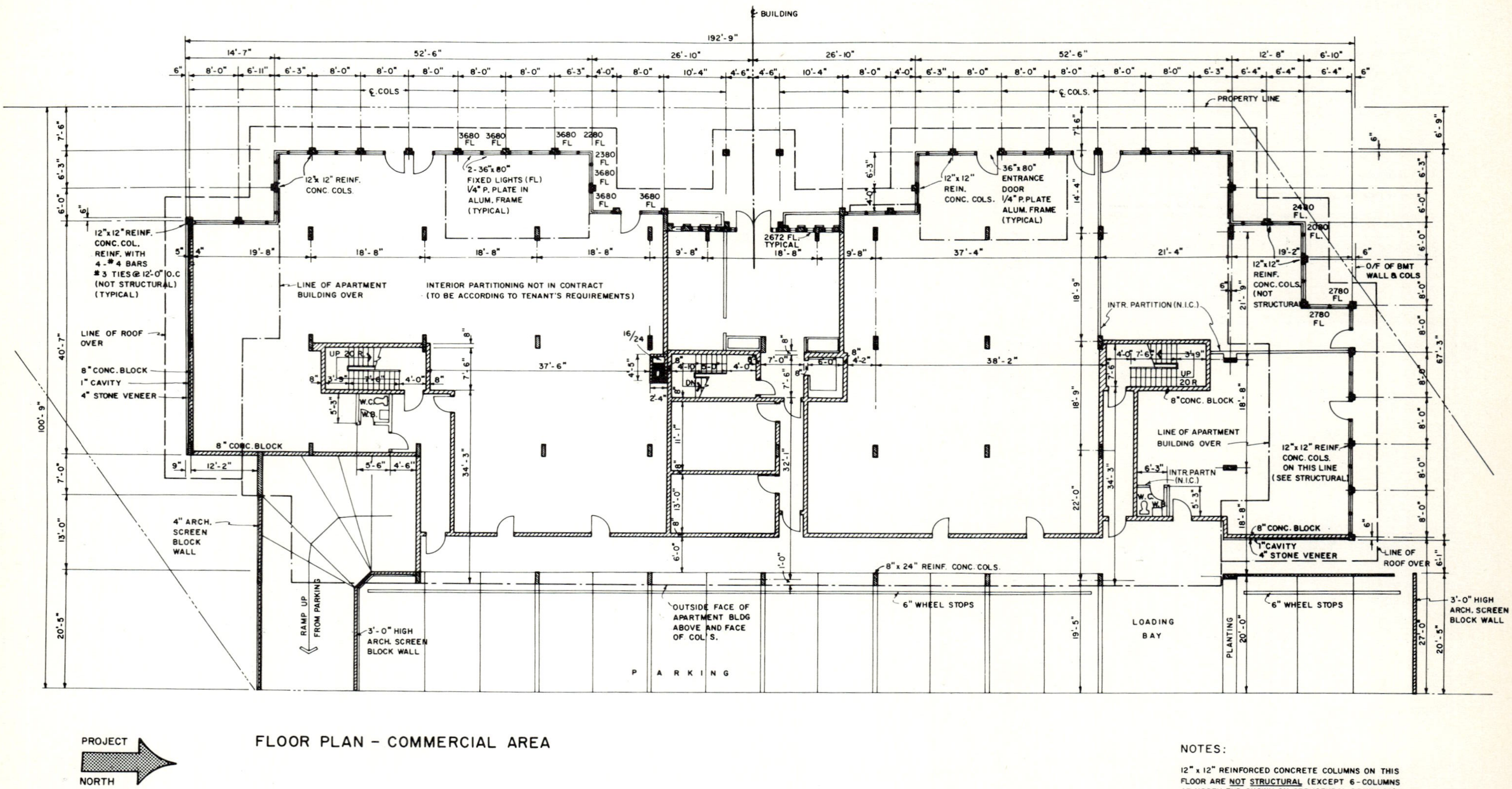

FLOOR PLAN – COMMERCIAL AREA

NOTES:

12" x 12" REINFORCED CONCRETE COLUMNS ON THIS FLOOR ARE NOT STRUCTURAL (EXCEPT 6-COLUMNS AT NORTH END SHOWN ON STRUCTURAL DRAWINGS) CHECK LOCATION OF COLUMNS WITH ARCHITECT BEFORE INSTALLING.

EXTR DOORS AT REAR OF RENTAL SPACES TO BE LOCATED TO SUIT TENANTS.

INTR PARTITIONS (OTHER THAN CONC BLOCK) ARE NOT IN CONTRACT.

APARTMENT BLOCK ②

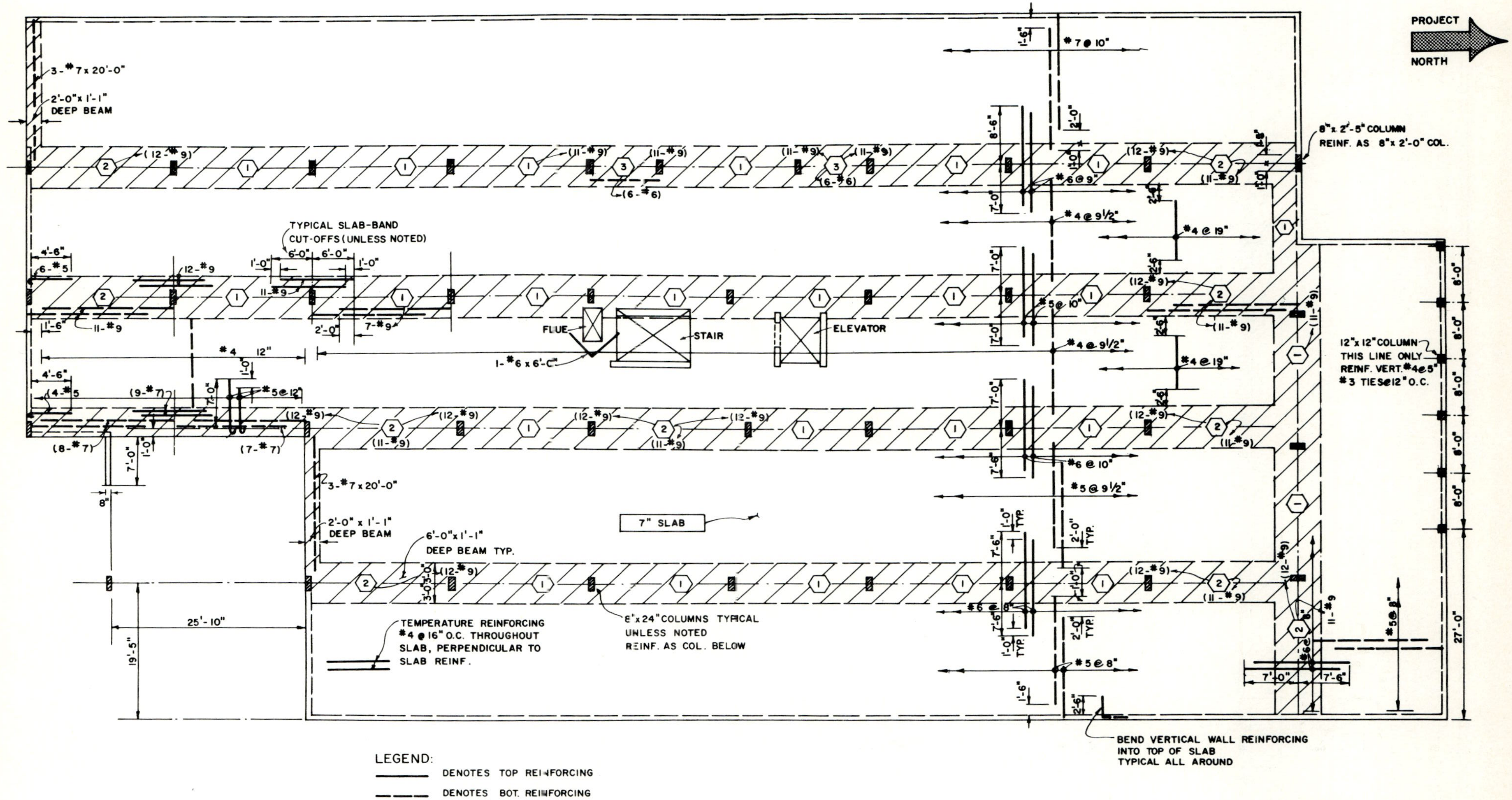

CEILING PLAN - PARKING STRUCTURE

APARTMENT BLOCK (3)

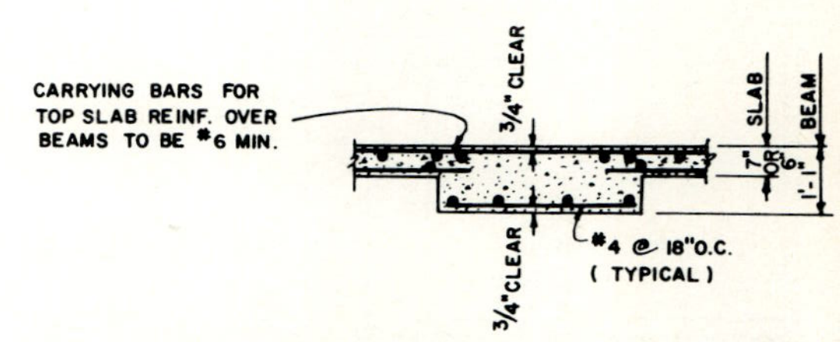

TYPICAL SLAB-BAND SECTION

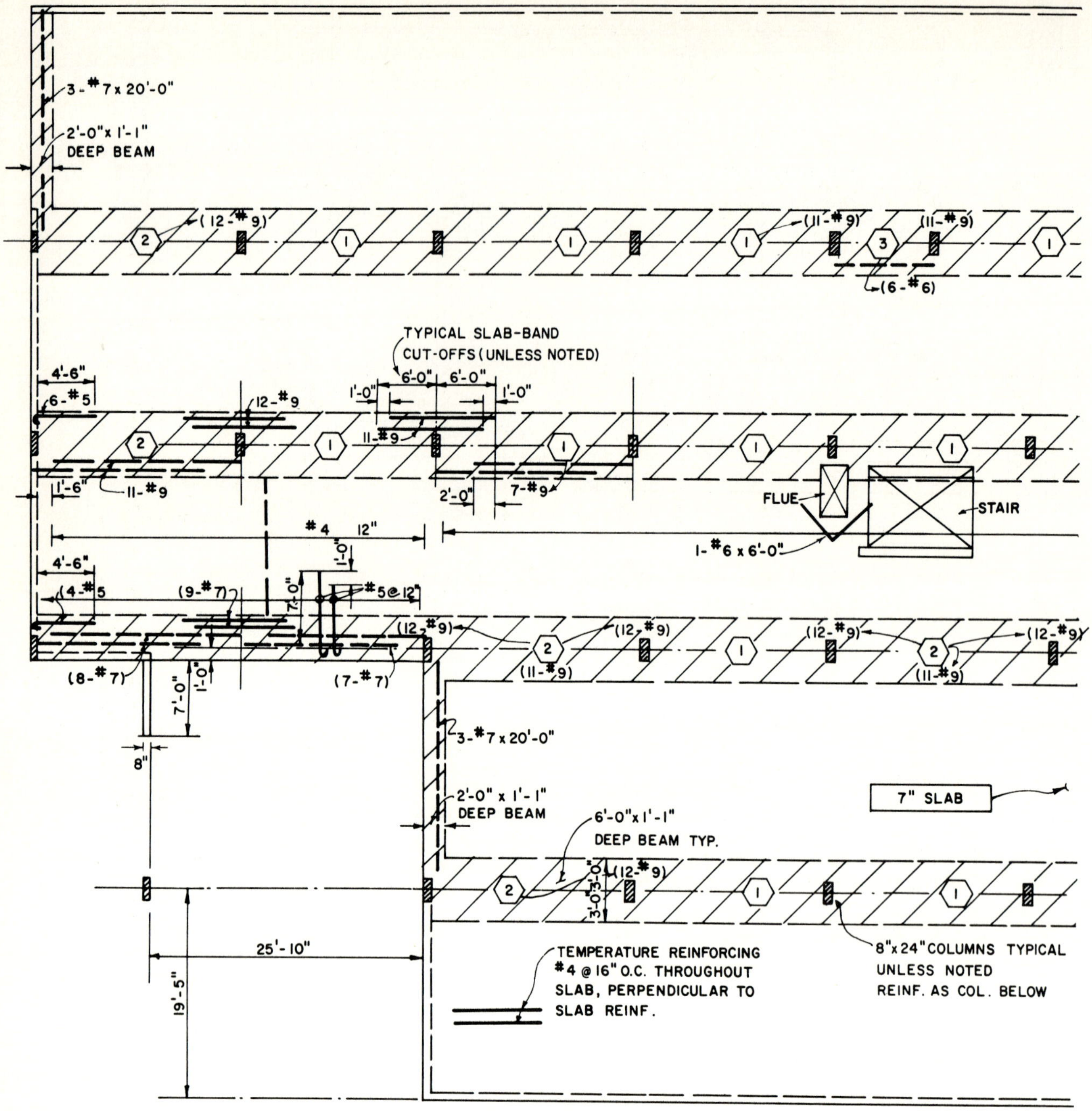

LEGEND:
——— DENOTES TOP REINFORCING
– – – DENOTES BOT. REINFORCING

CEILING PLAN - PARKING STRUCTURE

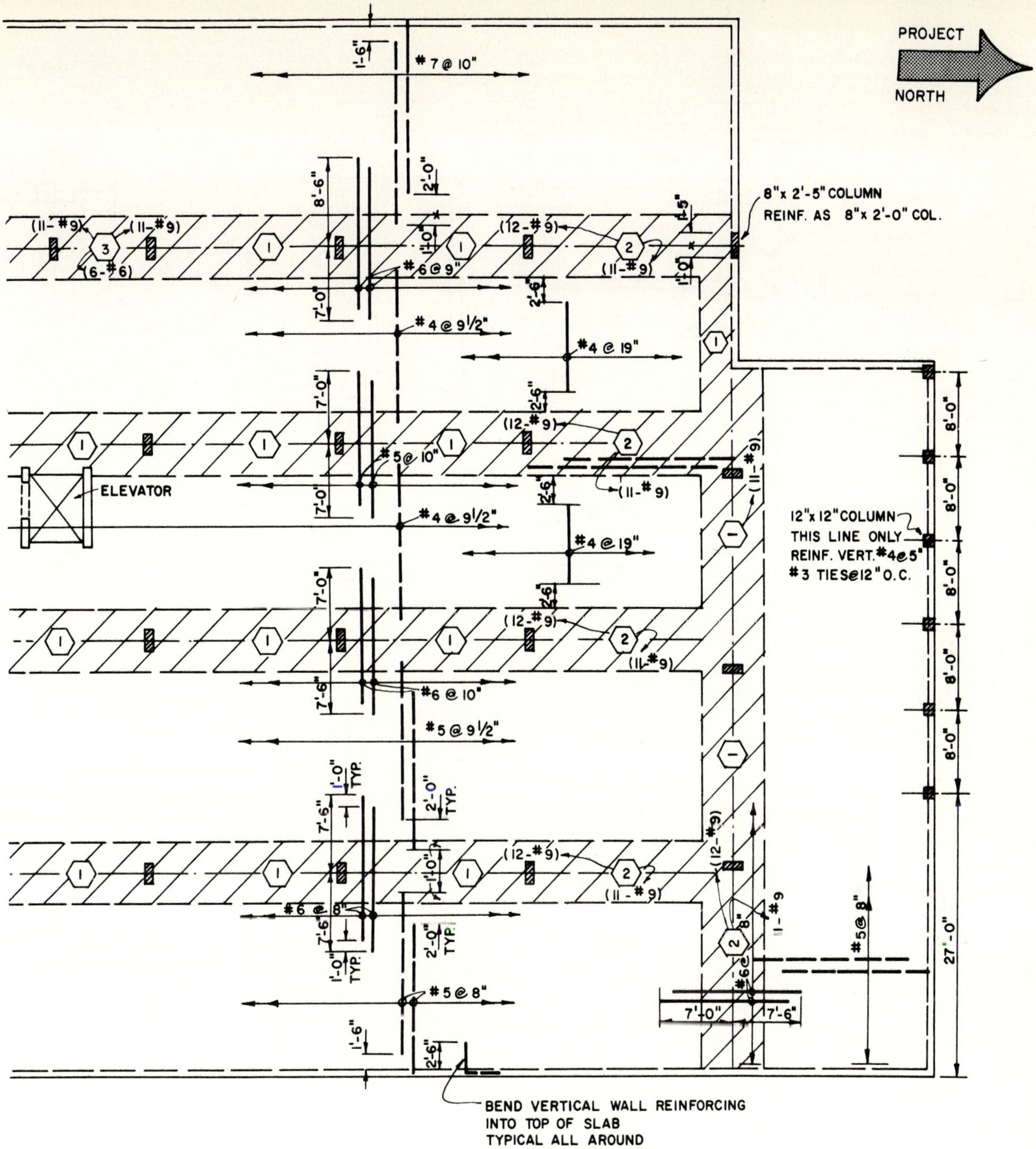

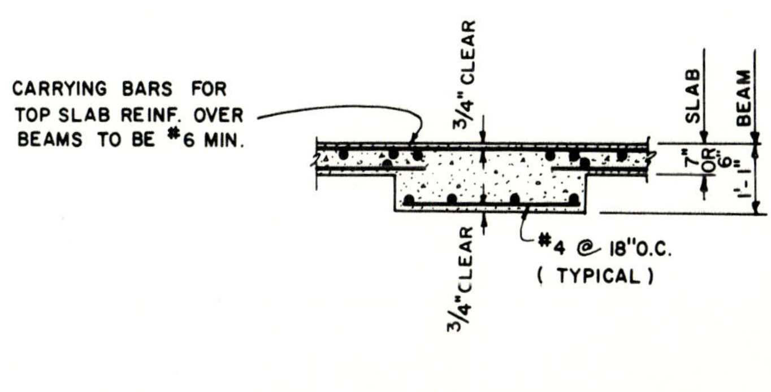

TYPICAL SLAB-BAND SECTION

APARTMENT BLOCK 3

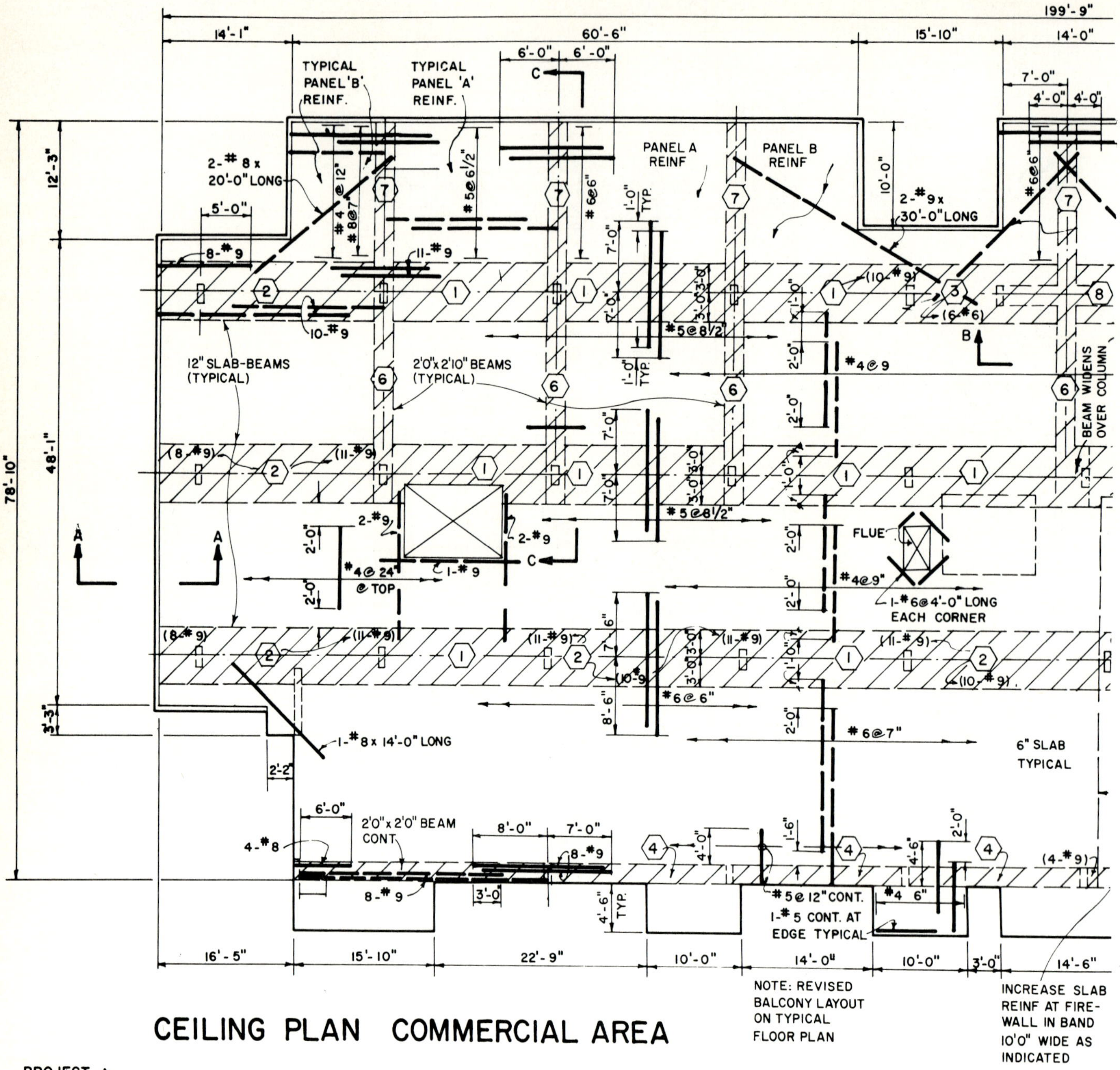

CEILING PLAN COMMERCIAL AREA

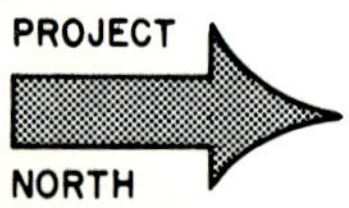

LEGEND:

——— TOP REINFORCING

— — — BOTTOM REINFORCING

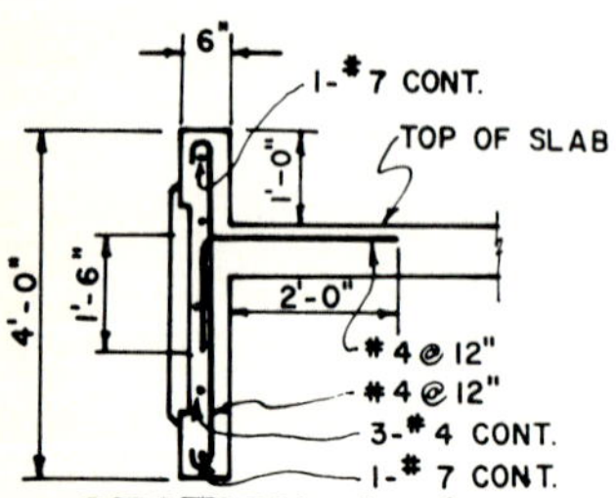

SECTION A-A

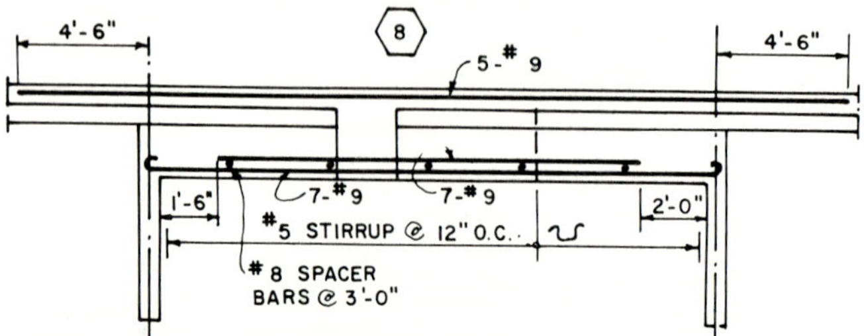

SECTION B-B

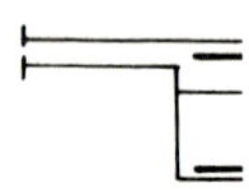

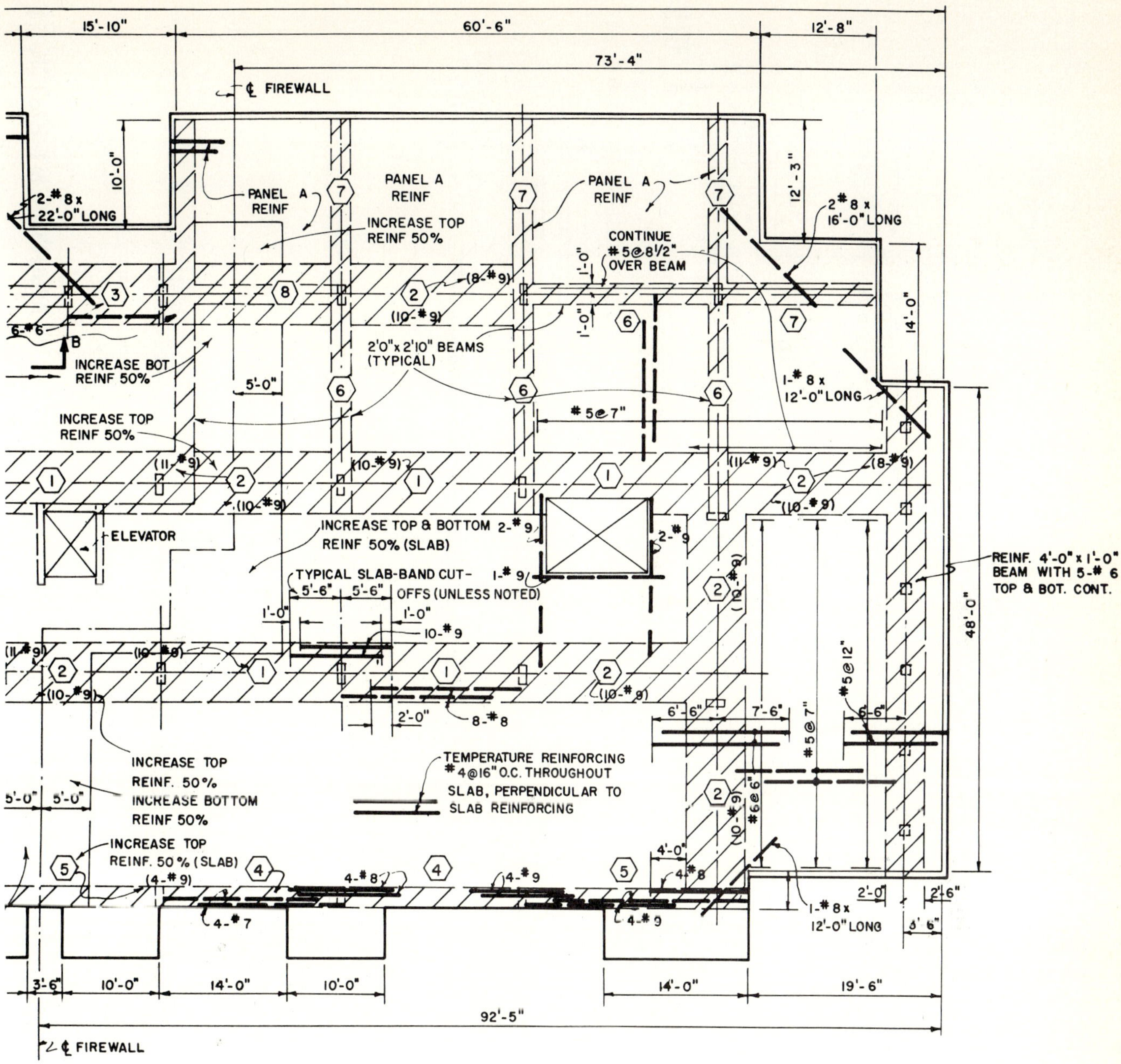

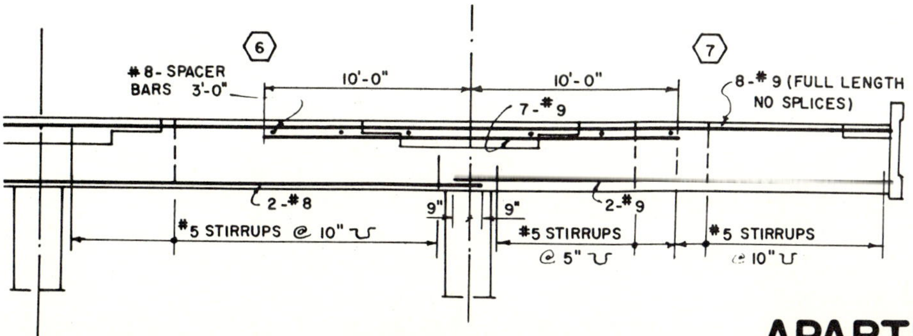

SECTION C-C

APARTMENT BLOCK 4

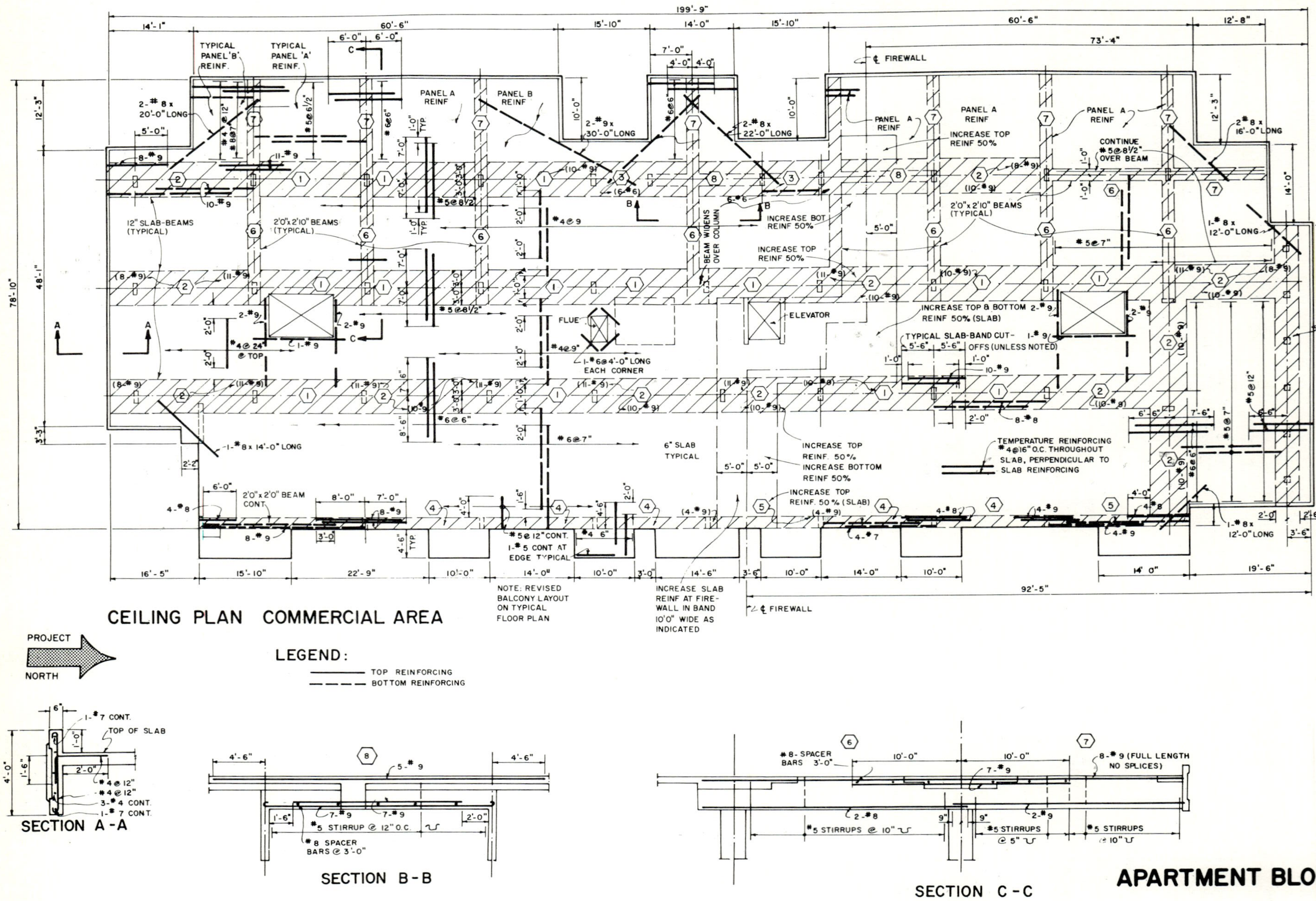

APARTMENT BLOCK 4

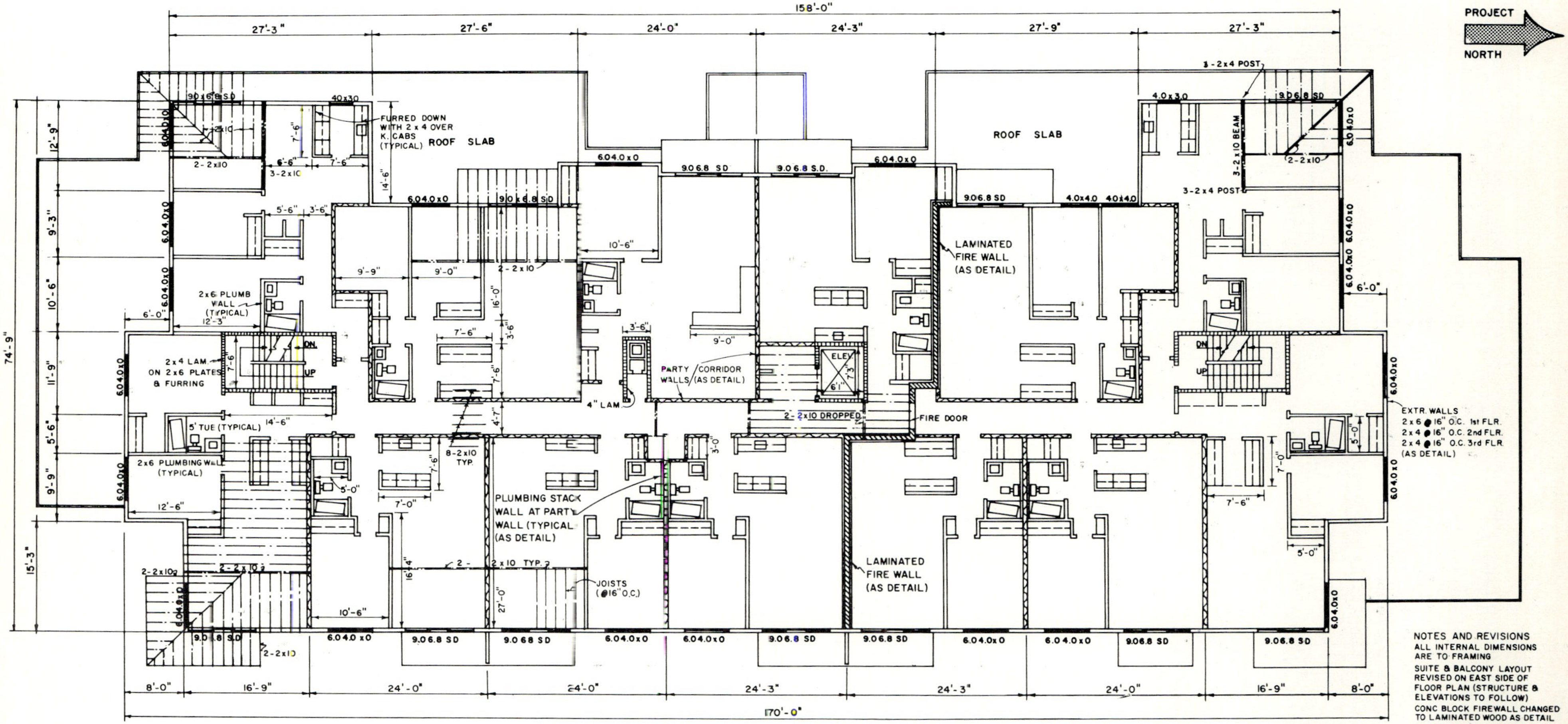

TYPICAL FLOOR PLAN

APARTMENT BLOCK 5

NOTES AND REVISIONS

ALL INTERNAL DIMENSIONS ARE TO FRAMING

SUITE & BALCONY LAYOUT REVISED ON EAST SIDE OF FLOOR PLAN (STRUCTURE & ELEVATIONS TO FOLLOW)

CONC BLOCK FIREWALL CHANGED TO LAMINATED WOOD AS DETAIL

CEILING JOISTS ONLY TO BATHROOMS ON 2nd AND 3rd FLOORS AND OMITTED ELSEWHERE

PROVIDE 2x4 FURRING OVER KITCHEN CPBDS. & TO TUB PANELS

PROVIDE CONT. 2x4/s AS REQUIRED TO SUPPORT WALLBOARD EDGES AT CORNERS OF WALLS & CEILINGS I.E. AT CEILING BOARD EDGES AT CORRIDOR WALLS

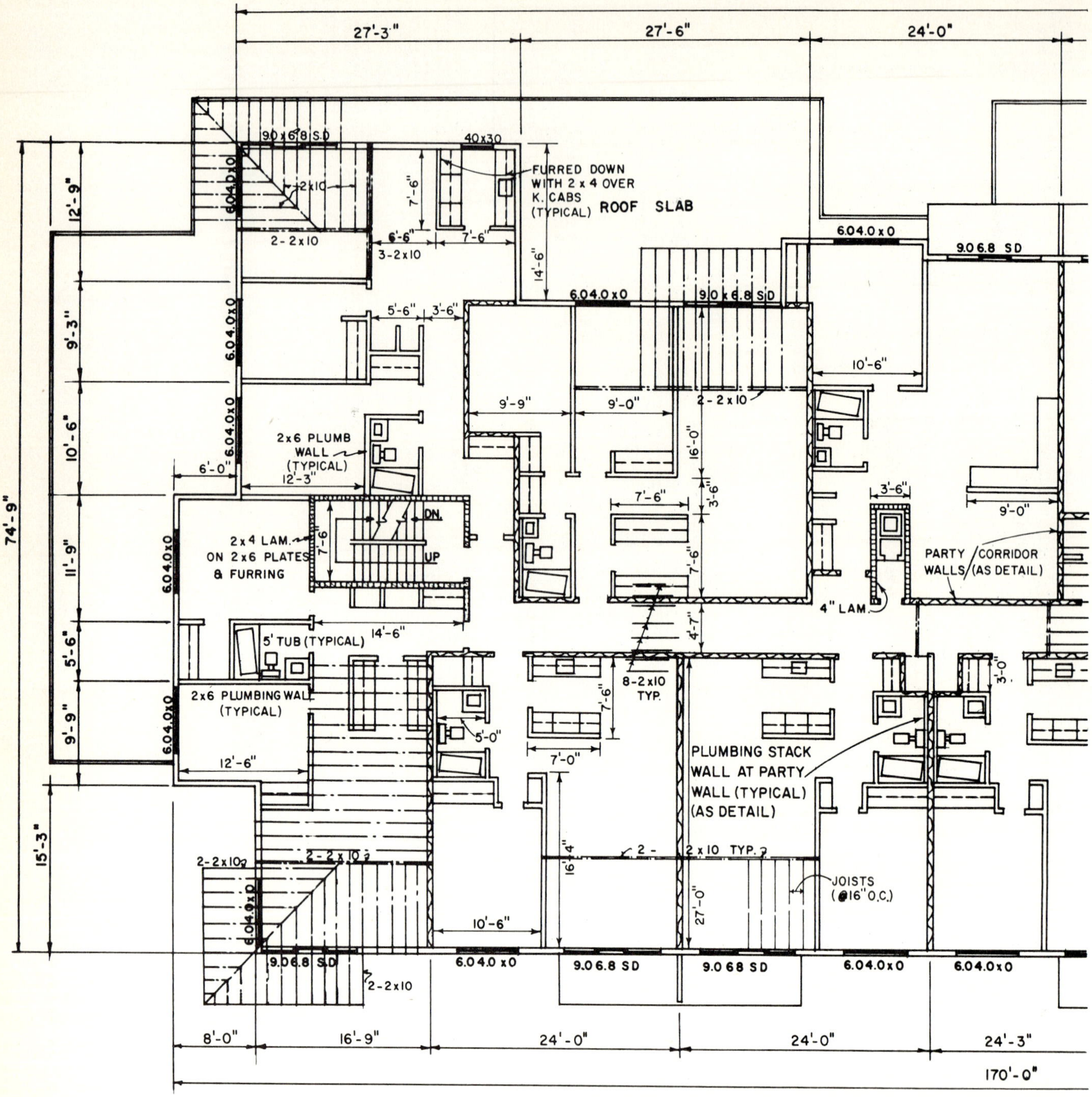

TYPICAL FLOOR PLAN

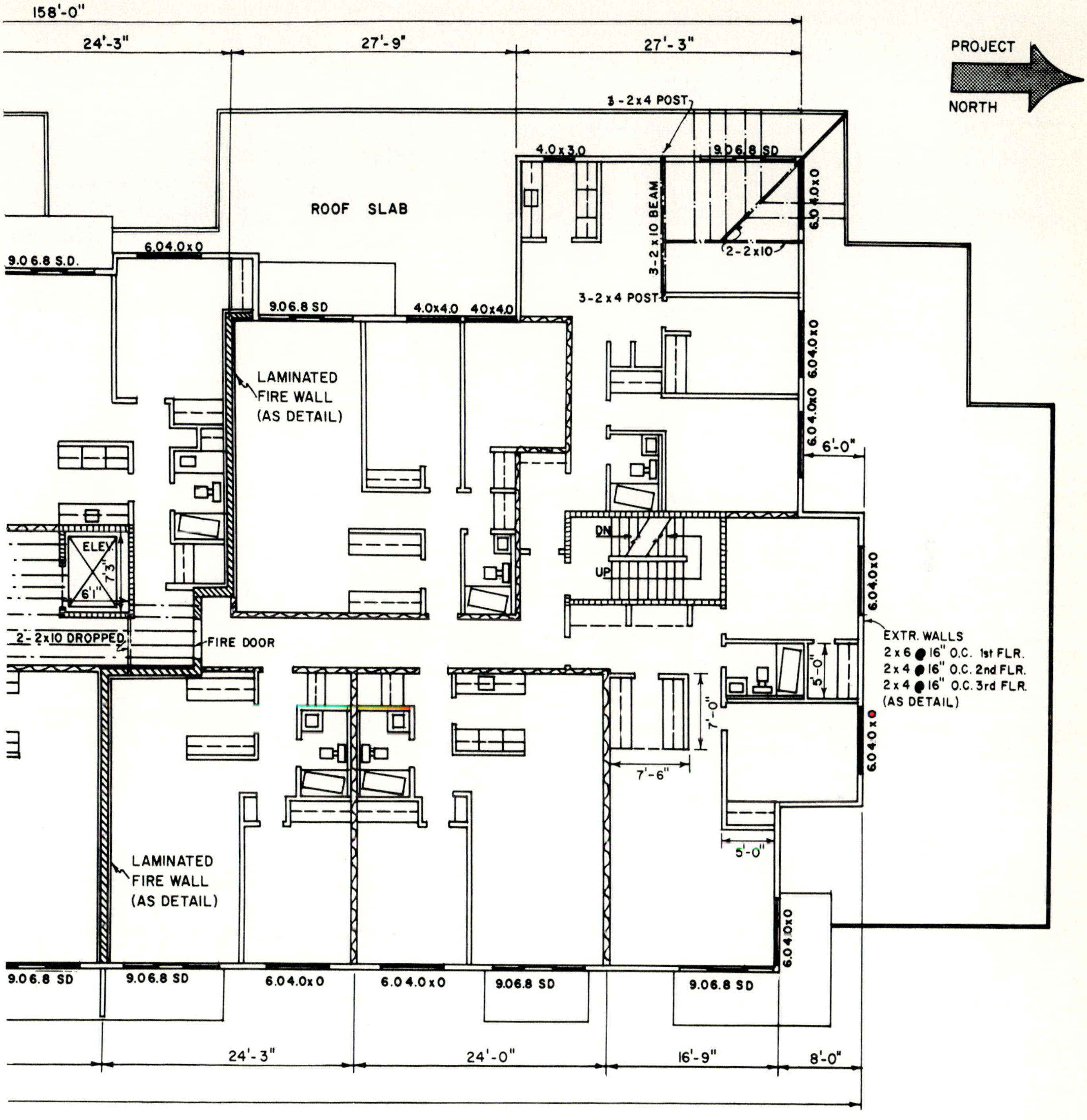

NOTES AND REVISIONS

ALL INTERNAL DIMENSIONS ARE TO FRAMING

SUITE & BALCONY LAYOUT REVISED ON EAST SIDE OF FLOOR PLAN (STRUCTURE & ELEVATIONS TO FOLLOW)

CONC BLOCK FIREWALL CHANGED TO LAMINATED WOOD AS DETAIL

CEILING JOISTS ONLY TO BATHROOMS ON 2nd AND 3rd FLOORS AND OMITTED ELSEWHERE

PROVIDE 2x4 FURRING OVER KITCHEN CPBDS. & TO TUB PANELS

PROVIDE CONT. 2x4/s AS REQUIRED TO SUPPORT WALLBOARD EDGES AT CORNERS OF WALLS & CEILINGS I.E. AT CEILING BOARD EDGES AT CORRIDOR WALLS

APARTMENT BLOCK 5

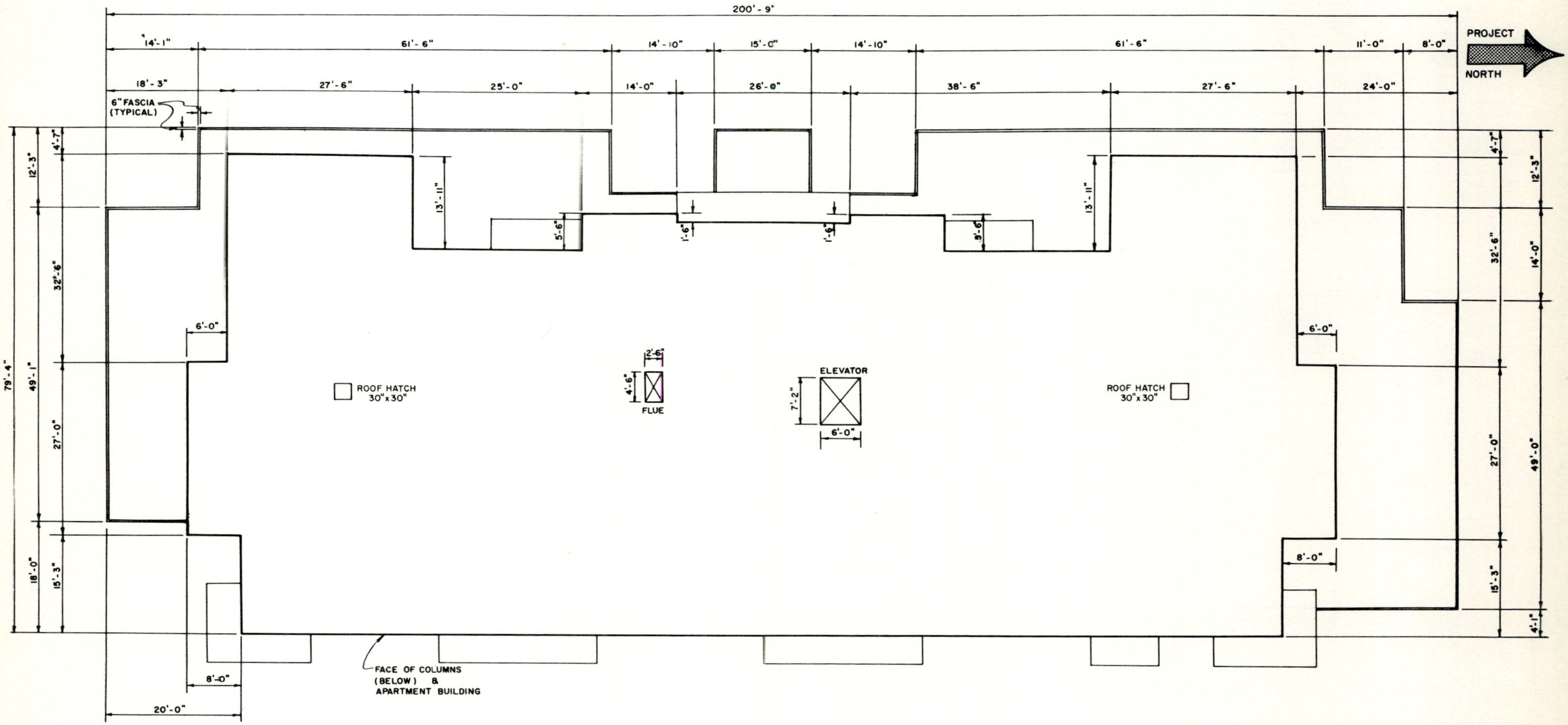

ROOF PLAN

APARTMENT BLOCK 6

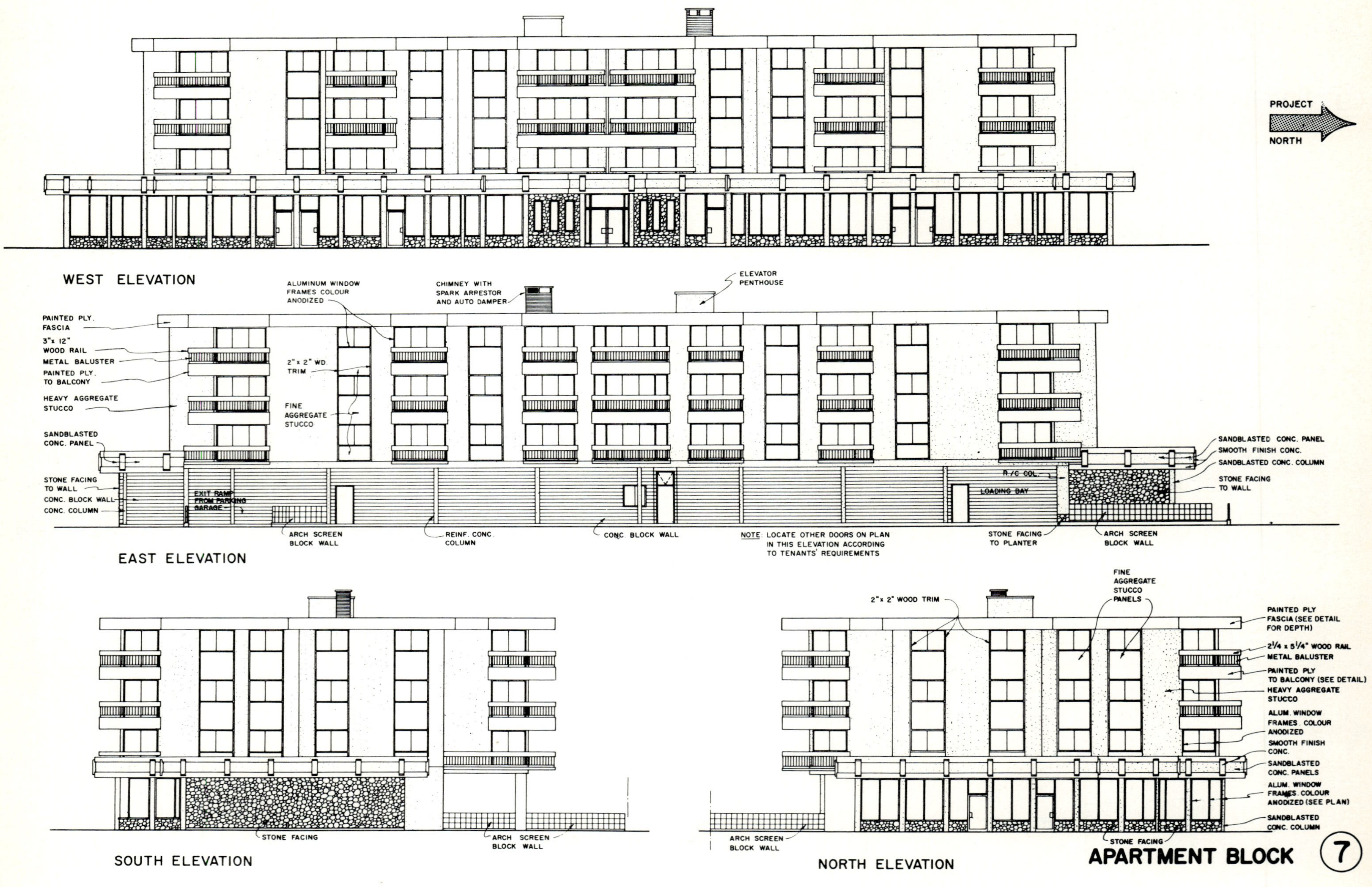
PROJECT
NORTH
WEST ELEVATION
ALUMINUM WINDOW FRAMES COLOUR ANODIZED
CHIMNEY WITH SPARK ARRESTOR AND AUTO DAMPER
ELEVATOR PENTHOUSE
PAINTED PLY. FASCIA
3"x 12" WOOD RAIL
METAL BALUSTER
PAINTED PLY. TO BALCONY
HEAVY AGGREGATE STUCCO
SANDBLASTED CONC. PANEL
STONE FACING TO WALL
CONC. BLOCK WALL
CONC. COLUMN
2" x 2" WD TRIM
FINE AGGREGATE STUCCO
EXIT RAMP FROM PARKING GARAGE
ARCH SCREEN BLOCK WALL
REINF. CONC. COLUMN
CONC. BLOCK WALL
NOTE: LOCATE OTHER DOORS ON PLAN IN THIS ELEVATION ACCORDING TO TENANTS' REQUIREMENTS
R/C COL.
LOADING BAY
STONE FACING TO PLANTER
ARCH SCREEN BLOCK WALL
SANDBLASTED CONC. PANEL
SMOOTH FINISH CONC.
SANDBLASTED CONC. COLUMN
STONE FACING TO WALL
EAST ELEVATION
STONE FACING
ARCH SCREEN BLOCK WALL
SOUTH ELEVATION
2" x 2" WOOD TRIM
FINE AGGREGATE STUCCO PANELS
PAINTED PLY FASCIA (SEE DETAIL FOR DEPTH)
2 1/4 x 5 1/4" WOOD RAIL
METAL BALUSTER
PAINTED PLY TO BALCONY (SEE DETAIL)
HEAVY AGGREGATE STUCCO
ALUM. WINDOW FRAMES. COLOUR ANODIZED
SMOOTH FINISH CONC.
SANDBLASTED CONC. PANELS
ALUM. WINDOW FRAMES. COLOUR ANODIZED (SEE PLAN)
SANDBLASTED CONC. COLUMN
ARCH SCREEN BLOCK WALL
STONE FACING
NORTH ELEVATION
APARTMENT BLOCK
7

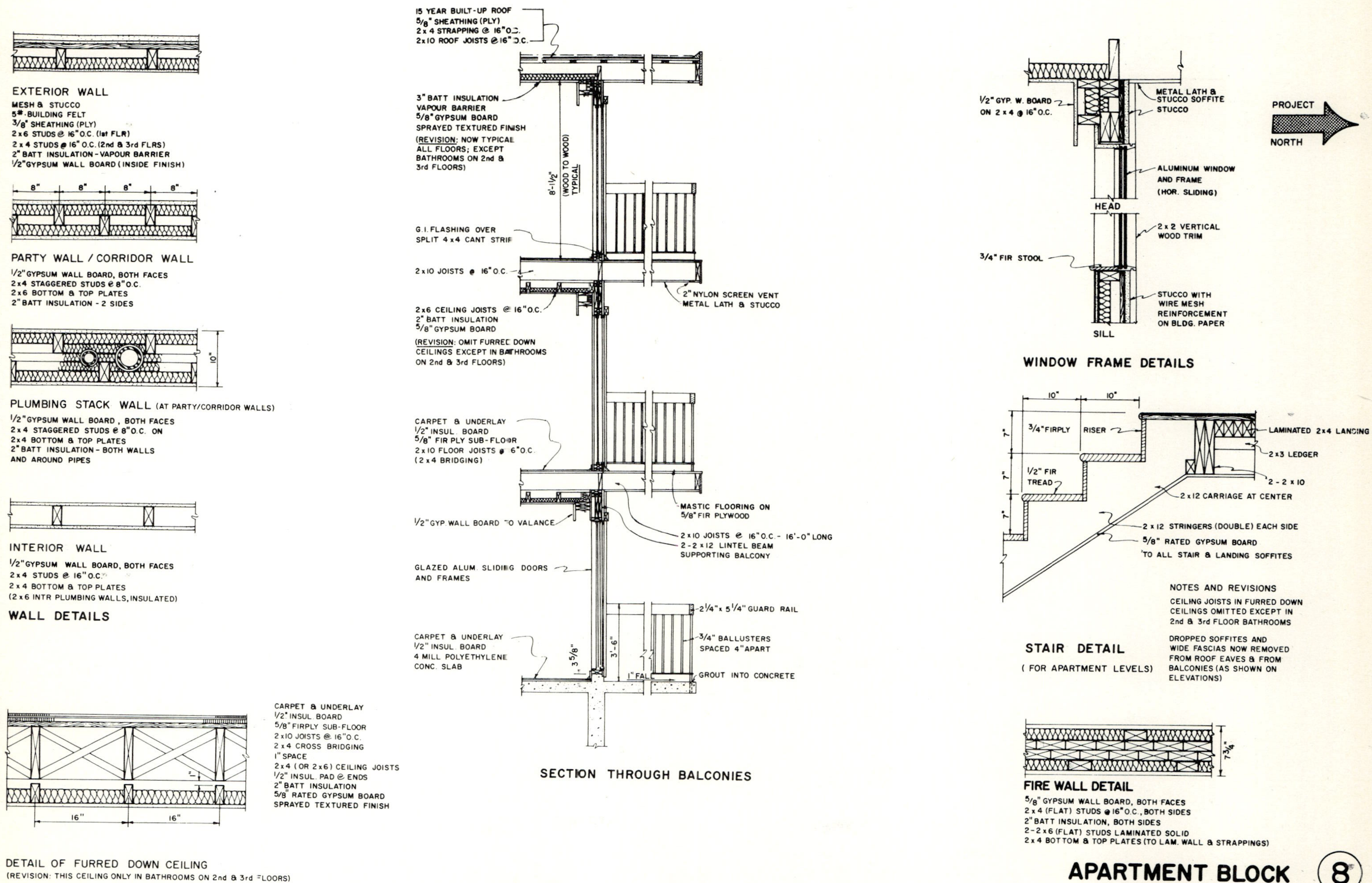

APARTMENT BLOCK 8

this class of building work go, these go further than many. In any case the inherent discrepancies should not hamper an estimator, especially one making a preliminary estimate; they require the kinds of decisions that estimators must make, especially in preliminary estimates.

One of the biggest problems in estimating costs of this kind of building arises from the use of yard lumber and plywood in formwork for concrete and then its reuse later as framing material. You should make reasonable assumptions, stipulate them, and get on with the estimate without too much concern as to their correctness; here it is academic, and you can compare your total estimated costs with those of others based on different assumptions and see their different effects on your total costs.

Framing is frequently undermeasured. Because this building was the subject of several court cases (in which I gave evidence as an expert witness) most of the trade work was measured more than once with care and special attention. For your guidance the following total quantities are given:

Concrete, all kinds,	1,520	CY
Steel rebar, all kinds,	112	TON
Formwork, all kinds,(contact area)	70,000	SF
Framing lumber, all kinds,	170	MBF
Ply sheathing, walls and decks,	47,000	SF
Exterior stucco, with soffits,	1,550	SY
Gypsum wallboard,walls, ceilings,	118,000	SF

In measuring for these quantities, openings were deducted from gross areas of walls and decks. The gross area of wallboard would be 5 to 10 percent higher. These quantities were for the complete complex, including the commercial floor, but that floor contained no partitions or fixtures.

Exercises

1. Make an elemental estimate of the costs of the apartment building excluding the commercial floor, as if it did not exist, using elements in the preceding list. (If you wish, substitute prices for another location and time).

2. Make a conceptual estimate of the apartment building (excluding the commercial floor) based on approximate quantities for as many trades as possible and, using outline sketches of the building traced or copied from the drawings in the book, represent the design information of a project provided in its earliest stages. Include only unusual drawing details and only the barest outline specifications.

3. Make a preliminary estimate of the apartment building (excluding the commercial floor) based on the building's area and using a selected cost manual. Compare your results with those of an elemental estimate and comment on the differences and similarities between them.

4. From an estimate of the apartment building based on approximate quantities, synthesize as many meaningful relationships as you can from among the several elements and items of work, expressed quantitatively and justified by argument as to their validity as useful data for other estimates of similar buildings (for example, the relationship between total floor area and total area of interior wallboard).

5. Discuss the following aspects of estimating costs relative to the apartment building:

(a) What is the significance of additional corners in the plan of such a building; how do they affect the costs; and how might an estimator measure the effects of corners?

(b) Describe an efficient way for an estimator to measure the costs of an additional (fourth) floor of apartments so as to arrive at an alternative cost for an apartment building of four floors of apartments instead of three.

6. (a) What is the reproduction cost of a single parking stall in such a parking structure as that shown for the apartment building at a given location and time?

(b) Synthesize individual prices for each kind of apartment in this apartment building, showing a detailed cost analysis.

(c) Show an analysis of the total costs of the apartment building based on the results of (a) and (b).

7. Establish costs of construction and of interest (for financing) and prepare a cash flow prediction for the construction of this apartment building relative to a particular time and location and to specified conditions.

8. From an estimate of costs based on trade work, prepare a suitable schedule for estimating the reduction in value of this apartment building due to physical depreciation, with proper attention to be given to the normal rates of depreciation of the several elements, subelements, and items of work in the building, appropriate to appraisals made by the cost approach. Provide suitable costs of new work and realistic annual depreciation rates. Describe briefly the reasons for the rates of depreciation.

5

Commercial Floor of an Apartment Block

The last chapter dealt with a common type of apartment building, one with three framed floors standing on a reinforced concrete (RC) parking structure; we deliberately ignored the RC commercial floor shown in the apartment block drawings and located above the parking floor and beneath the framed structure. In this chapter, we deal with that commercial floor—as an alternative feature of the apartment block. In other words we are assuming that the owner has not yet decided whether to include the commercial floor in his apartment development project.

PRACTICAL "STANDARDIZED" FORMS FOR CONTRACTORS
Form 514 MFD. IN U.S.A.

GENERAL ESTIMATE

BUILDING: 39-SUITE APARTMENT BLOCK
LOCATION: with COMMERCIAL FLOOR
ARCHITECTS:
SUBJECT: R. CONCRETE – SUMMARY 1/2

ESTIMATE NO. PRELIM #1
SHEET NO. SUMMARY 1/2.
ESTIMATOR: KC
CHECKER: –
DATE: 04-01-83

FRANK R. WALKER CO., PUBLISHERS, CHICAGO

DESCRIPTION OF WORK	NO. PIECES	DIMENSIONS	EXTENSIONS	EXTENSIONS	TOTAL ESTIMATED QUANTITY	UNIT PRICE M'T'L	TOTAL ESTIMATED MATERIAL COST	UNIT PRICE LABOR	TOTAL ESTIMATE LABOR COST
Ready-Mix Concrete (3000 psi)									
PARKING LEVEL					CY				
4" Slab on Ground (parking)					230				
Column Footings					85				
Columns (8"x24")					14				
Wall Footings		(16"x 8")			8				
Ditto		(12"x 8")			8				
Ditto		(x10")			13				
8" Walls					152				
					510 cy				
COMMERCIAL LEVEL									
7" Suspended Slab					367				
Slab Bands x 6" thick					77				
Slab Beams		(24"x13")			3				
Columns		(8" x 24")			20				
Columns		(8' x 29")			1				
Columns		(12" x 12")			14				
8" Walls					9				
					491 cy				

PRACTICAL "STANDARDIZED" FORMS FOR CONTRACTORS
Form 514 MFD. IN U. S. A.

GENERAL ESTIMATE

BUILDING 39-SUITE APARTMENT BLOCK
LOCATION WITH COMMERCIAL FLOOR.
ARCHITECTS
SUBJECT R. CONCRETE - SUMMARY - 2/2

ESTIMATE NO. PRELIM. # 1
SHEET NO. SUMMARY 2/2
ESTIMATOR KC
CHECKER -
DATE 04-01-83

FRANK R. WALKER CO., PUBLISHERS, CHICAGO

DESCRIPTION OF WORK	NO. PIECES	DIMENSIONS			EXTENSIONS	EXTENSIONS	TOTAL ESTIMATED QUANTITY	UNIT PRICE M'T'L	TOTAL ESTIMATED MATERIAL COST	UNIT PRICE LABOR	TOTAL ESTIMATE LABOR COST
Ready-Mix Concrete (3000 psi)											
APARTMENT LEVEL							CY				
6" Suspended Slab							276				
Slab Beams x 6" thick							66				
Slab Beams, (24"x18" & 24"x28")							72				
Slab Fascia Beams (6"x48")							27				
							441	CY			
PARKING LEVEL (sheet 1/2)							510 ✓				
COMMERCIAL " (" ")							491 ✓				
APARTMENT " (" 2/2)							441 ✓				
							1,442				
CONCRETE NOT SHOWN ON DWGS. (IN PAVINGS, STEPS, ETC.) =							+ 65				
							1,507 ✓				
WASTE (EXCESS DELIVERED OVER NET QUANTITIES MEASURED) =							+ 13				
TOTAL AMOUNT of CONCRETE							1,520	CY			
REBAR											
1507 CY x 150#/CY (APPROX. AVG.) =							113	TON			

There is only one way for an estimator to estimate the costs of this commercial floor properly; that is to measure it in detail—concrete, forms, and steel—and to price it with unit prices for the various items of work. This is not as formidable as it sounds. In order to give you an attainable and verifiable goal, the total concrete quantities of the different concrete members are provided, totaling, in volume, 1,520 cubic yards for both the parking and the commercial floors. It is suggested that the *concrete* be measured first, followed by measurement of the *formwork* (thus making an opportunity to review the concrete measurements), followed by *detailed measurement of the steel rebar.*

Do not try to measure from drawings immediately after a quick perusal; even experienced estimators need time to become familiar with a set of drawings. First, actively study the drawings, make notes, and use the drawings to assist you in visualizing the concrete structure in three dimensions. You cannot measure properly until you can do that. Forget about the apparent complexity of the steel rebar for a while and concentrate on the concrete. Once you can visualize the concrete, you can measure it, and once you can measure it, both the forms and the rebar are easy to measure.

Novices are usually put off by the complexity of detail in drawings, especially those of steel rebar. Even the best and most experienced estimators must measure items one after the other, just as the novice. The difference with the experienced estimator is that he is clearsighted and methodical. There are no tricks or secrets to measurement other than to know construction and to measure it well, following the precepts contained in this book. You may want to refer to them. To measure the RC work follow these directions.

Visualize the RC work below the framed apartments in two parts: (1) the parking, in the form of a closed RC box, with a slab on the ground (at the lowest level), perimeter walls, and a slab on top, which in most cases supports the framed apartment floors above; and (2) a RC commercial floor, in the form of a RC table (that is, a slab on columns), inserted between the RC parking floor and the framed apartment floors. You may want to measure only the concrete table, or only the concrete box, or both, depending on your objective (see the following exercises). The estimate sheets reproduced in this chapter show the quantities of concrete for the parking level, the commercial level, and the RC slab that supports the apartment floors, each separately.

First, measure the slab, then beams and columns, and finally any other concrete work. Then measure the formwork for each of the different concrete members, followed by the steel rebar in the same order. Always measure to the fullest possible extent of a member (do not measure piecemeal) and then go back to make any necessary adjustments.

Exercises

1. Measure from the drawings all the reinforced concrete work in the commercial floor and in the parking floor, keeping separate the quantities for the two floors, and compare your results with those given in the reproduced handwritten sheets.

2. Calculate from the results obtained in (1): (a) the weights of steel (per cubic foot of concrete) for each kind of concrete member (verify the total amount);

(b) the amounts of formwork (in contact area) per cubic yard for each kind of concrete member;

(c) the amount of concrete (in cubic feet) for each square foot of gross floor area of the two floors (parking and commercial). For each result, (a) through (c), discuss its possible application in *elemental estimates* of similar buildings. Indicate which results, if any, might be taken as reasonable constants for other elemental estimates and which results are not suitable for this purpose; explain your answers in detail.

3. Estimate the costs of the commercial floor for a given location and time and for given unit costs of labor, materials, and equipment rental. Present your results in the form of a report to an owner-client to indicate the estimated additional costs of this floor to the apartment project.

4. Discuss the significance of the distribution of costs among the three kinds of work: (1) concrete, (2) formwork, and (3) steel rebar, in this structure, in the light of general cost information about reinforced concrete buildings to (a) estimates, generally; (b) estimates of this kind of building, specifically; (c) the economics of this kind of construction (frame on concrete); and (d) the economics of reinforced concrete construction, generally.

5. Using a selected cost manual (with sufficient detailed information about reinforced concrete, such as that published by Means and listed in the Bibliography), compare your estimated costs for the commercial floor with those of alternative and structurally similar RC floors, and comment on the results of your comparisons. (If necessary for valid

comparisons, first convert all costs to a common base.)

6. Based on cost information that you have produced for the commercial floor, design a simplified but suitably accurate method for estimating the costs of this kind of RC structure suitable for use by appraisers and assessors making reproduction cost valuations.

6

A Warehouse with Office

The complete working drawings for this building are reproduced in this book to provide an opportunity for comparisons of estimates of different kinds. The warehouse's masonry, structural steel, and rough carpentry work were measured in detail in my earlier estimating book.[1] Serious students might consider measuring all the work in detail and pricing it, using a cost data book. Having done that, they might then try making alternative estimates by approximate quantities, by elements, and by building area, to compare their respective results and the efficacy of the different approaches. To those ends, some information about this project is proffered below.

The building was constructed in 1967, and the construction costs at that time were about $8 per square foot of gross area measured over both warehouse and attached office. It is of interest to note that in 1967 the Mean's cost data book showed under square foot costs a median cost per square foot for warehouses of $7.70 (national average costs). The data book is not specific about whether those warehouse costs included attached warehouse offices, but the increase in total costs because of an attached office is not usually great because

[1]Collier, *Fundamentals of Construction Estimating and Cost Accounting* (Prentice-Hall, 1974).

the proportion of office space is usually small, in this example, about 13 percent (20,000 square feet of warehouse and 3,000 square feet of office). Nevertheless, variations in the proportion of office space will cause variations in square foot costs. In the Mean's cost data book for 1983, the median cost for warehouses is shown as $22.90 per square foot.

The following approximate total quantities of trade work in the warehouse may be useful as a guide:

Site Work (warehouse and office together)

Strip topsoil, 6 inches deep	1,130 CY
Machine excavated trenches for footings	640 CY
Ditto for column bases	81 CY
Hand excavated bottoms, 3 inches deep	36 CY
Trim ground under gravel bed	22,640 SF
Pit run gravel fill, in place	1,800 CY

Concrete (warehouse and office separated)

Continuous strip footings (warehouse)	28 CY
Ditto (office)	9 CY
Column footings (warehouse)	14 CY
Ditto (office)	2 CY
Slab, 5 inches (warehouse)	310 CY
Slab, 4 inches (office)	41 CY
Walls, 12 inches and 8 inches (warehouse)	94 CY
Ditto (office)	14 CY
Walltie blocks, or deadmen,(2 x 2 x 2 feet)	5 CY
Sundry concrete sumps, etc	3 CY
Concrete sidewalks ands crossings	30 CY
Formwork to footings (contact area) (warehouse)	1,100 SF
Ditto (office)	635 SF
Formwork to walls (contact area—warehouse)	4,970 SF
Ditto (office)	1,110 SF
Form walltie blocks (contact area)	290 SF
Sundry forms and screeds to sidewalks, etc	allowance

Steel Rebar (warehouse and office separated)

Straight bars, #4, 5, 6 (warehouse)	3,000 LB
Bent bars, #3 (warehouse)	10 LB
Bent bars, #6, 7 (warehouse)	1,920 LB
Straight bars, #3 (office)	30 LB
Straight bars, #4, 5 (office)	2,000 LB
Bent bars, #6 (office)	1,050 LB

Sundry Items (warehouse and office separated)

3/4 inch- x 12 inch- bolts (warehouse)	NO
5/8 inch- x 15 inch- bolts (warehouse)	48 NO
Ditto (office)	24 NO
Steel trowel, hardener, cure slab (warehouse)	20,000 SF
Ditto (office)	2,650 SF
Polythene slab membrane (inc.laps) (warehouse)	21,500 SF
Ditto (office)	3,000 SF
Gravel bed under slab, 6-inches thick (warehouse)	370 CY
Ditto (office)	50 CY

Concrete Block Masonry (Reinforced) (warehouse and office separated)

Walls, 8 inches (warehouse)	2,140 SF
Ditto, small areas (office)	430 SF
Walls, 12 inches (warehouse)	8,780 SF
Walls, 4 inches at eaves (warehouse)	270 SF
Extra for 12 x 8 corner blocks	44 LF
Extra for bond beams, 8inches	157 LF
Ditto, 12 inches	594 LF
Extra for concrete fill in 12 inch blocks	26 NO
Extra for vertical ditto, with #4 rebar	190 LF
Extra for control joints in 12 inch walls	324 LF
8 x 16 piers filled with 2 #5 rebars	120 LF
8 x 24 piers, ditto	154 LF

Structural Steelwork
Total weight, 22,000 LB.

Rough Carpentry (warehouse and office together)

Roof deck, 2inches T. & G.	53,000 BF
Roof joists, 3 x 14	20,000 BF
Ditto, 2 x 12	4,800 BF
Ditto, x 8 1/2 feet long, cut to curved top edge	50 NO
Headers, 2 x 12 (dressed) next to last	130 BF
Blockings, 2 x 2 on top of joists	450 LF
Ledgers, 3 X 6	3,000 BF
Ditto, 3 x 8	180 BF
Plates, 3 x 4	100 BF
Plates, 3 x 6	150 BF
Cant strips 4 x 4	64 LF
Ply sheathing, 1/2 inch	2,800 SF
Ditto to gutter x 8 inches-girth	62 LF
Sundry framing	6,000 BF

Glued-Laminated Beams (warehouse and office together)

9 x 27 5/8 inches (9 no.)	12,400 BF
9 x 29 1/4 inches (9 no.)	19,200 BF
5 1/4 x 27 5/8 inches (1 no.)	840 BF
5 1/4 x 29 1/4 inches (1 no.)	1,275 BF
5 x 16 1/4 inches (1 no.)	240 BF
3 1/4 x 16 1/4 inches (3no.)	555 BF
Sundry bolts, straps, hangers, clips	allowance

These measured quantities are net, and allowances for waste[G] should be made in the unit prices (as previously described). Apart from dampproofing and roofing work, the areas of which are easily measured, most of the remaining work requires only enumeration. Painting in the offices can be covered by a cash allowance. In 1983, mechanical and electrical work should cost about $72,000 (at 1983 national average costs, as cited in Means' cost books), including sprinklers. That is based on actual updated costs, but it also fits with the current median percentage for M & E work in warehouses in Mean's 1983 cost data book. For all design costs, all overhead costs, and profit, you might add 20 percent to the

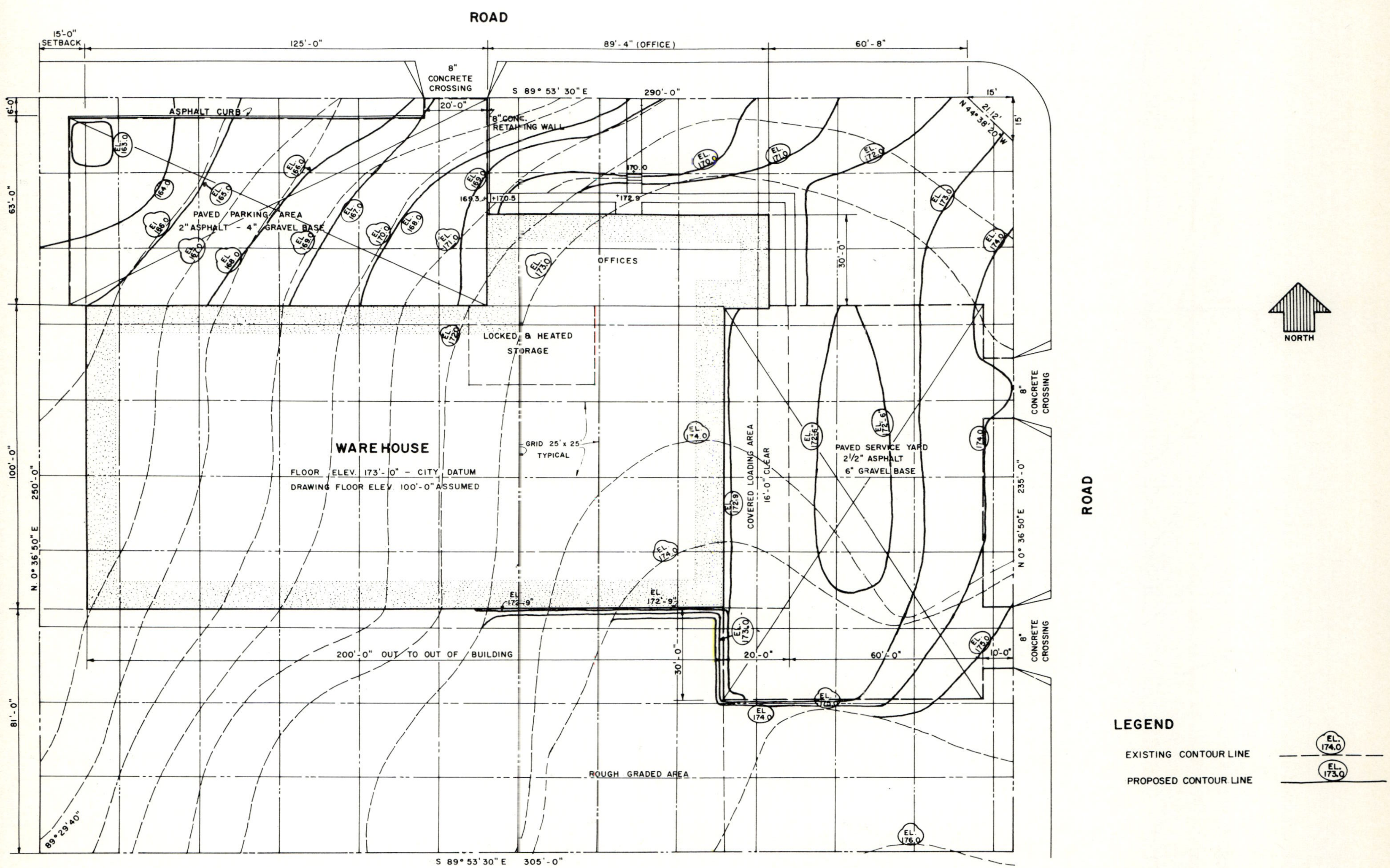

SITE PLAN

WAREHOUSE ①

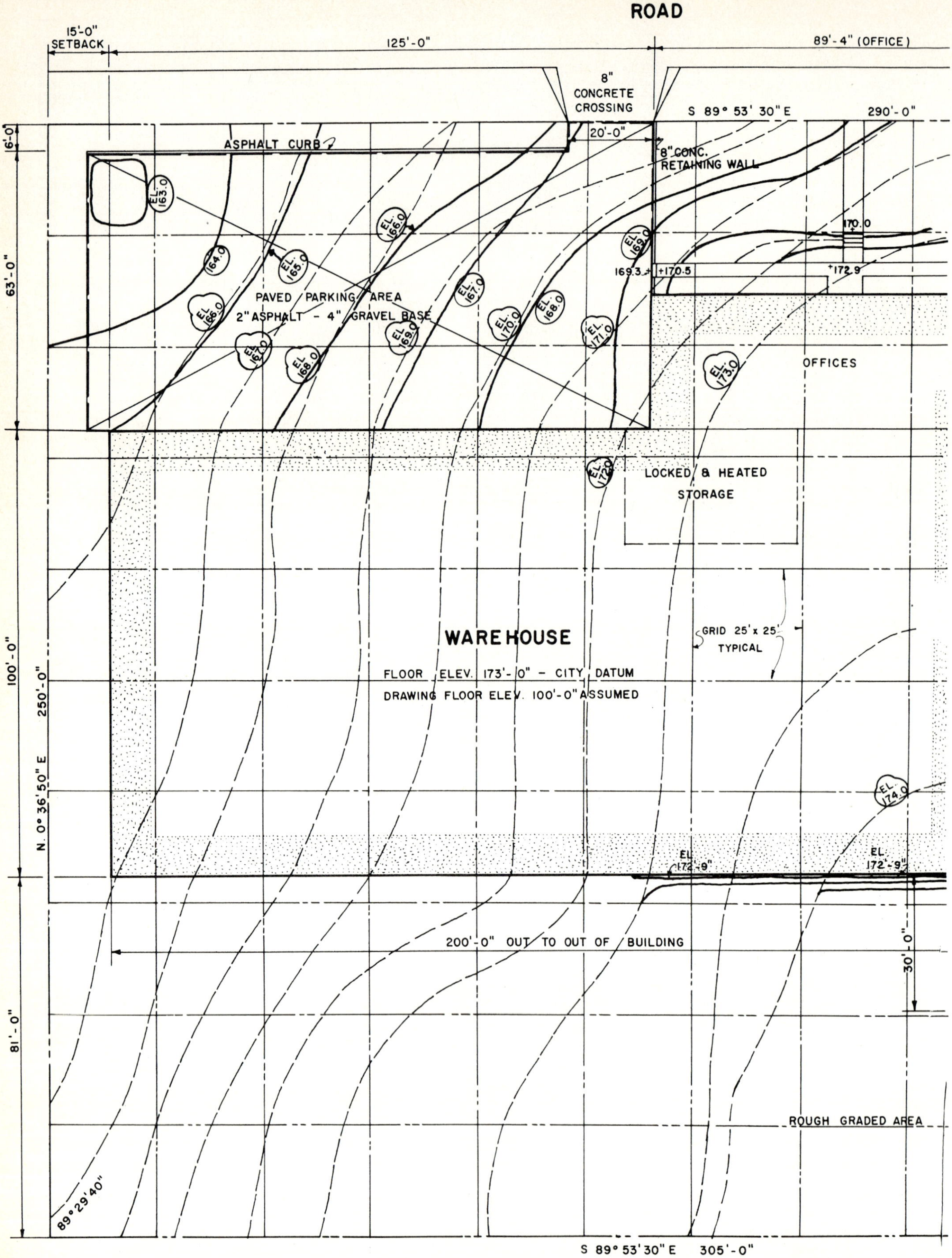
ROAD
15'-0" SETBACK
125'-0"
89'-4" (OFFICE)
8" CONCRETE CROSSING
S 89° 53' 30" E
290'-0"
16'-0"
ASPHALT CURB
20'-0"
8" CONC. RETAINING WALL
EL. 163.0
EL. 166.0
170.0
EL. 169.0
169.3
+170.5
+172.9
63'-0"
EL. 164.0
EL. 165.0
PAVED PARKING AREA
2" ASPHALT - 4" GRAVEL BASE
EL. 166.0
EL. 167.0
EL. 168.0
EL. 167.0
EL. 169.0
EL. 170.0
EL. 168.0
EL. 171.0
EL. 168.0
EL. 173.0
OFFICES
EL. 172.0
LOCKED & HEATED STORAGE
WAREHOUSE
GRID 25' x 25' TYPICAL
FLOOR ELEV. 173'-0" - CITY DATUM
DRAWING FLOOR ELEV. 100'-0" ASSUMED
100'-0"
250'-0"
N. 0° 36' 50" E
EL. 174.0
EL. 172'-9"
EL. 172'-9"
200'-0" OUT TO OUT OF BUILDING
30'-0"
81'-0"
ROUGH GRADED AREA
89° 29' 40"
S 89° 53' 30" E 305'-0"

SITE PLAN

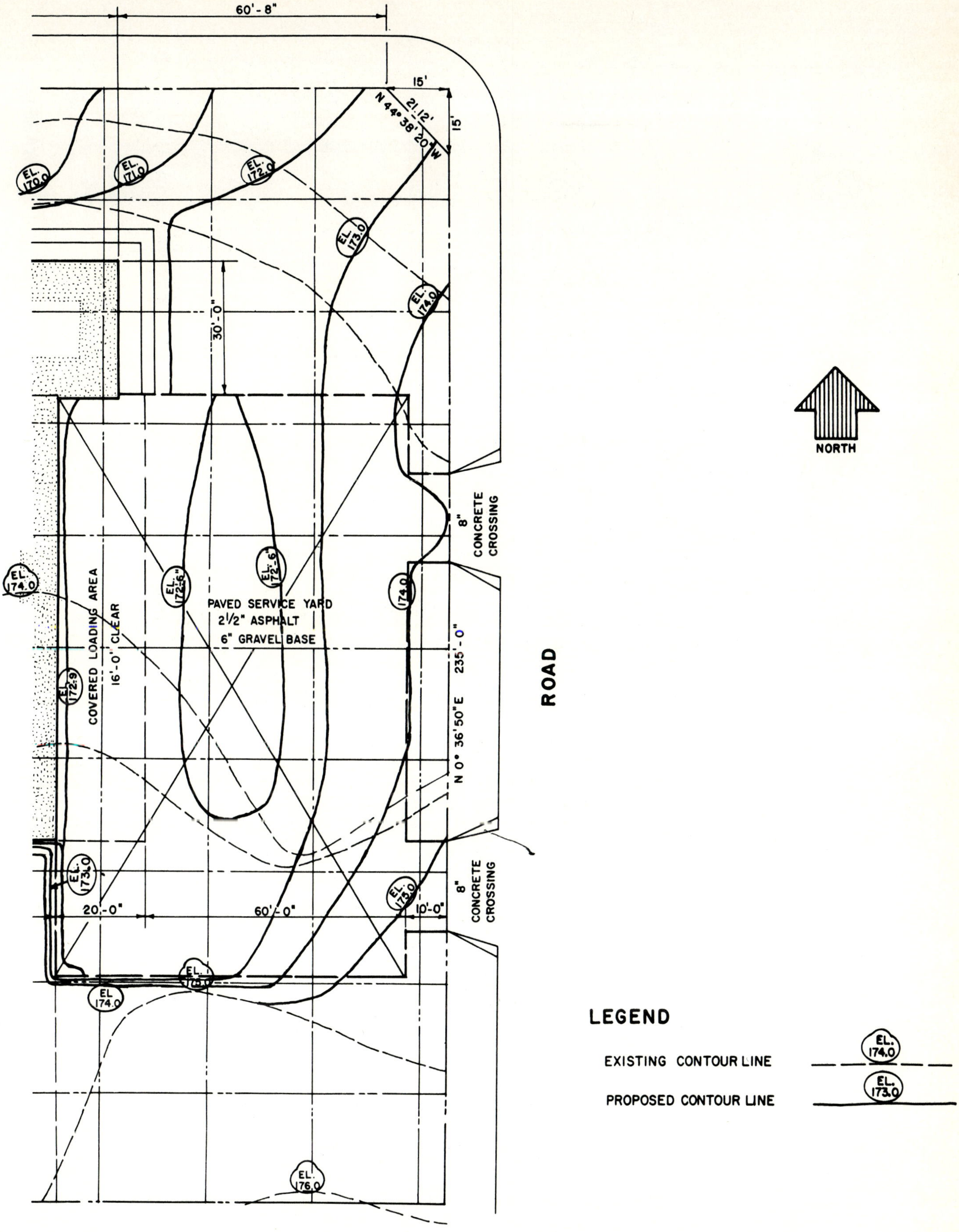

60'- 8"
15'
21.12'
N 44° 38' 20" W
15'
EL. 170.0
EL. 171.0
EL. 172.0
EL. 173.0
EL. 174.0
30'-0"
EL. 174.0
COVERED LOADING AREA
16'-0" CLEAR
EL. 172.9
EL. 172.6"
EL. 172.6"
PAVED SERVICE YARD
2 1/2" ASPHALT
6" GRAVEL BASE
174.0
8" CONCRETE CROSSING
235'-0"
N 0° 36' 50" E
ROAD
EL. 173.0
EL. 173.0
20'-0"
60'-0"
10'-0"
8" CONCRETE CROSSING
EL. 174.0
EL. 174.0
EL. 176.0
NORTH
LEGEND
EXISTING CONTOUR LINE
EL. 174.0
PROPOSED CONTOUR LINE
EL. 173.0

WAREHOUSE ①

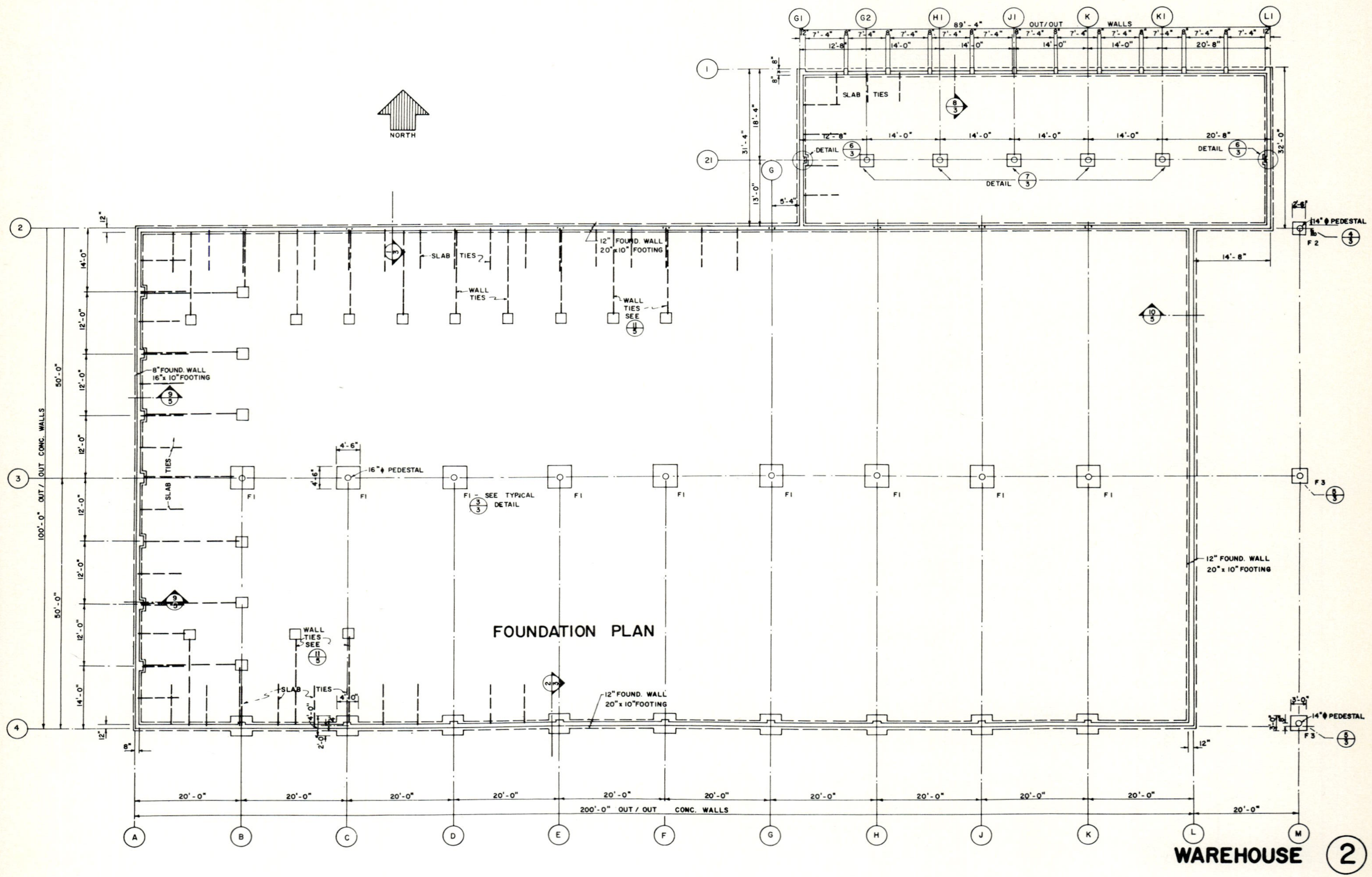
FOUNDATION PLAN
NORTH
200'-0" OUT / OUT CONC. WALLS
100'-0" OUT / OUT CONC. WALLS
89'-4" OUT/OUT WALLS
12" FOUND. WALL 20"x 10" FOOTING
8" FOUND. WALL 16"x 10" FOOTING
16" ϕ PEDESTAL
14" ϕ PEDESTAL
F1 - SEE TYPICAL DETAIL
SLAB TIES
WALL TIES SEE
DETAIL
WAREHOUSE
2

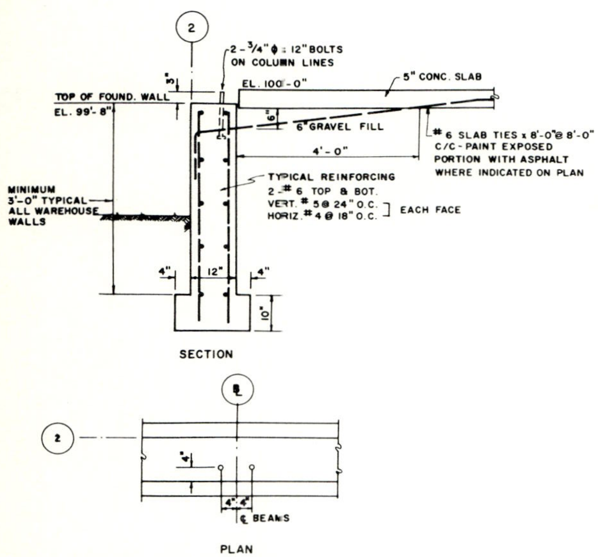

PLAN AND SECTION THROUGH NORTH FOUNDATION WALL (1/3)

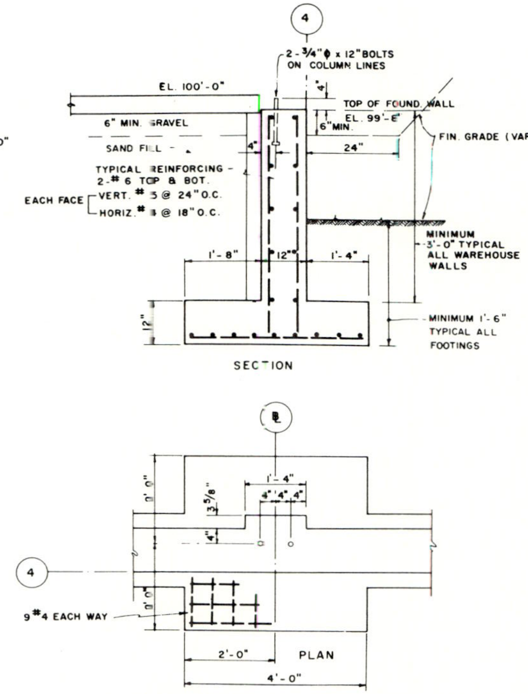

PLAN AND SECTION THROUGH SOUTH COLUMN FOOTINGS (2/3)

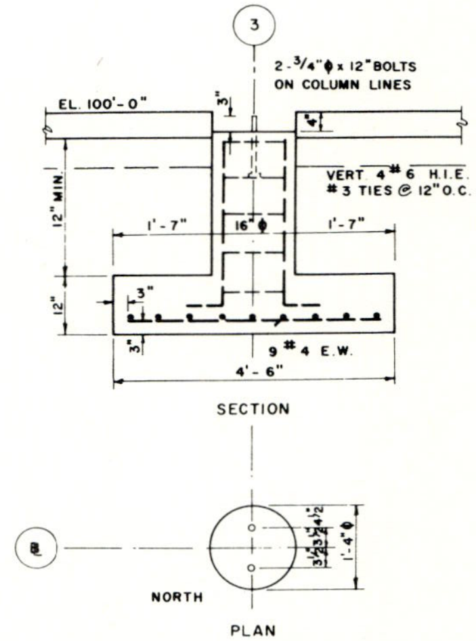

INTERIOR FOOTING TYPE F1 9 THUS (3/3)

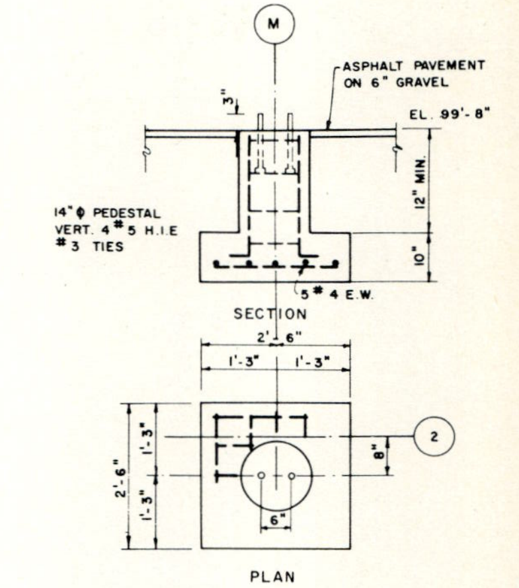

CANOPY FOOTING TYPE F 2 1 THUS NORTH WALL (4/3)

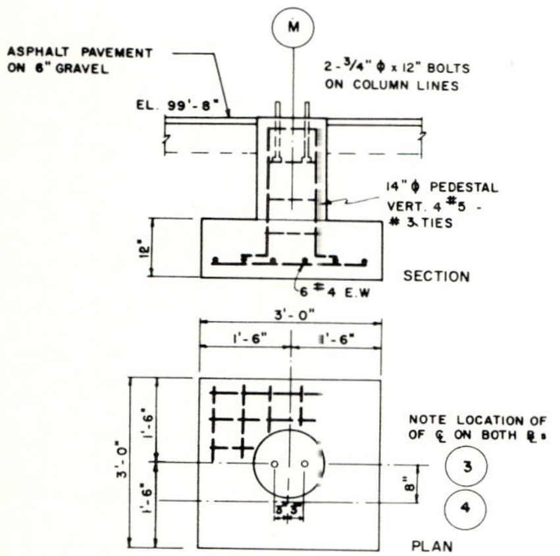

CANOPY FOOTING TYPE F3 (5/3)
2 THUS ON ℄ & SOUTH WALL

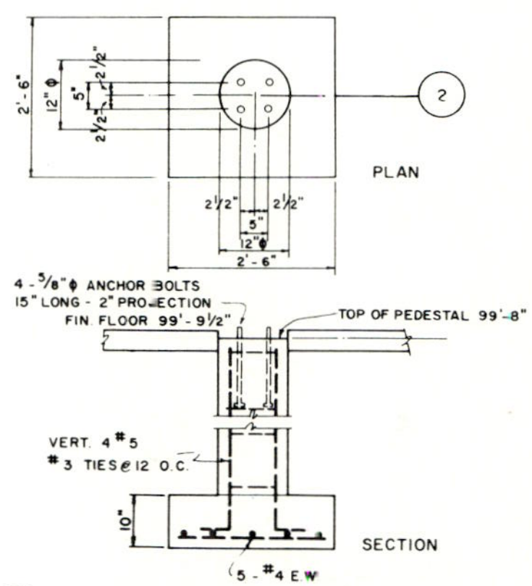

INTERIOR OFFICE FOOTING 5 THUS (7/3)

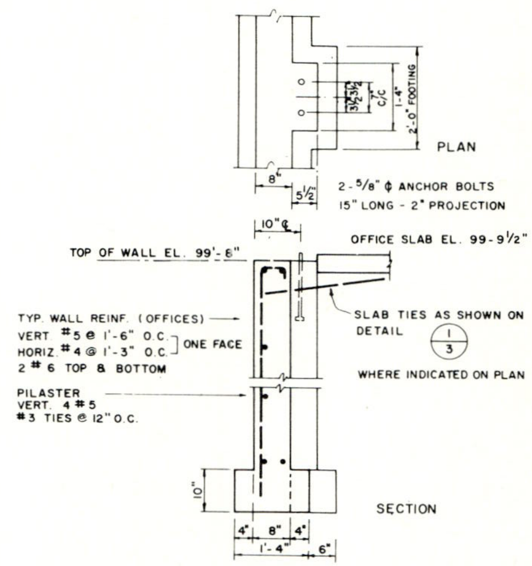

COLUMN PILASTER AT EAST AND WEST OFFICE WALL (6/3)

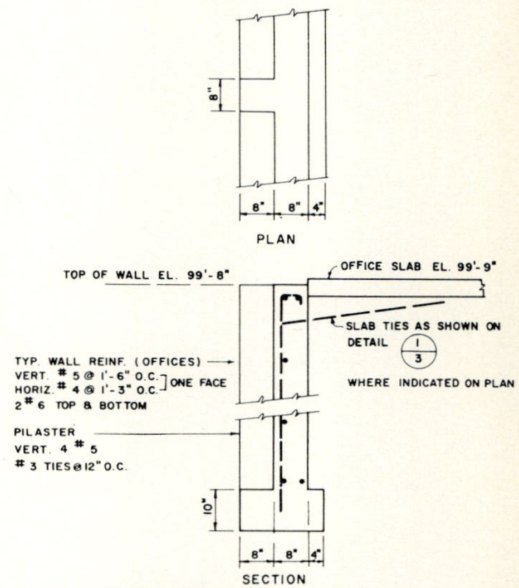

COLUMN PILASTER AT NORTH OFFICE WALL (8/3)

FOOTING DETAILS (SEE OTHERS ON SHEET 5)

WAREHOUSE 3

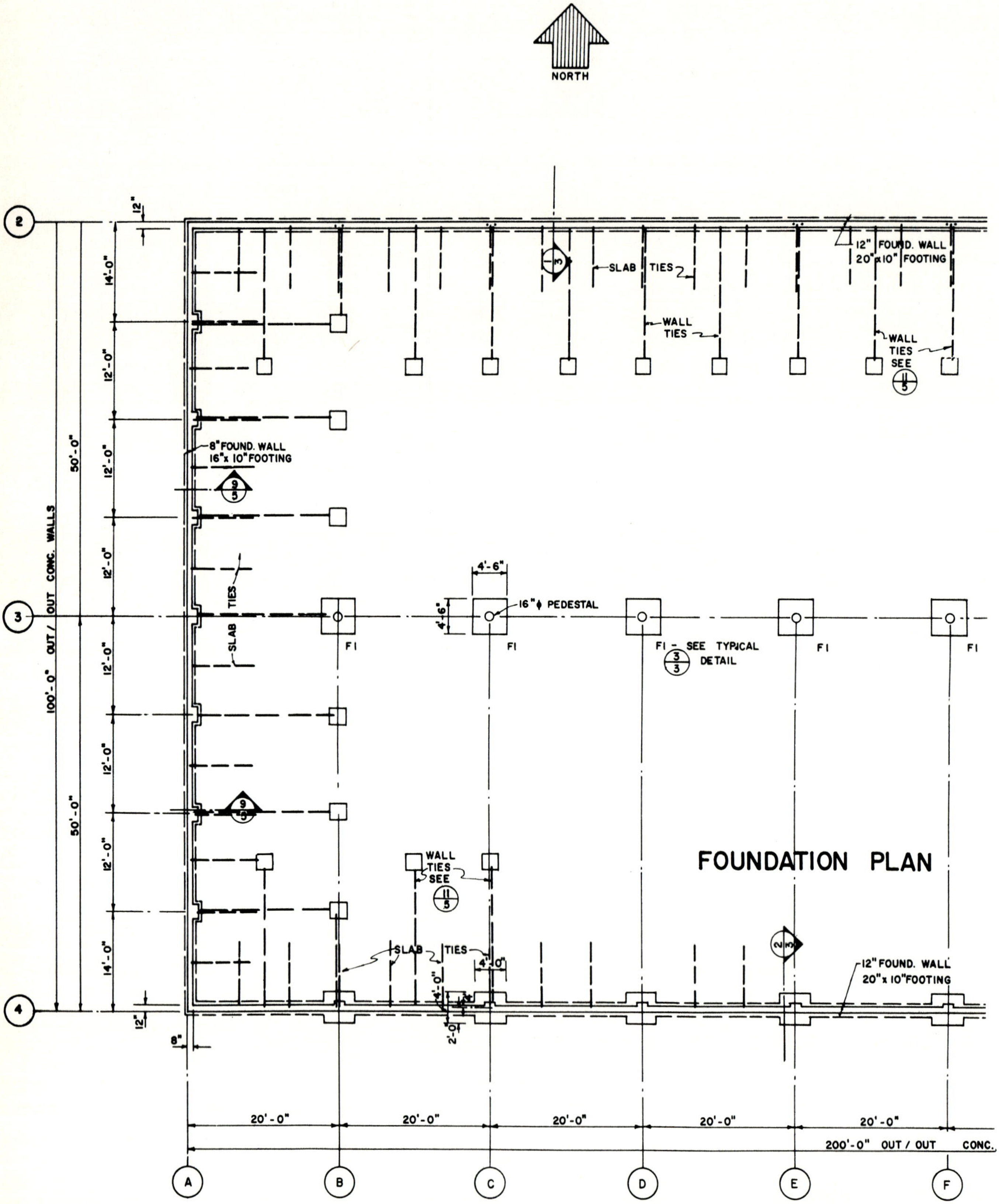

FOUNDATION PLAN

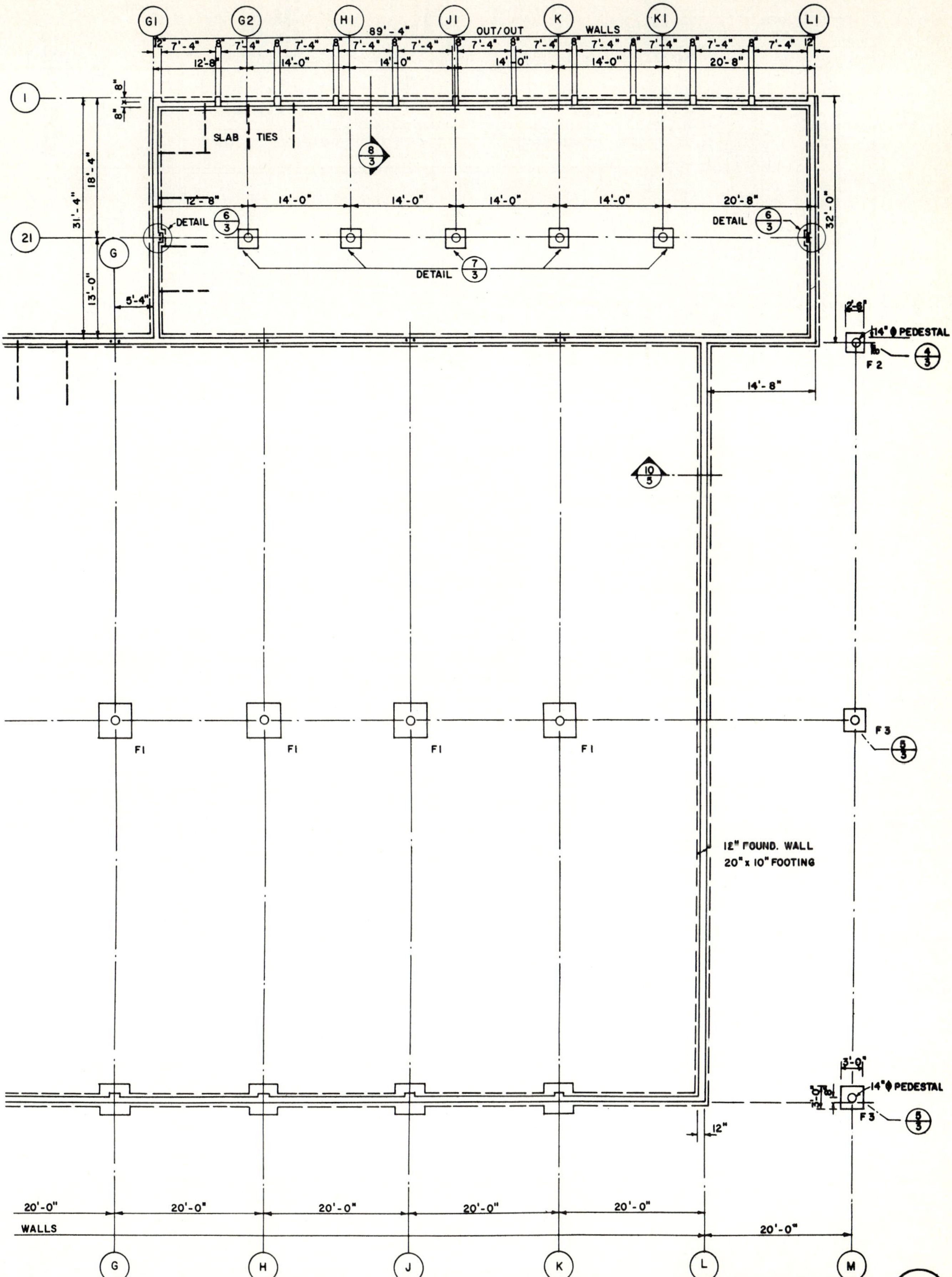

G1
G2
H1
J1
K
K1
L1
89'-4"
OUT/OUT
WALLS
12'-8"
14'-0"
20'-8"
SLAB
TIES
DETAIL
14'-0"
32'-0"
31'-4"
18'-4"
13'-0"
5'-4"
14" ϕ PEDESTAL
F 2
14'-8"
F1
F 3
12" FOUND. WALL
20" x 10" FOOTING
3'-0"
12"
20'-0"
WALLS
G
H
J
L
M

WAREHOUSE 2

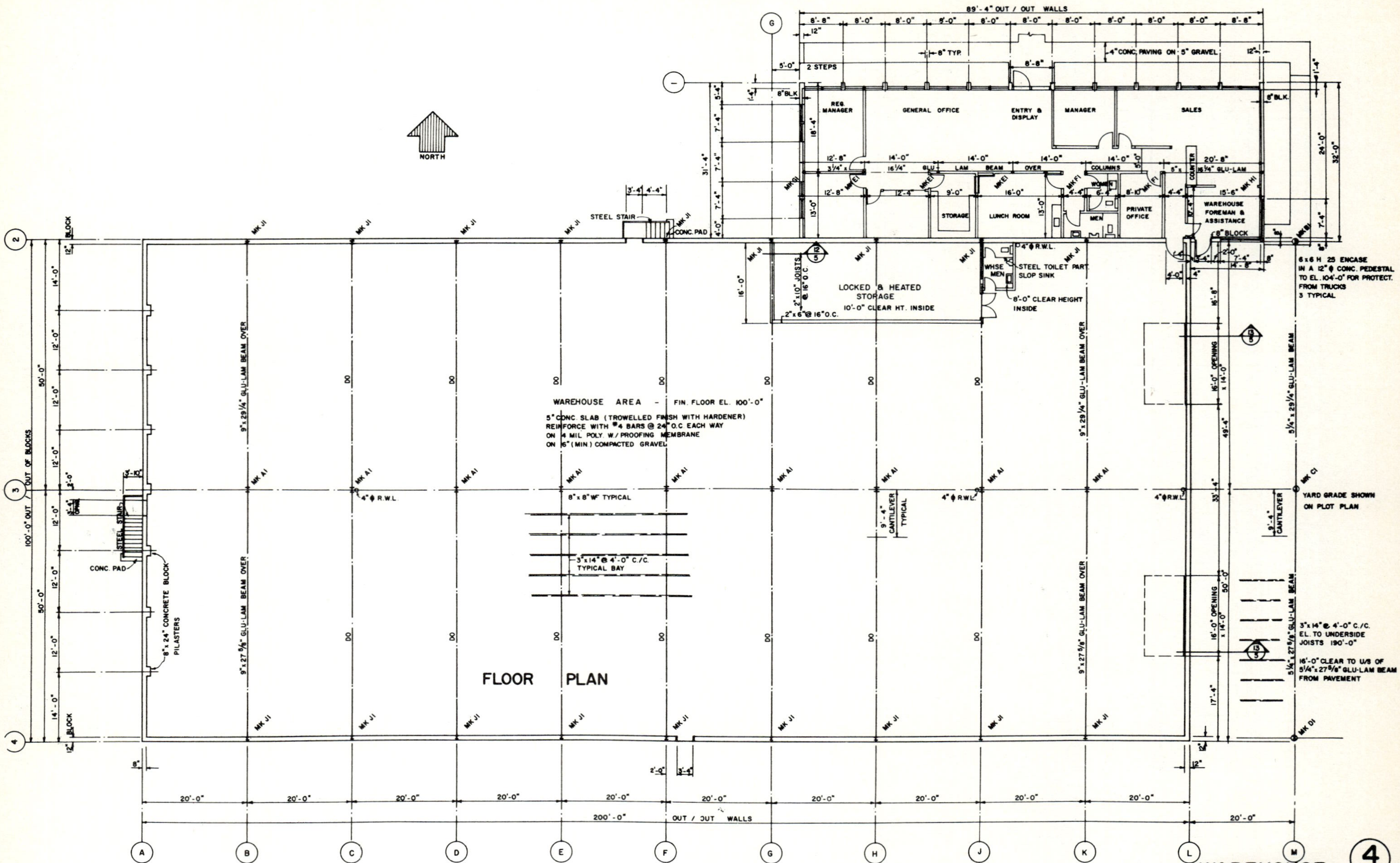

89'-4" OUT / OUT WALLS
4" CONC. PAVING ON 5" GRAVEL
2 STEPS
NORTH
REG. MANAGER
GENERAL OFFICE
ENTRY & DISPLAY
MANAGER
SALES
STORAGE
LUNCH ROOM
PRIVATE OFFICE
WOMEN
MEN
WAREHOUSE FOREMAN & ASSISTANCE
COUNTER
GLU LAM BEAM OVER
COLUMNS
STEEL STAIR
CONC. PAD
LOCKED & HEATED STORAGE
10'-0" CLEAR HT. INSIDE
2"x10" JOISTS @ 16" O.C.
2"x6" @ 16" O.C.
WHSE MEN
STEEL TOILET PART.
SLOP SINK
8'-0" CLEAR HEIGHT INSIDE
4" ϕ R.W.L.
6 x 6 H 25 ENCASE IN A 12" ϕ CONC. PEDESTAL TO EL. 104'-0" FOR PROTECT. FROM TRUCKS 3 TYPICAL
WAREHOUSE AREA - FIN. FLOOR EL. 100'-0"
5" CONC. SLAB (TROWELLED FINISH WITH HARDENER)
REINFORCE WITH #4 BARS @ 24" O.C. EACH WAY
ON 4 MIL POLY. W/ PROOFING MEMBRANE
ON 6" (MIN) COMPACTED GRAVEL
9" x 29 1/4" GLU-LAM BEAM OVER
9" x 27 5/8" GLU-LAM BEAM OVER
5 1/4" x 29 1/4" GLU-LAM BEAM
5 1/4" x 27 5/8" GLU-LAM BEAM
8" x 8" WF TYPICAL
3" x 14" @ 4'-0" C./C. TYPICAL BAY
CANTILEVER TYPICAL
YARD GRADE SHOWN ON PLOT PLAN
16'-0" OPENING x 14'-0"
3" x 14" @ 4'-0" C./C. EL. TO UNDERSIDE JOISTS 190'-0"
16'-0" CLEAR TO U/S OF 5 1/4" x 27 5/8" GLU-LAM BEAM FROM PAVEMENT
8" x 24" CONCRETE BLOCK PILASTERS
100'-0" OUT / OUT OF BLOCKS
200'-0" OUT / OUT WALLS
FLOOR PLAN

WAREHOUSE

4

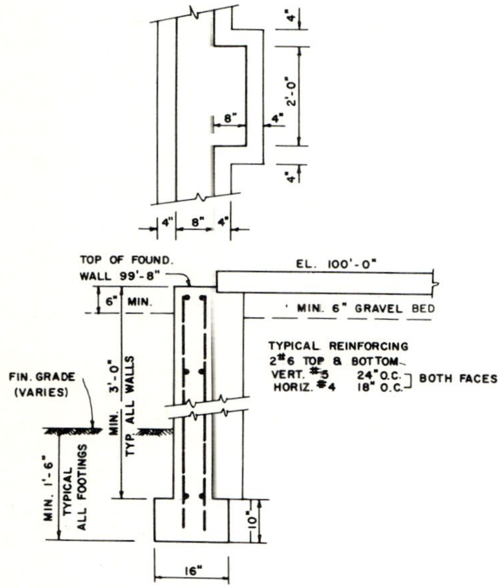

9/5 PLAN & SECTION THROUGH WEST COLUMN FOOTING

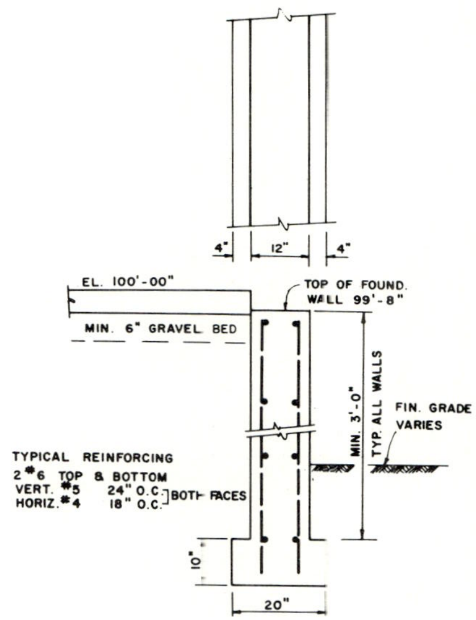

10/5 PLAN & SECTION THROUGH EAST FOUNDATION WALL

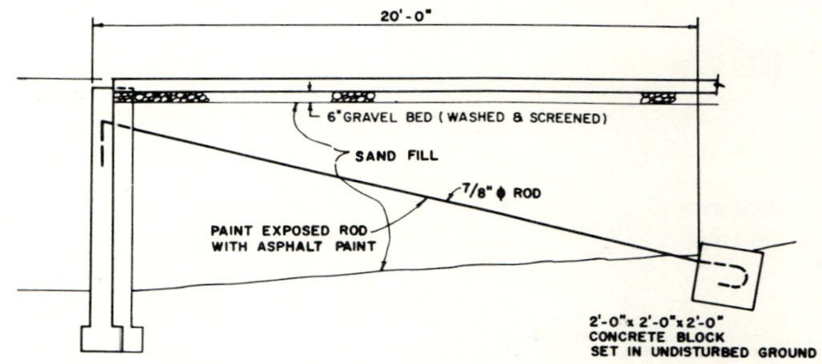

11/5 TYPICAL SECTION AT WALL TIES

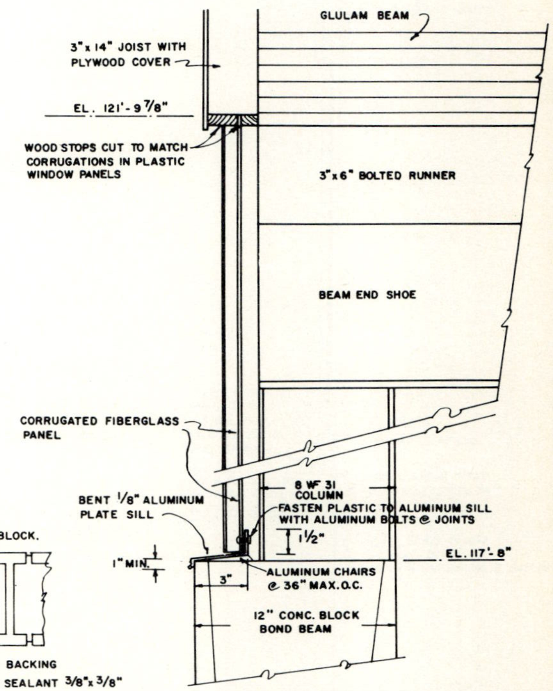

14/5 WINDOW HEAD & SILL DETAIL FOR WAREHOUSE

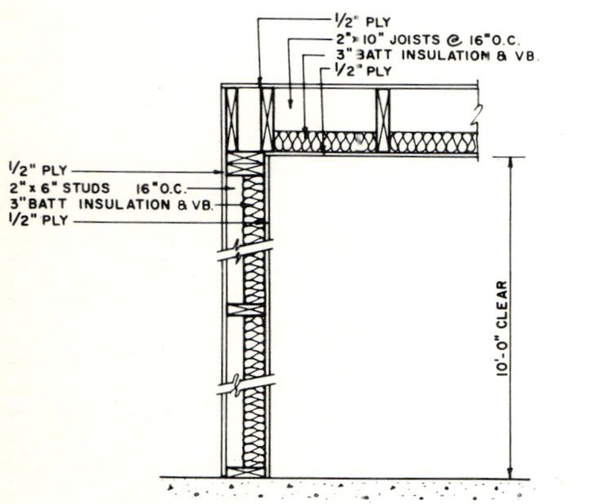

12/5 SECTION THROUGH WEST WALL OF STORAGE AREA (MENS TOILET SIMILAR)

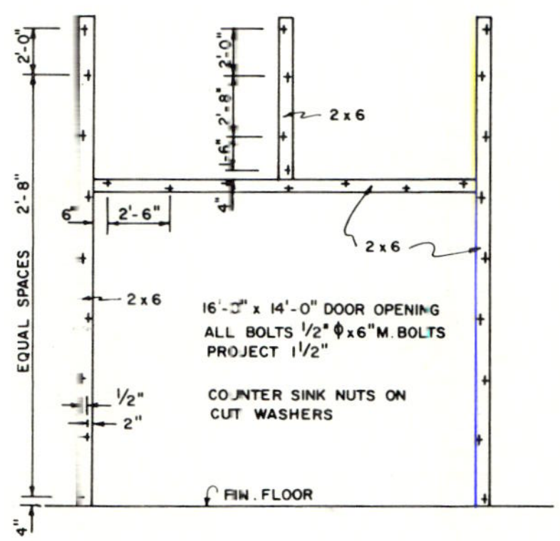

13/5 WOOD FRAMING FOR 2 - 16' x 14' HI-LIFT O.H. DOORS

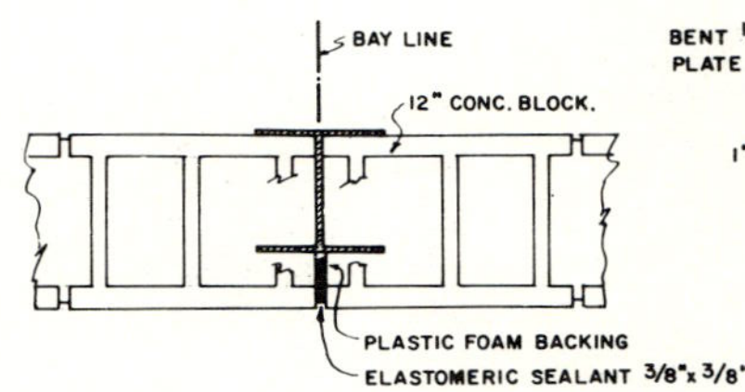

15/5 CONTROL JOINT DETAIL (TYPICAL @ EACH BAY LINE NORTH & SOUTH WALLS)

WALL SECTIONS & DETAILS (WAREHOUSE)

WAREHOUSE 5

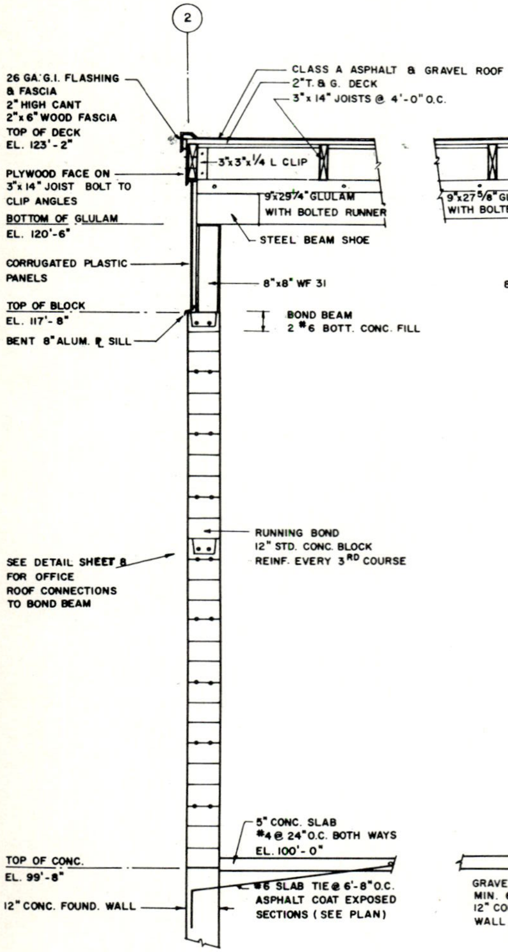

SECTION THROUGH NORTH WALL

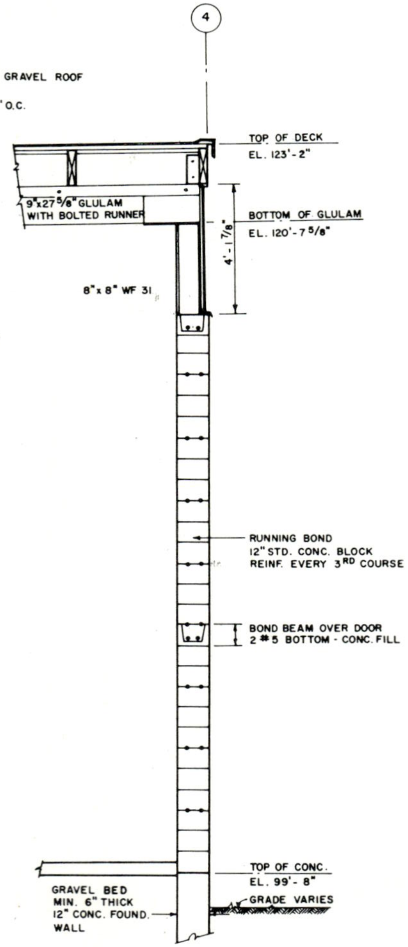

SECTION THROUGH SOUTH WALL

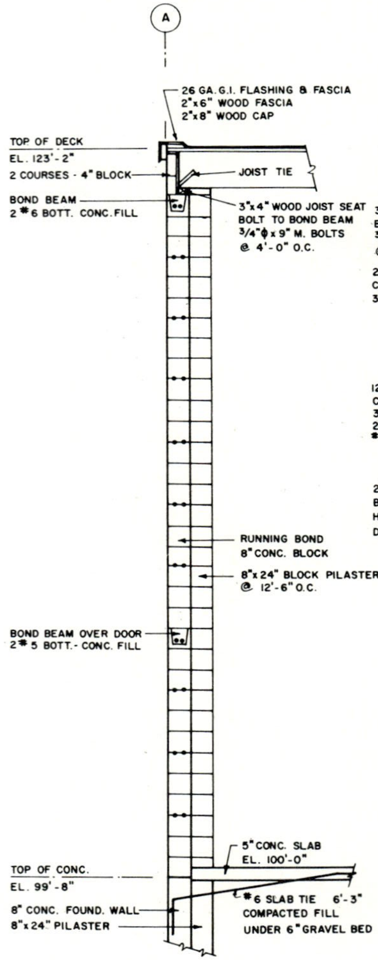

SECTION THROUGH WEST WALL

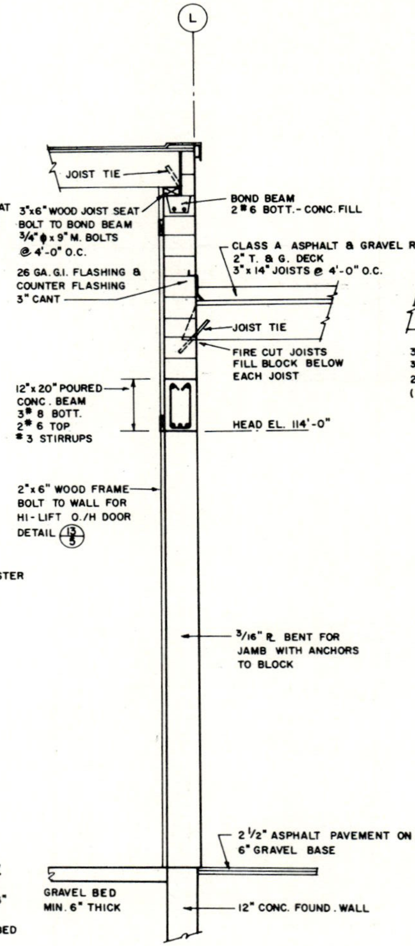

SECTION THROUGH EAST WALL

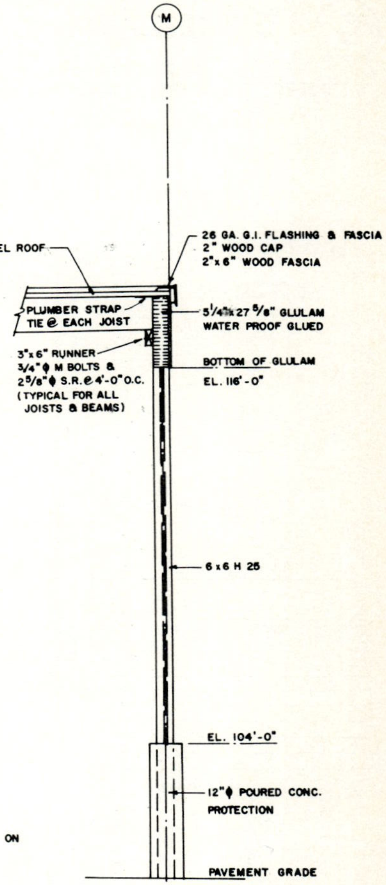

SECTION THROUGH CANOPY SUPPORT BEAM

WAREHOUSE ⑥

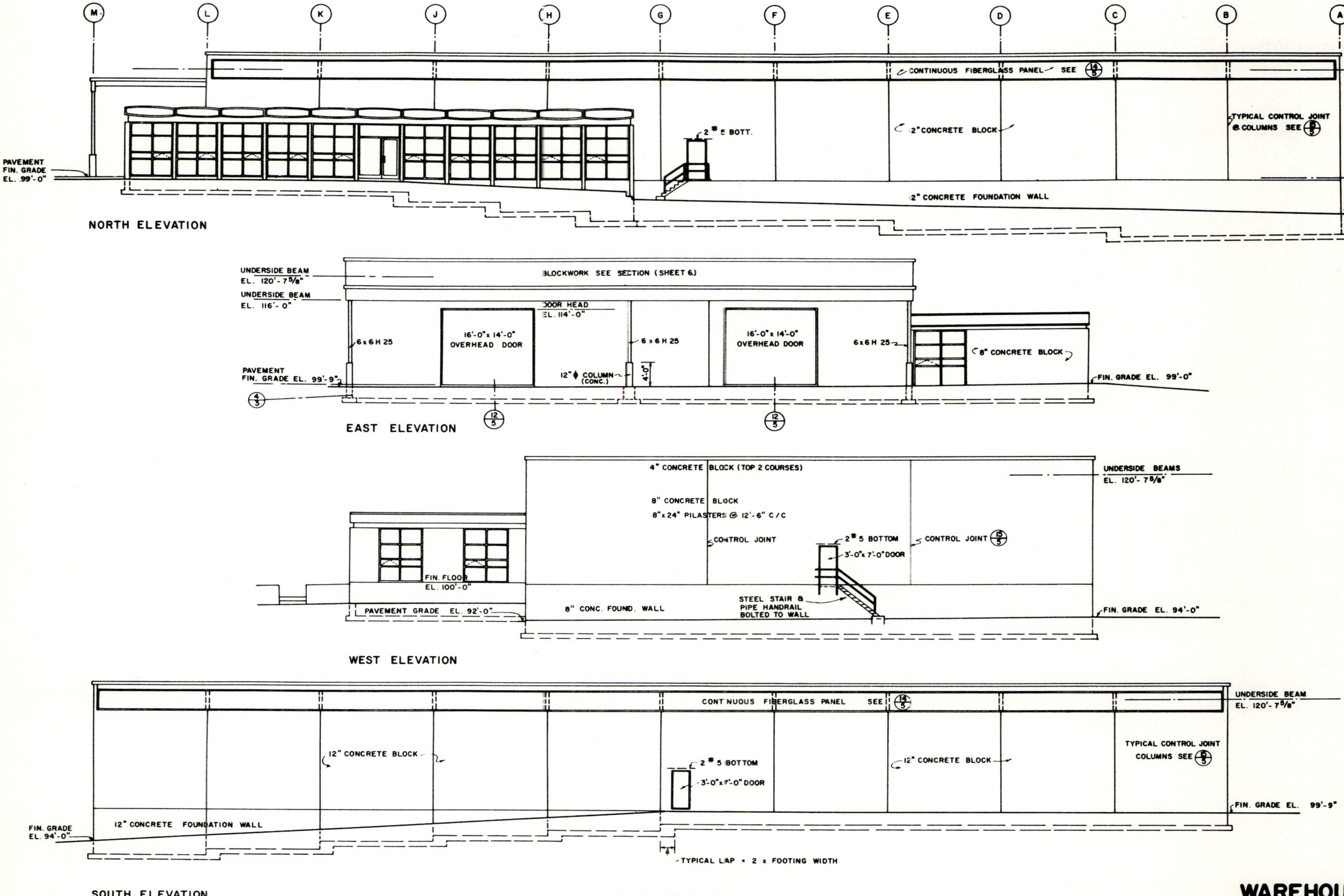

ELEVATIONS

WAREHOUSE 7

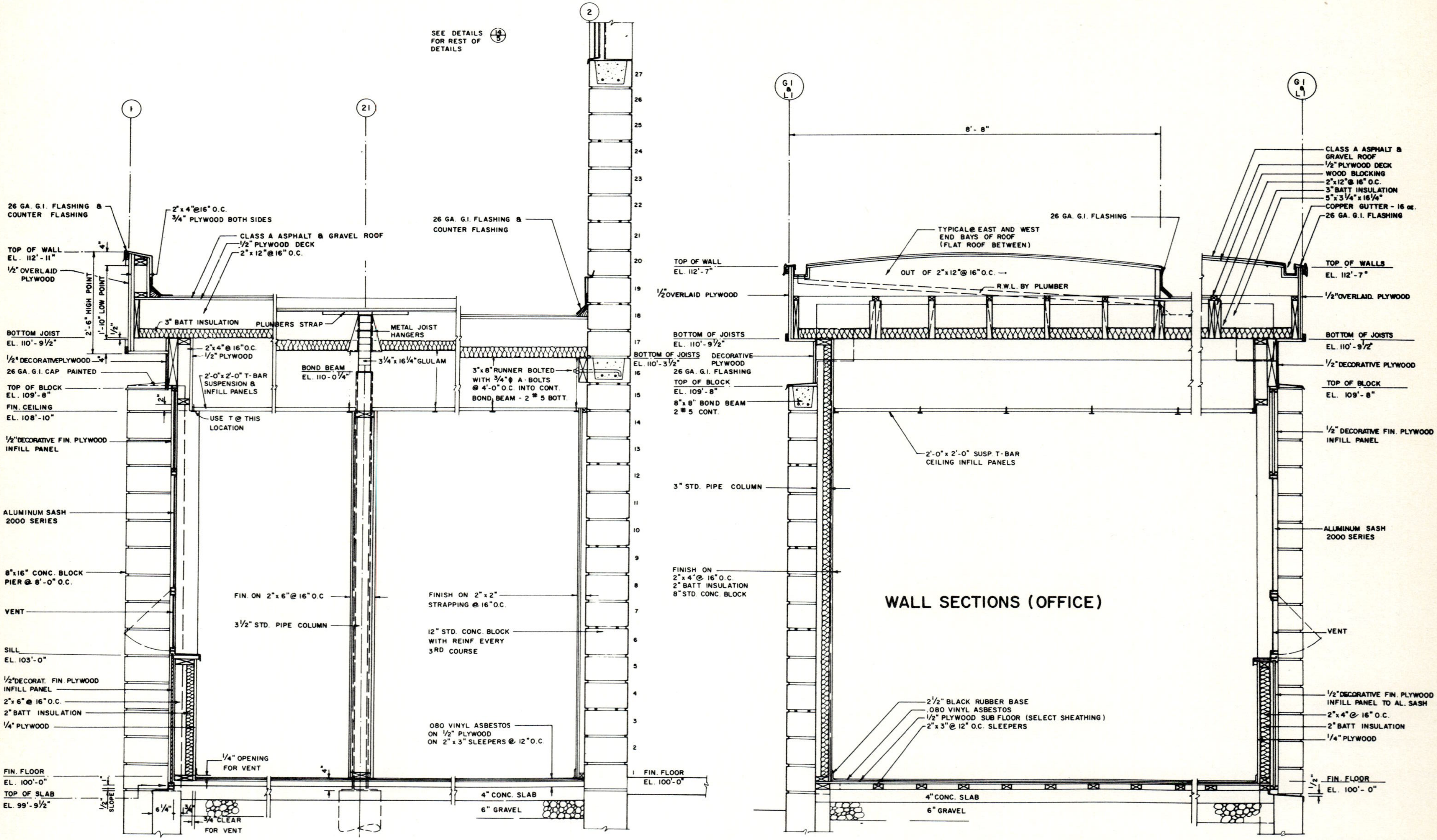

SECTION THROUGH NORTH WALL OF OFFICE

SECTION COLUMN IN OFFICE

SECTION THROUGH SOUTH WALL OF OFFICE

SECTION THROUGH EAST WALL OF OFFICE (TYPICAL FOR WEST WALL)

SECTION THROUGH WEST WALL OF OFFICE AND CURTAIN WALL

WAREHOUSE 8

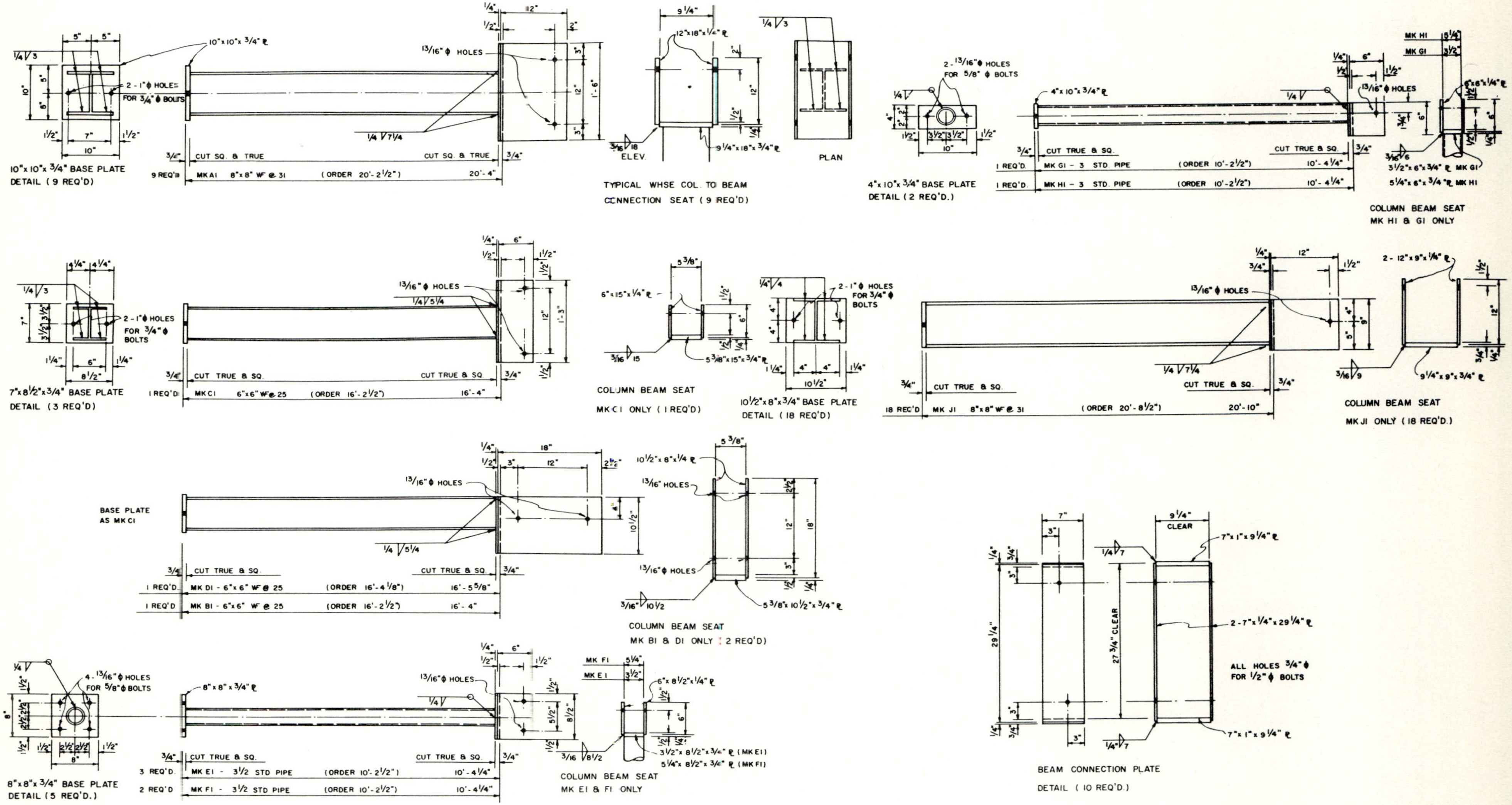
10"x10"x 3/4" BASE PLATE
DETAIL (9 REQ'D)
9 REQ'D MK A1 8"x8" WF @ 31 (ORDER 20'-2 1/2") 20'-4"
TYPICAL WHSE COL. TO BEAM
CONNECTION SEAT (9 REQ'D)
ELEV.
PLAN
4"x10"x 3/4" BASE PLATE
DETAIL (2 REQ'D.)
1 REQ'D. MK G1 - 3 STD. PIPE (ORDER 10'-2 1/2") 10'-4 1/4"
1 REQ'D. MK H1 - 3 STD. PIPE (ORDER 10'-2 1/2") 10'-4 1/4"
COLUMN BEAM SEAT
MK H1 & G1 ONLY
7"x8 1/2"x3/4" BASE PLATE
DETAIL (3 REQ'D)
1 REQ'D MK C1 6"x6" WF @ 25 (ORDER 16'-2 1/2") 16'-4"
COLUMN BEAM SEAT
MK C1 ONLY (1 REQ'D)
10 1/2"x8"x3/4" BASE PLATE
DETAIL (18 REQ'D)
18 REQ'D MK J1 8"x8" WF @ 31 (ORDER 20'-8 1/2") 20'-10"
COLUMN BEAM SEAT
MK J1 ONLY (18 REQ'D.)
BASE PLATE
AS MK C1
1 REQ'D. MK D1 - 6"x6" WF @ 25 (ORDER 16'-4 1/8") 16'-5 5/8"
1 REQ'D MK B1 - 6"x6" WF @ 25 (ORDER 16'-2 1/2") 16'-4"
COLUMN BEAM SEAT
MK B1 & D1 ONLY (2 REQ'D)
8"x8"x 3/4" BASE PLATE
DETAIL (5 REQ'D.)
3 REQ'D. MK E1 - 3 1/2 STD PIPE (ORDER 10'-2 1/2") 10'-4 1/4"
2 REQ'D MK F1 - 3 1/2 STD PIPE (ORDER 10'-2 1/2") 10'-4 1/4"
COLUMN BEAM SEAT
MK E1 & F1 ONLY
ALL HOLES 3/4" ϕ
FOR 1/2" ϕ BOLTS
BEAM CONNECTION PLATE
DETAIL (10 REQ'D.)

total direct costs. Also include the following cash allowances for the work[G] indicated:

Underground drainage	$5,000
Washroom accessories	600
Miscellaneous metals	4,000
Exterior and interior painting	7,000
Landscaping work	20,000
Asphaltic pavings	25,000
High-lift doors, installed	8,000
Finish hardware (materials only)	1,500

The following items should be allowed for, in addition to the 20-percent allowance for direct and indirect overheads and profit: rented construction equipment; survey and layout of building; site cleanup; temporary access road and temporary services; and a temporary site office.

The building site is located at the outskirts of a major city and is within a hundred yards of a freeway. The site is naturally well-drained and was free from any unusual features that would add to the costs of normal site works.

Exercises

1. Measure all the work in the warehouse and office, incorporating the measured quantities provided here. Price the work using Mean's cost data book (or other equally detailed data book); incorporate the cash allowances given, allow for the general-requirement items mentioned, and apply a 20 percent markup for job overhead costs, operating overhead costs, and for profit. To the fullest extent possible, separate the warehouse and office costs. Calculate and compare the overall unit cost per square foot of building with that cited earlier (making any necessary adjustments for time and place).

2. Describe and measure valid *building elements* and with these and the information provided, estimate the costs of the warehouse and office, using the same sources of cost information used in the detailed estimate described in Exercise 1. Compare the results of this and the last exercise and write a commentary on their similarities and differences.

3. Using a selected cost manual (such as Mean's, Boeckh's, or other similar source of cost information and data), estimate the costs of the warehouse and office, using the building areas as the basis, in order to make valid and instructive comparisons with the two estimates made in the previous exercises. Write a commentary in an essay style (of not less than 500 words) on your findings.

4. In detail, make an alternative estimate of the warehouse and office, substituting tilt-up reinforced concrete construction for concrete masonry, and write a commentary on your results as if to advise an owner who wishes to decide on the type of construction.

5. Advise an owner as to the probable costs of warehouse and office buildings similar to that shown in the drawings but of the following areas:

(a) warehouse: 22,000 square feet; office: 2,000 square feet; and

(b) warehouse: 27,000 square feet; office: 3,400 sqauare feet. Clearly show the reasoning behind all of your calculations. Write a report explaining your estimated costs to the owner.

7

A Government Office Building

With this example we see the successful use of the volume method; an estimating method with a poor reputation, a method that has been spurned for decades, but one that should not be allowed to disappear, any more than carpenters should allow the adze or the spokeshave to disapppear, because these tools have special uses that in these days may not arise often but are valid uses nevertheless. Similarly the volume method of estimating building costs also has its applications, and one of them is demonstrated here.

One reason presented earlier (in Part One) for the poor reputation of the volume method (or, as it is sometimes called in the United States, the *cubic foot method*, because we have not yet succumbed to the international metric system) was the lack of any predictable relationship between a building's volume and the total area of its superficial elements, those parts that give rise to most of a building's costs. This inevitable lack of a predictable relationship begins two-dimensionally with the *area* and the *perimeter* of buildings; in the volume method we see this weakness compounded from two to three dimensions. But, as we are also about to see, there are times when the volume method of making a preliminary estimate is both useful and successful.

The subject of this estimate is a reinforced concrete office building of one story with a partial basement. (No drawings are included as we are solely concerned with estimating method.) At first sight it looked rather like a military command post built to withstand attack; instead it was designed to withstand the extremes normal at a weatherstriken northern site. A similar building had been constructed about a year earlier at another site about 800 miles away, but there were significant differences between the two buildings, particularly in their mechanical and electrical services and, of course, between their respective site works. Because of the similarities in design and (what is more to the point) because of the brief time given to me to produce a preliminary estimate, I decided to risk using the volume method—with certain critical adjustments. The estimate proved reasonably successful.

Compare the estimate with the bids received, which (to the nearest $50) were:

Bids Received	
Company A	$899,400
Company B	$905,550
Company C	$957,000
Company D	$1,106,400
Company E	$1,132,300
Average Bid	$1,000,130
Estimate Amount	$906,000
Difference	$94,130 (= 9.4% of the average bid amount)

Classically among quantity surveyors this difference between the estimate and the bids is considered acceptable, especially with two bids lower than the estimate's amount and the estimate's amount within 10 percent of the average bid. The spread between highest and lowest bids may appear excessive at first sight, but the project site was remote.

Briefly the method of estimating was as follows: (1) to analyze the costs of the similar building already built after deducting the costs of atypical items of work, and then to calculate a cost per cubic foot for the typical building; (2) to apply that unit cost to the subject building's volume; and (3) to add the costs of atypical items and conditions related to the subject building. The handwritten estimate sheets illustrate how this methodology was followed.

Because the aim of the estimate is to be as close as possible to the average of the bids subsequently received, the average of the bids received for the previous building was used to calculate the cost per cubic foot, rather than the amount of the lowest bid, or the contract sum. From this amount ($780,000) were deducted the costs of the atypical features: site works, mechanical, electrical work, and also the cash allowances (mostly for atypical items). The residual amount was divided by the building's volume to give a unit cost per cubic foot of $4.35.[1]

Next this unit price was applied to the volume of the subject building, and the estimated costs of atypical items of work added: the costs of site works, mechanical, electrical work, and the specified cash allowances. Also added were the costs of other atypical work for a house and a garage. Finally a subjective judgement was made as to the additional costs likely because of location, and an allowance of 5 percent was added to give a total estimated amount of $906,250. This was then rounded to the nearest thousand dollars.

It would be interesting to have seen just how much the various bidders added to their bid estimates for the location, and one cannot but wonder if it was not at that point in their estimates that the bidders created the major part of the differences among their bids.

To estimate the costs of atypical items of work, approximate quantities were measured from working drawings (contract drawings in the case of the earlier building and almost completed working drawings for the subject building) and priced with unit prices, or in the case of the earlier building, by cost information from the contract *schedule of values*.[G] In order to show this procedure, transcriptions of the estimate sheets for this analysis are reproduced here (sheets G1 through G4). The deductions of atypical items from the earlier building were more or less a matter of fact, and their inclusion in the example serves only to make it complete, but the addition of atypical items for the subject building may deserve some comment.

Site works. Details have been omitted from the example as we have already seen enough about excavation, fill, and pavings.

Atypical features. The subject building contains a chimney stack, a clerestory with windows, a luminous ceiling, and double-glazed windows. The mechanical and electrical services were priced by building area except for the plumbing work (details omitted).

Was it worthwhile deducting and adding the atypical items? Or could equal results have been obtained by simply using one unit price per cubic foot of building? Let us consider the *analyses of costs.*

Of the earlier building, $780,000 was divided by 85,750 cubic feet equal $9.10 per cubic foot.

Of the subject building, 91,000 cubic feet was multiplied by $9.10 to equal $828,100 (estimated total costs.)

[1]As mentioned, all costs in these examples have been increased by a factor selected for each estimate so that they are closer to today's costs.

SHEET G-1

(1). Analysis of Costs of Basic Structure of Similar Building

Building project's total costs (average of bids)		$ 780,000
Deduct: Site Works; Non-typical features; Mechanical and Electrical; and Cash Allowances		-407,000
Divide total costs of basic structure by volume 85,750cf		$ 373,000
Cost per cubic foot of basic structure	=	$4.35

(2). Estimate of Costs of Subject Building

Basic Structure's costs: 91,000cf @ $4.35	=	$ 395,850
Add: Site Works; Non-typical features; Mechanical and Electrical; and Cash Allowances	=	384,300
Add: Ancillary Buildings (residence and garage)		82,950
		$ 863,100
Add: Extra costs due to different location (+5%)		43,150
Total Estimated Costs of Project		$ 906,250

(3) Comparison with Bids Received

Average of five bids received*		$1,000,130
Estimated Costs of Project		906,000
Difference		$ 94,130
	=	9.400%

*Lowest—$ 899,400
Highest— 1,132,500

The difference between the estimated costs and the average of the bids received is just within acceptable tolerances (usually 10%). The estimated costs of $906,000 was about $7,000 more than the lowest bid. The relatively large difference may be due to:

(a) The remote locations of the two buildings compared and the distance and difference between them

(b) The method of estimating employed.

SHEET G-2

Analysis

Calculation of the costs of a Basic Structure from a previous project's costs for a similar but slightly smaller office building:

Average of bids received for previous project			$ 780,000
Deduct: Site Works; Non-typical Features; Mechanical and Electrical Work; and Cash Allowances:			
1. Site Works			
Excavation and drainage	$22,500		
Irrigation system	17,450		
Asphalt pavings	16,390		
Landscaping and misc.	87,000		
	=	143,340	
2. Non-typical Features			
Precast concrete facings	21,750		
Additional insulation for electrical heating	10,650		
Canopies, installed	18,000		
Related structural work	21,000		
	=	71,400	
3. Services			
Mechanical work	39,000		
Electrical work	+69,160		
	=	108,160	
	322,900		
Mark-up for overhead and profit	+32,300		
	355,200		
4. Cash Allowances	+51,800	−407,000	
(carried forward)			$373,000

This estimated total ($828,100) is less than the lowest bid and about 17 percent less than the average of bids; not a successful result. Even if the 5 percent allowance for location is added, the result is still less than the lowest bid.

Thus the volume method of estimating is at least vindicated in part—with certain important provisions: that it be used only where costs of similar buildings are available and that atypical items of work are kept separate from the actual application of the building volume to costs. In this example the earlier building was also situated far from any large towns or cities, and therefore the costs derived from it were already influenced by a remote location. The 5 percent added to the costs of the subject building was for *additional costs* due to location. The fact that

SHEET G-3

Net Costs of Basic Structure (brought forward)	=	$ 373,000	
Gross Volume of Building (divider)		÷85,750 cf	
Cost of Basic Structure: per cubic foot	=	$4.35	
Estimate			
Building's Basic Structure: 91,000cf @ $4.35	=		$ 395,850
Add: Site Works; Non-typical Features; Mechanical and Electrical Work; Cash Allowances; and Ancillary Buildings:			
1. Site Works			
Excavation, fill, drainage (as detail sheets)	=	$ 15,000	
Pavings & landscaping		110,000	
	=		+125,000
2. Non-typical Features			
Lightweight topping roof 4,000sf @ 75 cents	=	3,000	
Masonry stack: 281f @ $100	=	2,800	
cleanout, cap, footing	=	200	
Clerestory			
8×72 RC Beam 921f @ $60	=	5,520	
skylights: 6 No @ $900	=	5,400	
Interior cladding 450sf @ $2	=	900	
Exterior cladding 450sf @ $6.50	=	2,930	
Flashings 2001f @ $6	=	1,200	
Luminous ceiling: extra cost	=	3,000	
(carried forward)		$24,950	$520,850

SHEET G-4

(brought forward)		$24,950	$520,850
Windows: extra costs of insulating type		9,000	
Cabinetwork (extra)		6,300	
		40,250	
Mark-up, as before		4,050	
		=	$44,300
3. Services			
Mechanical work (H & V.) 8,000sf @ $7.50	=	60,000	
Electrical work 8,000sf @ $6.00	=	48,000	
Plumbing (as detail sheet)	=	29,000	
		=	137,000
4. Cash Allowances (as specified)		=	78,000
5. Ancillary Buildings			
Residence: 1120sf @ $60	=	67,200	
Garage: 750sf @ $21	=	15,750	
			=82,950
			$863,100
Add: Extra Costs of Project due to Location and additional distance to site (5%)		=	43,150
			906,250
Total Estimated Costs of Project		=	$906,000

both buildings were not only similar but also built in remote places was one more reason for the successful use of this estimating method. It would have been more difficult to make an accurate estimate using approximate quantities and to obtain numerous unit prices that were valid for a remote site.

Exercises

1. Find drawings and cost information for a building similar to one illustrated in this book (or find such information for any two similar buildings) and emulate the estimating method described in this case study. Do not overlook cost differences due to location, time, and different construction markets and make estimated allowances in your costs accordingly. For atypical items, use whatever estimating method appears effective and expedient.

2. (a) With an analysis of the costs of a known building (possibly from a schedule of values or from another example) for which you have drawings and specifications or to which you have access, first measure the building's volume and then apply the volume to the total costs for each trade or division of work in the building.

(b) Write a brief report on the significance of your findings, showing the unit cost per cubic foot (cubic meter) for each trade or division, and your opinion on the significance of this method of analysis to estimating generally.

(c) Compare the unit prices with unit prices per

square foot of building area for each trade and division of work and comment on their significance in estimating.

3. (a) List, describe, and briefly discuss each pro and con relative to the estimating method used in the actual case study presented in this chapter, as it applies to (1) the particular building and (2) estimates generally.

(b) Describe what you consider the primary weaknesses in the estimate presented in the example.

4. Answer the following questions about the case study in this chapter:

(a) Which atypical items of work were not considered?

(b) Why do you think those atypical items were not considered?

(c) What are the main things about location that affect construction costs? Can these be accurately estimated by adding to an estimate's total one amount calculated by a percentage figure? Explain your answer.

5. Other than approximate quantities what other estimating basis might have been used in this case? List and briefly describe the pros and cons involved.

6. Compare the unit costs per cubic foot of building volume for selected classes of buildings to the costs per pound of selected commodities, such as prime beef, or pork bellies, or another basic commodity, over a recent period of at least 5 years, and report on your findings.

7. Select a suitable cost index for building construction costs that can be easily obtained from common sources and applied in construction cost estimates. Justify the efficacy of your selected index.

8

Estimating Costs of Structural Steelwork

How many new estimators and students of estimating have struggled to translate a building's structural steel drawings into measured quantities of work ready for pricing? How many have been mystified by the absence of information on drawings and in specifications about steelwork bases, connections, bracings, and other details of structural steelwork and have wondered how to measure what is not shown? For the inexperienced estimator, measurement of structural steelwork is often a problem, but even the experienced cost consultant and designer do not escape entirely.

How many have wondered if there is not a better way to make preliminary estimates of steel structures, other than so many pounds of steel per square foot of building as the only apparent alternative to a detailed takeoff of steelwork quantities? The Canadian Institute of Steel Construction (CISC) together with the Canadian Steel Construction Council (CSCC) has come to their aid with a booklet entitled "A Project Analysis Approach to Building Costs," which shows a rational and efficient method of estimating steelwork costs for budget and design purposes. Because the booklet's methodology is so streamlined and well explained, even the neophyte can quickly grasp the fundamentals of steelwork economics and thereby acquire a better understanding of steelwork in general.

It is not, primarily for these reasons however, that I have selected from this booklet on steelwork costs an example of estimating and have reproduced from it an actual case study.[1] Rather it is because the estimating method shown is a model of methodology for estimates of costs of many other kinds of construction work, a method that will become increasingly important as newer methods of contracting replace the traditional, as applied construction economics become a normal part of the design process, and as the costs of a project are normally seen as continuing throughout the life of its buildings and infrastructure.

The CISC/CSCC booklet in describing its project analysis approach to building costs does in fact refer to not only capital costs but also to operating and maintenance costs, life cycle costs, and financing costs, including mortgaging, construction funding, rental income, and capitalization. Section A of the booklet deals with the project analysis approach to costs. In Section B is a working example that compares the cost of a steel-framed structure with a reinforced concrete structure for the same building; in Section C the estimating method already cited is explained in detail; in Section D is an explanation of the cash flow analysis used in the comparison of steel with concrete; and finally in Section E is information on other CISC/CSCC publications and computer programs on steelwork design and costs.

Even the most traditional contracting methods recognize that it is not always necessary for a bidder to have the kind of details that one expects to find in shop drawings[G] in order to bid for a stipulated-sum contract or subcontract. In bidding for other kinds of contracts, the design information (translated by estimators into economic information) can take forms different from that required by stipulated-sum contracts. For example, for maximum cost-plus-fee contracts[G] the contract sums (the maximum costs) may be established by estimating methods similar to those used in preliminary estimates cited in Part 1 and illustrated in this Part 2; although the CISC/CSCC estimating method is intended for use by designers to select an economical structural system, in principle it has other applications.

The CISC/CSCC method of estimating steelwork costs refers to:

1. Steel member length (L). From grid dimensions (center to center of members).

2. Net steel mass (NSM). Member length x nominal mass per unit length.

3. Connection factor (Confac). To allow for material and labor for connections between members.

4. Gross steel mass (GSM). Obtained by applying a Connection factor (Confac) to the Net steel mass (NSM) for each category of member.

5. Cost factored mass (CFM). Obtained for each steel category by applying a Cost factor (Cosfac) to measure the influences of *design complexity* and *raw material price.*

6. Metric cost index (MCI). Calculated at time intervals for different regions from cost estimates for a sample building; this is a base cost upon which to build an estimate of steelwork costs, as illustrated in the reproduced project example, for a specific building in a specific place at a specific time.

7. Bay size modifier (BSM). Because the MCI is calculated for a building with a standard bay size, different bay sizes and the resultant differences in sizes of members and numbers of connections must be allowed for in an estimate.

8. Estimated steel cost (C). (Expressed in dollars per metric ton) for each steel member is, Member length x Nominal mass per unit length (in kilograms) x Confac x Cosfac x ((MCI x BSM) divided by 1,000). (Or, NSM x Confac x Cosfac x $\frac{\text{MCI x BSM}}{1{,}000}$).

Connection and cost factors are included in the CISC/CSCC booklet. A metric cost index figure is available at a CISC regional office (in Canada only).

With the definitions given, the reproduced example should be self-explanatory; it also illustrates the conciseness and simplicity of the method. But to appreciate fully the method it is necessary to see all the content of the CISC/CSCC booklet on which I have largely relied in this chapter. I have, however, only outlined its contents.

[1]With the permission of the publishers to whom acknowledgment is due also for most of the content of this chapter.

CISC/CSCC METHOD OF ESTIMATING STRUCTURAL STEEL
PROJECT: ManuLife Place, Edmonton, Alberta

Member Classification	Net Steel Mass *(NSM)* -t	X	Connection Factor *(Confac)*	=	Gross Steel Mass *(GSM)* -t	X	Cost Factor *(Cosfac)*	X	CISC/CSCC Metric Cost Index *(MCI)*	=	Estimated Steel Cost *(C)* -$
Beams											
C, M, or S to 20 kg/m	1.61		1.05		1.69		2.5		1 100		4 650
W-Shapes to 50 kg/m	46.18		1.05		48.49		1.2		1 100		64 010
W-Shapes to 50 kg/m (Stub Girder Purlins) - Gerber construction - *Cosfac* = 80% of 1.55	366.17		1.05		384.48		1.24		1 100		524 430
W-Shapes 51 - 240 kg/m	195.08		1.05		204.83		1.0		1 100		225 320
Stub Girders	335.43		1.4		469.6		1.05		1 100		542 390
WWF-Shapes	27.5		1.05		28.88		1.15		1 100		36 530
Cambering-(No. of Members x $ $\left\{\frac{MCI}{30}\right\}$ each)			728 x $ $\left\{\frac{1100}{30}\right\}$								26 690
Columns											
W-Shapes to 50 kg/m	4.77		1.15		5.49		1.15		1 100		6 940
W-Shapes 51 - 285 kg/m	84.19		1.15		96.82		1.05		1 100		111 830
WWF-Shapes	797.67		1.15		917.32		1.2		1 100		1 210 860
Spandrels											
W-Shapes to 50 kg/m	126.06		1.05		132.36		1.3		1 100		189 280
W-Shapes 51 - 240 kg/m	345.47		1.05		362.74		1.15		1 100		458 870
Trusses - (HSS Chords)	13.04		1.15		15.0		1.8		1 100		29 690
Ancillary Steel	132.83		1.2		159.4		2.3		1 100		403 270
TOTALS					2827.1 t						$3 834 760

$$\text{Average Unit Price} = \frac{\$3\ 834\ 760}{2827.1\ \text{t}} = \$1\ 356.43/\text{t}$$

This method and example of estimating by CISC/CSCC illustrates well many of the principles and precepts for construction economics and estimating—especially for preliminary estimates—cited in this book: It shows the need for estimating methods to be based on design fundamentals; it demonstrates the principle of economic utility; it places capital costs in realistic relationship with the other costs of developing and owning buildings; it shows the importance of cash flow to a project; and it shows how construction economics should be applied as part of the design phase of construction; and it shows generally how more estimates of costs will be made in the future.

Additional material on estimating steelwork costs is included in Part 3.

Exercises

1. From your study of the steelwork estimate reproduced here, write an essay describing the estimating method used and its underlying principles; explain in detail why this method of estimating is or is not suitable for bidding purposes.

2. Referring to the cost factors (Cosfac) listed in the estimate example, write your comments on the significance of these numerical factors relative to the costs of steel members and to an entire estimate.

3. Because steel members' lengths are measured from grid lines on the steelwork drawings, their lengths are described in the booklet as *theoretical.* Explain why theoretical lengths of steel members

are used and why these lengths can be used in estimates.

4. Obtain from a structural steel contractor copies of the drawings for a steel structure together with the total costs of the steelwork. Make an estimate of the costs using the CISC/CSCC method of estimating and compute the equivalent of a cost index, or base cost, for steelwork appropriate to that project.

5. Obtain a copy of the CISC/CSCC publication and study the methodology used in comparing steel and reinforced concrete structures and the costs related to them. Apply a similar comparison to the structure of a selected building of known cost.

9

An Elemental Analysis

This example comes from the files of Hanscomb Consultants, Inc.,[1] an international company of quantity surveyors, engineers, and construction consultants. The company has offices in many parts of the world, and the elemental analysis reproduced here is from the *Hanscomb Newsletter* of July 1982. (That same issue also reported on Hanscomb Inc.'s twenty-fifth anniversary in Canada. The company's first office in the United States was opened in Chicago in 1969, which shows that quantity surveyors are not strangers in North America.)

Hanscomb "provided cost management and value analysis services" for the Alberta Agricultural Building project, cited here. "The value analysis program, which was continuous from project inception to contract award, generated 97 value analysis proposals [for the owner], of which 44 were accepted, contributing \$1.9 million in capital cost reductions. The contract award was approximately \$2.0 million below budget."

[1]With whose permission the cost analysis in this chapter is reproduced and to whom thanks are due for most of the material in this chapter.

I have included this example to show how building elements are used in practice by professional cost consultants. The cost analysis shown is not strictly an estimate but rather a cost account and the final result of several preliminary estimates. But, of course, an elemental estimate might take the same format. The costs shown (in Figures 9-1 and 9-2) are based on measurements in square meters (1 square meter = 10.7638 square feet; divide the costs by 11 to get an approximation per square foot). In particular, your attention is drawn to the following points.

The first sheet of the cost analysis, shown in Figure 9-1, contains the physical data of the building. Notice particularly the areas and volumes, the ratios, and the capacities that are calculated. These are all quantitative representations of features of the building's design that facilitate the comparison and rationalization of building costs.

The second sheet, shown in Figure 9-2, contains the cost analysis proper, listing elements, element costs, element amounts, and their unit costs. The list of elements on which the costs are based is essentially the same as that standard list of elements reproduced in the Appendix. (More is said about *elemental estimates* in Part 3.)

Look at any cost analysis and anyone might be bemused by all the figures. It is necessary to learn to read this kind of document. (The format used here is not uncommon.) A useful precept is to go always from the general to the particular when faced with a complex array of detailed information. (The same approach is useful in studying estimate summaries.) First, read through the building's outline specifications to get a general idea of its features. Then find the gross floor area (GFA): in this case, 30,240 square meters (325,500 square feet, approximately). The total costs are $20,981,000, and the total unit cost $693.81 per square meter (=$64.62 per square foot). Now we can begin to get a handle on it.

Looking at the second sheet (Figure 9-2), we see that among the major elements, *100 Foundations* cost $436,960, the unit cost is $14.45 per square meter ($1.34 per square foot), and it represents 2.08 percent of total costs. Similarly, for *200 Building Shell*, the total costs are $6,426,350 and the unit cost $212.51 per square meter ($19.74 per square foot), representing 30.63 percent of total costs. Let us look further into the makeup of these figures for the building shell in the analysis—in which the breakdown is shown *below* the totals. Subelements numbered 210 through 240 are shown, each with its area (except for Stairs), unit rate, total cost, and unit cost.

210 Structure is shown with a total area of 37,674 square meters, yet the total floor area (GFA) is given as 30,240 square meters. How can this be? The difference is made up by adding the roof area and deducting the total area of the stairs. In practice I should want to see a third sheet of information, including the total volume of concrete or the total weight of steel for the structure, which information is assuredly in the cost analysist's file. This structure (210) cost $3,182,310 (the total amount of subelements 211 through 213), or $9.78 per square foot (= $105.24 per square meter). (You must become accustomed to reading the totals above the constituent parts, the reverse of the usual order for written addition.)

This cost analysis, because it is published in a newsletter, does not give all the information that one might otherwise expect, but it gives a good idea of the general approach to elemental analysis and to the use of quantitative expressions for design features of buildings. Studying it, I have the feeling that we have scratched only the surface of this kind of analysis and that we shall be able to express quantitatively the features of buildings with much greater precision and effect. The North American construction industry and its contracting companies in particular did not take to the use of contracts with quantities[G] or bills of quantities[G] as contract documents when they were tentatively introduced by firms such as Hanscomb, and these are rarely ever seen in North America (although they are common enough elsewhere).

As purchasers of construction become more aware of the possibilities of applied construction economics it seems probable that some kind of more intensive quantitative analysis will become necessary. Metaphorically we might view this example of a cost analysis as an iceberg, with only about 15 percent of its bulk showing above the surface. Below the surface we may imagine (and see if we look for them, through more analysis) much more detail, with each element and subelement consisting of items of work,[G] and each of these containing several unit prices for material costs, labor costs, equipment costs, job overhead costs, and so on.

Exercises

1. From the information given, calculate possible overall dimensions for the building (shown in the sketch) and test their appropriateness and correctness (mathematically) against the given areas and volumes and ratios until you are satisfied with them. Demonstrate your calculations, tests, and results.

2. Using a cost manual (such as Means' *Construc-*

Building type: OFFICE BUILDING	Classification No. 315
Project: ALBERTA AGRICULTURE BUILDING	Location: EDMONTON, ALBERTA
Owner: ALBERTA GOVERNMENT Architect: FORBES LORD FELDBERG SCHMIDT CROLL General Contractor: P.C.L. CONSTRUCTION LTD.	Tender date/completion: Oct. 1980/Dec. 1982 Market conditions: Competitive (6 bidders) Cost index: 230.8

Description of building Low rise 3 floors above grade with atrium, and one basement. Rectangular on plan with perimeter glazed fire exit corridor and four stairs. Basement parking for 232 stalls with cafeteria, library, printing, storage and loading dock on main floor. Open plan offices on 2nd and 3rd level. Penthouse mechanical room.

Sketch

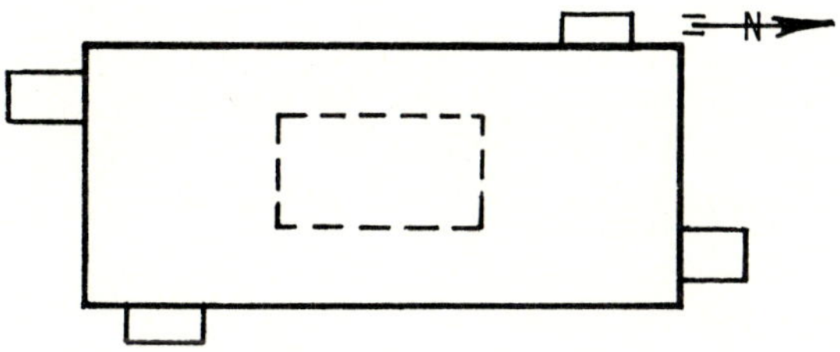

Outline specifications

100 R.C. spread footings, col. bases and fnd. walls
210 R.C. bsmt. slab & susp. slab on steel beams to main level, struct. steel & composite conc. steel deck electrified & fireproofed beams
220 4-ply built-up roofing, double glazed atrium skylight
230 Brick, R.C. walls; coated steel spandrel panels, insulation, precast conc. back-up; profiled metal siding & insul. to penthse
310 Alumin. doors & windows, concrete block, drywall coated steel spandrel panels to atrium walls. Tempered glass to corridor
320 Concrete floor hardener, quarry tile, carpet, glazed ceramic, VAT - paint, susp. DW, susp. T-bar, acoustic tile, Parline reflective & non-reflective ceiling - paint, face brickwk, vinyl covering, glzed mosaic
330 Kitchen equipment and built-in fittings
400 Hydraulic passenger and freight elevators
510 V.A.V. system
520 Electrified metal "Q" deck flooring, daylighting control
900 External parking, roads, soft and hard landscaping

Cost per m²: $ 615.04	per m³: $149.52	per - $ -

Areas and volumes: Incl. Basement Parking

Gross floor area (GFA):	30,240 m²
Net floor area:	23,442 m²
Volume:	124,392 m³
Exterior cladding area:	5,914 m²
Roofing area:	7,757 m²
No. of storeys above grade:	3 No.
No. of basement levels:	1 No.
Accommodation units:	680 Staff

Ratios : Incl. Basement Parking

Complexity factor	–	2.02:1
Net floor area/GFA	–	0.78:1
Volume/GFA	–	4.11:1
Exterior cladding area/GFA	–	0.20:1
Roof area/GFA		0.26:1
Length of partitions/GFA	–	0.14:1

Capacities : Excl. Basement Parking

Percent exterior wall glazed:	35%
Soil characteristics:	Lacustrine Clay Glacial Till
Density plumbing fixtures :	1 per 1,737 SF
Heating capacity :	33.6 BTU/Hr per SF
Cooling capacity :	793 SF/Ton
Ventilation capacity :	0.69 CFM/SF
Lighting intensity :	70 ft. candles

Floor area (by level)

Level	No.	Gross Area	%
Basements	1	8,133 m²	27
Ground floor	1	7,707 m²	26
Upper floors	2	14,046 m²	46
Penthouse etc.	1	354 m²	1
Totals	5	30,240 m²	100%

Floor area (by type)

			%
Office Area		12,933 m²	43
Basement Parking		7,257 m²	24
Storage		1,128 m²	4
Kitchen & Servery		386 m²	1
Other		1,738 m²	6
Net usable/rentable floor area:		23,442 m²	78%
Circulation	3,105 m²		
M & E spaces	1,552 m²		
Toilets	302 m²		
Walls, Partitions etc.	1,839 m²		
Unassignable/unusable floor area:		6,798 m²	22%
Gross floor area		30,240 m²	100%

Analysis No: N.13

Elemental Category	Element cost: Quantity	Element cost: Unit Rate	Element amount: Sub	Element amount: Group	Cost per m^2: Sub	Cost per m^2: Group	%
100 FOUNDATIONS	8,363 m^2	52.25		436,960		14.45	2.08
110 Normal foundations	8,363 m^2	40.15	335,760		11.10		
120 Basement excavation	30,224 m^3	3.18	96,200		3.18		
130 Special foundations	–	–	5,000		0.17		
200 BUILDING SHELL	m^2			6,426,350		212.51	30.63
210 Structure	37,674 m^2	84.47	3,182,310		105.24		
211 Lowest floor construction	8,363 m^2	33.76	282,370		9.34		
212 Upper floors construction	21,242 m^2	116.21	2,468,430		81.63		
213 Roof construction	8,069 m^2	53.48	431,510		14.27		
220 Roof finishes	7,757 m^2	92.95	721,050		23.84		
230 Exterior cladding	5,914 m^2	413.20	2,443,690		80.81		
231 Basement walls) 232 Walls above grade)	3,694 m^2	542.55	2,004,200		66.28		
233 Windows	2,093 m^2	196.39	411,050		13.59		
234 Entrances & storefronts	127 m^2	223.94	28,440		0.94		
240 Stairs	25 flights	3172.00	79,300		2.62		
300 INTERIORS	–	–		5,156,250		170.51	24.58
310 Partitions and doors	14,636 m^2	165.92	2,428,380		80.30		
320 Interior finishes	–	–	1,882,040		62.24		
321 Floor finishes	28,401 m^2	32.53	924,000		30.56		
322 Ceiling finishes	20,492 m^2	25.57	524,000		17.33		
323 Wall finishes	16,868 m^2	25.73	434,040		14.35		
330 Specialties & equipment	–	–	845,830		27.97		
331 Specialties	30,240 m^2	16.67	503,960		16.67		
332 Equipment	30,240 m^2	11.30	341,870		11.30		
400 CONVEYING SYSTEMS				206,000		6.81	0.98
410 Elevators	8 stps	25750.00	206,000		6.81		
420 Moving stairs and walks							
500 MECHANICAL & ELECTRICAL				4,466,000		147.69	21.29
510 Mechanical	30,240 m^2	–	2,866,750		94.80		
511 Plumbing and drains	–	12.00	363,000		12.00		
512 Fire protection	–	11.68	353,200		11.68		
513 HVAC	–	71.12	2,150,550		71.12		
520 Electrical	30,240 m^2	–	1,599,250		52.89		
521 Distribution	–	24.82	750,590		24.82		
522 Lighting	–	24.19	731,340		24.19		
523 Special systems	–	3.88	117,320		3.88		
600 OVERHEADS & PROFIT	–	63.07		1,907,360		63.07	9.09

NET BUILDING COST: $ 18,598,920 $ 615.04

Elemental Category	Quantity	Unit Rate	Sub	Group	Sub	Group	%
900 SITE DEVELOPMENT	45,733 m^2	52.09		2,382,080		78.77	11.35

TOTAL COST: $ 20,981,000 $ 693.81

Analysis No: N.13

tion Cost Data,1983) establish the index for Edmonton, Alberta, in 1983, and convert the costs shown in the example cited to national average costs. With this average cost information, calculate the following: (a) the total length of interior partitions in the building and their average cost per linear foot; (b) the total number of plumbing fixtures in the building and each average cost; (c) the average cost per ton of cooling; (d) the average cost of ventilation per CFM and; (e) the number of elevators and their average cost each. Show how these costs that you have calculated compare to those cited in the cost manual.

3. (a) Suggest possible meanings (supported mathematically) for the complexity factor of 2.02:1 indicated in the example; (b) suggest another ratio that might be used to express numerically a significant characteristic of buildings.

4. Explain in detail the following terms used in the cost analysis example illustrated in this chapter: (a) *gross floor area;* (b) *net floor area;* (c) *unassignable/ unusable floor area (walls, partitions, etc.);* (d) *lowest floor construction* and (e) *cost index.*

5. Make an itemized and descriptive list of all other information about the building illustrated in the example that is not shown and which you believe would be included in the file of such an analysis.

10

A Comprehensive Approach

In this final chapter of Part Two we examine something more than a method of estimating; we examine an approach to construction costs and building economics that may be a model for others and a reason for a new and different view of estimating. We have reviewed life-cycle costing, the economic view of all costs over the life of a particular real property development, including the costs of owning and using improvements,[G] costs arising not only from the original construction but also the costs arising from the use of a property throughout its economic life. Here we examine a formal method for determining at the outset of a project what the economical capital costs of building construction *should be*. Combined with life-cycle costing, this "capitalized income approach to project budgeting" (CIAPB), as it is called by its originators,[1] should give an owner a complete view and understanding of the economics of a development.

[1]U.S. General Services Administration (GSA), with whose permission this material is taken from Document PBS P 3425.11, *Capitalized Income Approach to Project Budgeting*, July 29, 1981.

The Principles and Objectives Involved in CIAPB

It is normal prudence for an owner-developer[G] to measure his or her construction budget according to the expected return on investment in the property. Profitable development of real property requires that this be done, and overdevelopment may cause greater losses than underdevelopment. Economically speaking, at least in theory, each parcel of land has (at different times) its highest and best use, and each has (at different times) a theoretically ideal development that will produce an optimum return. Underdevelopment will not reach the optimum because all of the property's resources are not utilized (e.g., an icecream stand erected on top of a potential oil well); overdevelopment causes losses because of the initial capital overexpenditure on something with the characteristics of a white elephant, followed by the costs of caging and feeding the beast.

It is significant of this CIAPB that it has been formalized by the U.S. federal government for its own purposes and, in a more general sense, that it puts the establishment and estimation of capital construction costs into a proper perspective; that is, within the life stream of a property beginning with a need and the initiation of development and ending with the inevitable economic death of the improvements. The land itself has a kind of immortality and remains to be redeveloped again and again.

Generally, real property owners have two principal attitudes toward the costs of developments, depending on their intentions for the developed property: (1) If it is to be sold immediately, building development costs should be minimized to maximize the profit from the sale; or (2) if the developed property is to be held to provide an income stream or a facility for the developer, then the prudent developer considers the costs in use and designs and builds to minimize replacements, repairs, and maintenance and to maximize the return. Understandably the General Services Administration (GSA) has the second kind of attitude and in its publication on CIAPB refers to its objectives, including the need:

1. To lower the costs of federal building construction;

2. To establish at the outset of a project the relationship between costs and income (or its equivalent) for each federal project;

3. To establish project cost controls to prevent overdesign and overconstruction;

4. To facilitate the handling of each federal building project as an individual cost center; and

5. To identify, reaffirm, or waive if necessary, federal requirements for projects that exceed the requirements of the private commercial sector.

An overview of the CIAPB process includes:

1. Perceiving a need for building space that initiates the process.

2. Surveying, *as though by a private investor*, the community in which the need is apparent and its market for building space to obtain current economic information about local building development.

3. Capitalization of the net income that would probably be derived from a specific building development to indicate the amount that a prudent private investor would invest in such a project.

4. Determination of construction costs by deducting from the amount of the investment (a) the costs of a site, and (b) all other costs not directly attributable to construction, including costs of design and engineering. These construction costs are then expressed in terms of a *conceptual estimate* to establish the levels of quality for design and construction of the project. The estimate's purpose is to verify that a building as described can be built for the estimated amount using "design criteria ... that a private developer would use."

5. A second conceptual estimate (which may be based on the first) that allows for all federal government criteria, requirements, and policies regarding federal buildings and their use. (It is expected that the total amount of this second estimate will be greater than that of the first estimate.)

6. Comparison of the two estimates to identify, explain, and justify their cost differences where possible and recommend waivers of government criteria, requirements, and policies where these cannot be justified.

7. Finally, control of the procurement of building space to ensure that the new building includes only those elements that have been justified and approved by this process.[2]

Having reviewed the principles and objectives of this approach to building costs, we shall examine the steps in the approach in more detail. At the end of this chapter we shall discuss some of the implications.

---

[2]I have made brief, paraphrased quotations from the GSA publication on CIAPB, and readers are urged to purchase and examine the document for themselves.

Determination of the Maximum Investment Amount for a Project

In Chapter 2, page 1 of its publication on CIAPB to which I refer, GSA indicates the three classical methods of valuing real property to which we have previously referred: (1) the cost approach (2) the market approach and (3) the income approach. It describes the *income approach* as based on the thesis "that value is the present worth of future rights to income" and selects this approach for its own purposes in CIAPB.

In Chapter 2, page 1 of its publication GSA proposes that:

$$\text{"project value} = \frac{\text{net income}}{\text{capitalization rate}}\text{"}$$

The publication then describes in detail in Chapter 2, page 1 through 6 how this formula should be applied in their "capitalized income approach to budget estimating."

GSA (and I am still referring to its publication, Chapter 2, page 3) rejects the classical band of investment theory as a means for synthesizing a capitalization rate (the critical figure in determining value from income) and turns instead to a modified band of investment theory expressed by this formula:

weighted capitalization rate
= ratio of equity x equity dividend rate
+ ratio of loan x amortization constant

GSA continues by saying (in Chapter 2, page 3), "This formula simply states that most properties are heavily mortgaged and that what the investor wants is coverage of his mortgage amortization plus an adequate return on his equity investment."

Using the following definitions,

K. . . . . The ratio of mortgage amount to total project cost,
ROR. . . . . The rate of return on equity,
CRF. . . . . The capital recovery factor (the amount required to retire a one-dollar mortgage with interest over a specified term at a constant annual payment).

GSA recommends for those "unfamiliar with the intricacies of more complex capitalization methods" the use of this formula:

$$\text{estimated value} = \frac{\text{net income}}{(1\text{—K})\,(\text{ROR}) + \text{K(CRF)}}$$

GSA also recommends the alternative use of more advanced techniques and the use of capitalization rates from the market but only under the guidance of a professional appraiser. As our immediate interest is primarily with the implications of CIAPB to *estimating* and the part that conceptual estimating plays in CIAPB, we may at this point leave the determination of the maximum investment amount.

Conceptual Estimates for Comparison of an Investor's Requirements for a Building Project with Those of GSA

GSA describes (and here again I paraphrase from Chapter 3) the estimator's procedures in making both of these conceptual estimates (one estimate to express the total amount of the derived construction costs in terms of a private investor's requirements for a building project; the other to do the same for the GSA's requirements):

1. Generally formulate an interpretation of what is required by each of the two developers (government and private) for a building project and forecast the costs of supplying those requirements.

2. Specifically provide a concept and a design solution for each of a project's requirements.

3. "Leave a traceable path identifying the assumptions upon which each estimate has been structured."

4. "Provide a budget for each concept that, if incorporated in a prospectus, could be achieved if adhered to during design."

5. In the absence of more specific market information, use the information about building quality, standards, and other criteria provided in the publication by GSA.

The CIAPB publication then proceeds to detail the methodology of making a conceptual estimate along the same lines as that described in this book but in a more specific and formalized way directed toward specific classes of government buildings. The persistent reader, however, might learn much about general methods for conceptual estimates from the publication. The astute reader may notice GSA's direction to the estimator about a traceable path and the assumptions made in an estimate. GSA has (in Appendix 3-E of the publication) its own standard list of building elements and their items, entitled, "Uniformat Element/Item," which may be worthy of more general application.[3]

[3]The list contains 12 major elements, from *01 Foundations* to *12 Site Work*, each with items under sub-elemental headings.

Implications in CIAPB for Estimating in General

The last two Chapters (chapters 4 and 5) of the GSA publication on CIAPB deal with maximum project cost and procurement, and as these are mainly concerned with government objectives, we need not discuss them here in detail. The former deals with the comparison to be made between the hypothetical private developer's project and that of GSA and the justification or elimination of the extra features usually found in a GSA project. (Justification is required to be based on life-cycle costing and on cost-benefit analysis.[G]) The last chapter, "Procurement," discusses federal government funding for space procurement, the possibilities of leasing space, and—of more immediate interest to us—the subject of design by the A-E (architect-engineer), about which GSA says in Chapter 5, page 2): "The project cost plan must be rigorously enforced through the design and construction process" and "The A-E will continue to be required, by contract, to redesign the project at his or her own expense if the bid cost exceeds by 5 percent the construction cost target."

The general implications of this comprehensive approach to building costs are several, and so far as our subject, estimating, is concerned, the implications are significant. The most significant implication is that apparently precise and detailed measurement of construction work, particularly as it has been formalized by the quantity surveying profession over the last few decades, is perhaps of less importance to the *overall economic view* of building development than we have assumed. Yet, at the same time, standard methods of measurement[G] are desirable and necessary as a common vocabulary for precise comparisons of costs, among other things. Perhaps several different standards are needed for different applications.

Another implication of CIAPB touches on the types of construction contracts used in building development. In looking at the bigger economic picture in the light of the life-cycle costs of buildings and of cost-benefit analyses of building elements, the significance of the initial capital costs of construction is reduced. The subsequent costs of maintenance, repairs, and replacement of elements and subelements during the life of a building often loom far larger. Is it sensible to pay so much attention to the initial costs of buildings while at times practically ignoring the later and often greater costs? (Yet we cannot estimate the later costs unless we we first estimate the initial costs.) And is it sensible to seek the lowest possible initial cost through competitive bidding for lump-sum contracts and almost to guarantee a contract to the lowest bidder[G] in the light of all the effects of construction economics? Therefore, should we not as construction economists be looking for more practical and more effective and more efficient methods of measuring and estimating the costs of work, and should owners not more often use something other than the traditional building contracts and contracting methods?

Questions and Topics for Discussion

1. Is a national government in a different position from a private investor when seeking to develop building space? Explain your answer in detail (and in not less than 1,000 words).

2. What are the principal considerations of private investors when investing in real property?

3. Explain briefly the modified band of investment theory used by GSA for income capitalization.

4. Explain in detail what you understand by a traceable path to be left by an estimator while making an estimate?

5. In the light of building economics, including life-cycle costing, discuss at length the subject of accuracy in project estimates.

6. Could conceptual estimates replace detailed estimates as the most common form of estimate in construction? Explain your answer in not less than 1,000 words.

7. Define the following terms used in this chapter: (a) *net income*; (b) *capitalization rate*; (c) *income approach*; (d) *return on equity investment* and (e) *capital-recovery factor*.

8. Using as a subject the Kuckein house (in Chapter 1), determine the costs of building, owning, and occupying this house over an economic life of 50 years in a specific location and over a specific period of time. In the light of your results, establish a rental rate for this house that would return to an owner-investor 15 percent on his equity investment. State all the assumptions upon which your studies are based.

9. Using as a subject the apartment block (in Chapter 3), excluding the commercial floor, determine the costs of building, owning, and renting the apartments over an economic life of 50 years, in a specific location, and over a specified period of time. In your results establish monthly rental rates for

each type of apartment that would return to an owner-investor 15 percent on his equity investment. State all assumptions on which your studies are based.

10. Design for yourself a house containing 1,500 square feet of interior living space, plus a two-car garage of 500 square feet, on a lot of 7,500 square feet (60 feet frontage), built to be occupied for at least 100 years in a specified place and built at a specified time; the house is to provide maximum benefits for a minimum cost over its lifetime. Sketch the house design, including primary details of construction, and make an outline specification. State all assumptions on which your economic study is to be based, and show all costs and calculations, including the net cost per year to you, the owner.

PART THREE

1

Proposal Estimates

In Part Three we turn to precepts and practices of estimating, and in particular, of *conceptual estimating*, for that is our main topic, even though it can be described as standing on the base of *bid estimating*. You may well ask, Why show examples before discussing methods? And I must answer, To spotlight the final goal before picking our way along the pathway to it.

In Part One we said that in order to understand *conceptual estimates* it was necessary to understand *bid estimates*,[G] and because most bids are for stipulated-sum contracts,[G] usually most bid estimates are detailed estimates with all the items of work shown in the drawings, specified, or both, and measured and priced in detail. The exceptions are bids for contracts other than those for a stipulated sum (i.e., for some kind of cost-plus-fee contract) for which less detailed estimates are usually made, namely, estimates that are more like preliminary estimates; often, conceptual estimates.

The title of this chapter, "Proposal Estimates," indicates a broader classification of estimates than bid estimates, and it includes all estimates made as a basis for proposals[G] by a contracting company to an owner, including those for stipulated-sum contracts and also proposals for any other kind of contract, even estimates made in preparation for and as a basis of negotiations leading to a construction contract, estimates that are not usually as detailed as those made for bids for stipulated-sum contracts and that in many respects are similar to conceptual estimates. For example, the kind of estimate that is made from incomplete design drawings by a contracting company for a proposal for a maximum cost-plus-fee contract.[G]

This chapter is not intended as a short course on construction estimating for estimators in contracting companies or for bidders. Instead it is a review of the precepts and practices of their kind of estimating, together with an examination of *estimate summaries* as a way to a better understanding of construction costs and how they are dealt with in proposal estimates, with special attention paid to *job overhead costs.*[G]

Many estimators say that job overhead costs are the most difficult to estimate. Certainly this appears true in conceptual estimates in which job overhead costs are frequently underestimated; in large part because these costs are not represented by specific items of work, hence they are less obvious. Job overhead costs are problematic because they are nebulous and dependent on unreliable things such as the weather, deliveries of materials, and all the other things that affect a project's duration; and job overhead costs are particularly dependent on a project's duration. Because they do depend on project duration and on the rest of the work they are usually found in an estimate's summary, which is why we shall examine estimate summaries.

Proposal Estimates and Cost Accounting

Estimating is not something that can be done in isolation. It is only one part of a continuous cyclical process, as illustrated in Figure 1-1. The cycle, as shown, has three main parts:

1. Estimating
2. Planning
3. Controlling the work.

Figure 1–1 shows and briefly describes the construction functions comprising each of these three main parts of the process of producing construction work (Notice that we do not call it *construction,* because construction also includes *design.*)

Estimating cannot be done properly without cost accounting,[G] which precedes estimating in the cycle because it is needed to provide an estimator with cost data from previous projects. (The cycle as shown does not apply only to one project in isolation but to all the operations of a construction company that is continuously seeking work by making bids and that strives to have a number of concurrent projects, each probably at a different stage.) The fact that the cost-accounting arm of many construction companies is underdeveloped while they can continue to do business does not belie the fact that cost accounting is necessary or the fact that a company that has a properly developed cost-accounting arm will probably be more successful than one without. A successful construction company is one that makes at least the estimated profit on almost every project that it performs—not one that survives through a succession of losing projects, each financed by the last.

Precepts and Practices of Estimating and Cost Accounting

These precepts were first set down in another book,[1] and they have been condensed here for brevity's sake. They are the precepts of:

1. Purpose,
2. Veracity,
3. Verification,
4. Measurement, and
5. Accuracy.

(1) The purpose of estimating is to predict the costs of construction work before the work is done, with the aid of cost accounting. *The purpose of cost accounting* is to discover the real *costs of work,*[G] to verify or facilitate the correction of estimates in the light of experience, and to provide *cost information* and *cost data*[2] for future estimates. The relationship between cost accounting and successful estimating is unchanging and extends to all kinds of estimating, including *conceptual estimating.*

This is not the place to discuss cost accounting in detail, and the reason for its introduction is to point out that all estimating, including conceptual estimating, requires information and data from other projects. General accounting is essential to all construction projects for general business purposes

[1]Collier, *The Fundamentals of Construction Estimating and Cost Accounting.*

[2]The difference between *information* and *data* in estimating is described in the Glossary and discussed in Chapter 1 in Part 4. The more general term is *information.*

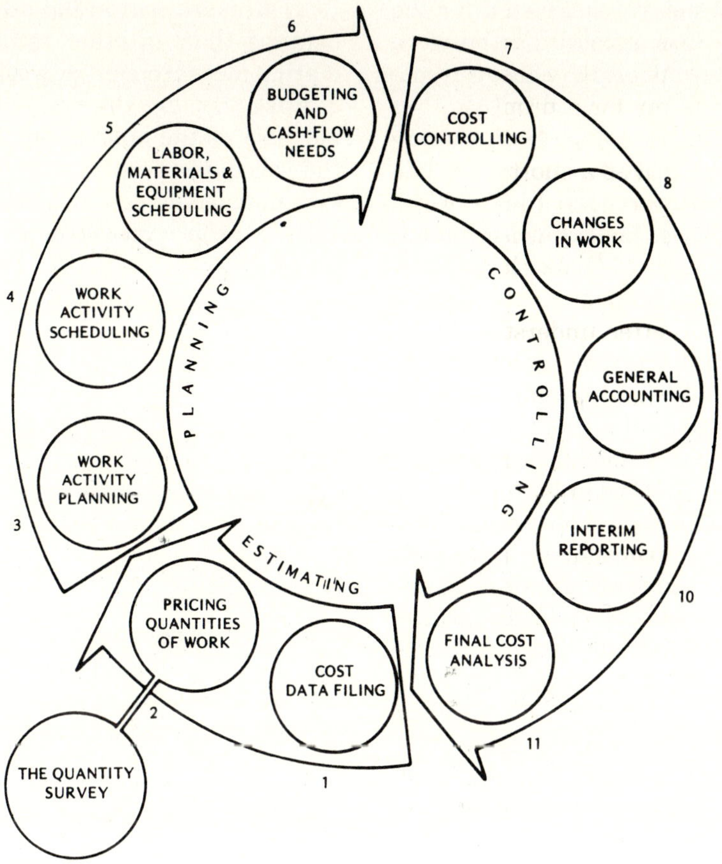

Estimating

(1) *Cost data filing:* an estimator draws on information obtained from other jobs in making an estimate

(2) *Pricing quantities of work:* this completes an estimate and bid and may lead to a construction contract

Planning

(3) *Work activity planning:* construction work is analysed into parts (activities) which are then arranged in logical sequence

(4) *Work activity scheduling:* estimated times for activities are included in a schedule for the project

(5) *Labor, materials, and equipment scheduling:* amounts of resources required at different times throughout a project are estimated and scheduled

(6) *Budgeting and cash flow needs:* these are calculated and scheduled to find out the amounts of financing required throughout the project

Controlling

(7) *Cost controlling:* daily costs of major items are obtained and compared with the amounts of work performed; resultant unit rates are compared to those in the estimate; unsatisfactory production rates are discovered and corrections in work methods may be made

(8) *Changes in work:* these require costs to be negotiated and recorded

(9) *General accounting:* including payrolls and payments for materials and equipment rentals, etc., and for cost accounting requirements

(10) *Interim reporting:* on actual costs and progress periodically made and the schedule adjusted accordingly

(11) *Final cost analysis:* compares actual costs and times spent on work activities with those in the estimate and schedule to provide information for future projects.

Construction management is a cyclical process of input and feedback; a constant learning process; a process requiring continuous revision and adjustment as actual results vary from predictions and estimates; it is a steering process towards a goal.

FIGURE 1.1 Cycle of Construction Management

(payrolls, determination of profit), and some confuse this with cost accounting. The latter is fundamentally different in that it requires the *quantities of work*[G] actually done but not yet completed for the costs actually expended. Cost accounting therefore extends throughout the controlling phase of construction work and its results are used in the other two phases, estimating and planning, to provide information. For these purposes, costs are meaningless unless accompanied by accurate quantities of work; which is why cost accounting requires measurement and technical knowledge and why general accountants and cost accountants uneducated in construction cannot do construction cost accounting.

(2) Veracity in estimating and cost accounting includes the avoidance of delusion and the related virtue of objectivity. The wrong allocation of costs is a common but serious problem that distorts the future estimates and plans that utilize the unreliable information. Careless or deliberately wrong allocations of costs or other information often begin with the deficient notion that an estimate has but one purpose: to make a bid and to get a job. In fact, an estimate should be intended for more: as a source of information for the management of construction work, including planning and scheduling, purchasing, and the control of costs. Part of the remedy for wrong allocations of costs is a better understanding of contracts and costs and submission to the general rule: That which gives rise to a cost must carry the cost. For example, the costs of using a masonry saw are part of masonry costs, not part of job overhead costs.

(3) Verification in estimating and in cost accounting is the outcome of comparing their results; one verifies the other. An estimate's worth remains unproven until a project is completed and the total costs are known and accounted for and until the costs are allocated to the many different items of work and overhead. Then the correctness of the estimate becomes apparent (assuming that the accounts are correct). But we should not wait until a project is completed to do this, only until a specific item of consequence is partially completed, because, in fact, costs can be allocated to the weightier items of work (those creating larger shares of the costs) before the total quantity of such an item is completed. This is how costs are controlled during construction. However, one cannot be completely sure about the costs of an item until it is quite complete. (With major items, production rates often increase as work progresses because the work is done more efficiently after the workers have overcome initial difficulties, learned the job, and found a rhythm of work.) Verification of an estimate is the same as calling the shots, the shot being the estimated costs of each item of work. Calling the shots is as essential in estimating as it is in shooting at a target.

(4) Measurement in bid estimates is usually more detailed than in other estimates, and in bid estimating measurement of work creates the quantities of work that are the basis not only of the estimate but also of the subsequent bills of materials[G] and the schedules used in purchasing and in construction management. For bid estimates the suggested rules of proper measurement include:

1. Measuring all work *net* as installed in place.
2. Allowing for waste,[G] shrinkage, and swell[G] in unit prices, *not* in the quantities of work.
3. Measuring *items of work*,[G] not just the materials.[G]
4. Measuring work consistently in accordance with some *standard method of measurement*.[G]
5. Not substituting and measuring one item of work for another.
6. Adequately describing all items of work and of job overhead for proper pricing.
7. Measuring work to a practical degree of accuracy according to its nature and the type of estimate.

Even in bid estimates these rules are sometimes broken, although the result may be to make proper estimating and cost accounting subsequently impossible. Reflection on the rules will show you why. For preliminary estimates it is necessary to know and understand these rules of measurement for bid estimates so that in *preliminary estimates* they may be judicially broken. For example, in a conceptual estimate[G] it may not be possible to describe an item of work fully; nevertheless, as full a description as possible at the time is desirable. It is not sufficient to say (as some will do, even in a bid estimate), Concrete foundations. At least say, Concrete (3,000 psi/20 MPa) in footings, or whatever is appropriate. The extent to which the rules of measurement are followed in preliminary estimates (and bid estimates) is a matter of common sense and expediency; the rules always remain as a practical ideal rather than an absolute law.

(5) Accuracy in estimating and cost accounting as in everything else is relative, and it should be commensurate with the kind of estimate and the item of work measured. Some estimators have delusions of accuracy and measure dimensions to several places of decimals, as if to write in accuracy. (Three inches [75 mm] is close enough for bulk excavation; one-half inch for most concrete work.) Published

standard methods of measurement are useful as guides in measurement, and in their absence an estimator should draw up his own. As for estimates generally, research into accuracy has produced inconclusive results. While examining estimating precepts and practices in the next three chapters appropriate degrees of accuracy are indicated for different kinds of estimates.

Custom says that usually it should be possible to make a preliminary estimate to within 10 percent of the average of the subsequent bids for the work in a stipulated-sum contract. There are too many possible exceptions and conditions to declare a hard and fast standard. For most proposal estimates the ultimate test of accuracy is cost accounting.

Bid Estimate Summaries

In this section we refer to bid estimates and to their summaries as embodying ideal standards of quality for all kinds of proposal estimates. Bid estimates for lump-sum contracts are the most common and should be the most detailed and the most precise.

The best way to review any estimate is to look first at its summary and then at the details that support it. The format of an estimate is something like that of a pyramid, with a wide base of details and with a summary at the apex. But because the detail comes first and the summary last we must stand the pyramid on its apex, upside down:

In an estimate:

1. *The last line*—Total Estimated Costs of Work
2. *The preceding line*—addition of Profit and Operating Overhead (either separately or together)
3. *The preceding line*—Total Estimated Net Costs of Work
4. *The preceding line*—addition of Job Overhead Costs
5. *The preceding line*—total of Labor, and Materials, Equipment, and Subcontractors' Work
6. *The preceding line*—gross Labor Costs, and gross Material Costs plus Total of Equipment Costs plus Total of Subcontractors' Work
7. *The preceding line*—addition of Small Tools to Labor Costs
8. *The line preceding labor costs*—addition of Indirect Labor Costs to Direct Labor Costs of the items of work.

An example of a bid estimate summary is shown in Figure 1–2. It is set down in a typical two-column sheet with totals brought forward at the top for *direct labor costs*[G] and for *materials' costs*[G] for the measured and priced *items of work.*[G] Immediately below these are the entries for an estimate summary.

Summaries vary in format according to the work and the methods of different estimators and different companies. Some prefer a three-column sheet to keep subcontractors' work separate from materials' costs. In some estimates subcontractors' work may be entered on separate sheets and the total carried to the final summary sheet. The same may apply to equipment costs. The essentials are (a) a total of all direct costs, except for the job overhead costs, just before adding the latter (because the total direct costs influence the amount of job overhead required), and (b) a total of all direct costs (including job overhead) to which operating overhead and profit are added. If an overall contingency allowance[G] is to be added (for whatever reason), it is better added after the profit and overhead, but there is no strict rule about this, except that it is better to follow customary practice (whatever it may be) as close as possible so that useful comparisons among estimates and project costs can be made. Therefore it makes more sense to add an unusual item such as a general contingency allowance right at the end.

Cost Categories in Bid-Estimates

Let us examine in detail the contents of a *bid estimate* under these typical headings:

1. Work by contractor's own forces
2. Subcontractors' work
3. Construction equipment
4. Job overhead items
5. Operating overhead[G] items
6. Profit.

Work by Contractor's Own Forces

Most contract documents[G] do not inhibit a contractor in making subcontracts,[G] but some public works contracts require a contractor to do a specified minimum amount of the total work of a contract to preclude contractors who act primarily as brokers and subcontract out all work. Private owners can, of course, do the same if they wish. The extent of

PRACTICAL "STANDARDIZED" FORMS FOR CONTRACTORS
Form 514 MFD. IN U. S. A.

GENERAL ESTIMATE

BUILDING: HIGH SCHOOL RENOVS
ESTIMATE NO. 1
SHEET NO. 14/14
LOCATION: PRESWICK, IL.
ESTIMATOR: JB
ARCHITECTS: MCDOUGAL & INNIS CO.
CHECKER: DA
SUBJECT: FINAL SUMMARY (BID ESTIMATE)
DATE: 05/31/74

FRANK R. WALKER CO., PUBLISHERS, CHICAGO

DESCRIPTION OF WORK	TOTAL ESTIMATED MATERIAL COST	TOTAL ESTIMATE LABOR COST
TOTALS (brought forward) GROSS MATERIAL	137350 ✓	—
DIRECT LABOR	—	107050 ✓
ADD: INDIRECT LABOR COSTS (+30%)	—	32100 ✓
		139150 ✓
ADD: SMALL TOOLS' ALLOWANCE (5%)		7000 ✓
TOTAL GROSS LABOR COSTS		146150 ✓
ADD: GROSS MATERIAL COSTS		= 137350 ✓
		283500 ✓
ADD: SUBCONTRACTORS' WORK (from Summary)		514100 ✓
TOTAL LABOR, MATERIALS, SUBCONTRACTORS' WORK		797600 ✓
ADD: JOB OVERHEAD COSTS (from Summary)		56000 ✓
TOTAL ESTIMATED NET COSTS OF WORK		853600 ✓
ADD: PROFIT AND OPERATING OVERHEAD (15%)		128000 ✓
TOTAL ESTIMATED COSTS OF WORK		$ 981600 ✓

FIGURE 1.2 Summary of an Estimate

the work done by a so-called general contractor[G] (now usually an inaccurate title) and that done by subcontractors[G] varies according to local custom, the type of project and work, and an owner's requirements. Work by a contractor's own forces (and the same applies to the work that a subcontractor does with his own forces, as distinct from that work that he subcontracts to a sub-subcontractor[G]) is properly estimated by detailed measurement and analytical pricing using cost information from previous projects.

But even for bid estimates some will use methods that are more like those of preliminary estimates—especially in a busy sellers' market, in which plumbing companies, for example, may estimate costs of plumbing work in dwellings at so much a fixture, or even at so much a bathroom or a dwelling unit. Whether this practice is commendable depends on whether it is successful in obtaining work at an acceptable price for an owner and at a profitable price for the bidder.

Subcontractors' Work

Most trade contracting companies[G] submit sub-bids to all the primary bidders[G] for a building project, but sometimes a sub-bidder decides not to give his sub-bid to a particular bidder, and nobody and nothing can force a person to bid to another if he does not want to. As a result an excluded bidder may be obliged to make a more than adequate allowance for that subtrade work[G] in his bid (with hopes of getting a lower sub-bid later), and consequently his own bid may be inflated. Conversely a bid may be the lowest bid because it contains a more competitive sub-bid that was not available to the other bidders. Which then is the "true total price" for the work?

Trade contracting companies may bid according to the market rather than according to actual costs: in a buyers' market in order to get a job by bidding low or in a sellers' market by bidding high in order to gain a larger profit should the bid be successful. Again, which is the "true price"? We can say only that construction costs are created in the market, whatever they may be. The right price for an owner is the lowest price that he can get in the market, all other things being equal. An owner's cost consultant may calculate accurate direct costs of work and add what he believes is a reasonable markup[G] for operating overhead costs and profit, but that does not mean that his estimated price is the "true price." In a sellers' market a price may be the right price for the owner but not for a contractor—if he can get a higher price. A cost consultant must consider the market in estimating costs, the same as a contracting company. (Examples of estimating costs of trade work are found in Chapter 3, Part 3.)

A perennial problem with sub-bids lies in misunderstandings among bidders and bid-recipients as to the precise content and scope of sub-bids. (Does a roofer's bid include for all sheet-metal work in the job, or only for flashings?) Local custom is a great aid to mutual understanding, but customs vary from place to place. Sometimes a project requires work that is not customary. Detailed descriptions of the scope of sub-bids is essential. Specification writers can help by writing appropriate sections,[G] but this problem of communication persists in construction and causes much trouble. Construction people are not much given to verbalizing, it seems.

Construction Equipment

The costs of construction equipment[G] (and of plant,[G] should there be any on a site) often defy precise classification and allocation, and these costs are often the bane of estimators; they are one of the primary causes of inadequate conceptual estimates because many do not understand equipment costs or their importance.

To avoid confusion follow this golden rule: That which gives rise to a cost must bear that cost. When applied to equipment the rule means, for example, that a masonry saw's costs are part of masonry costs, which is easy to follow. But problems arise with large general-use equipment such as hoists and tower cranes. This equipment kind is best dealt with in a separate section of an estimate with the total costs taken to the summary of direct costs. (See Chapter 3, Part 3 for some examples.)

Small tools[G] are often a single item in a summary, a percentage addition to the labor costs based on past experience and on cost accounting in similar projects. Greater accuracy may be achieved if small tools are calculated separately as a percentage of labor costs for each trade.

Job Overhead Items

Job overheads were introduced in Chapter 3, Part 1, where an example from an actual project was examined. We shall discuss in detail some of the common job overhead items in building projects to show how job overhead costs[G] are estimated in bid estimates as a guide to estimating them for other kinds of estimates.

Field Supervision: A competent superintendent[G] (or, in a CM project,[G] a supervisor[G]) will probably cost $3,000 to $4,000 per month for wages, plus fringe benefits[G] amounting to about another $1,000 per month, which amounts to almost $100,000 for a typical 18-month long project. A problem arises when a project is of such size or complexity that it requires more than one superintendent—but not two. Similarly some smaller projects hardly need a full-time superintendent, but usually one is required by the contract. (This requirement is found in all nationally published standard construction contracts.) You cannot have half a superintendent on a project, but sometimes arrangements can be made for two jobs to share a superintendent. Some contractors try to save costs by surreptitiously withdrawing a superintendent once a project is about two-thirds completed, but usually they are in breach of contract by doing so. Usually field supervision costs are the biggest among job overhead costs, often amounting to about 50 percent of total job overhead costs, therefore they are critical.

Contractors may be required by law to employ on a site a full-time qualified first-aid man, or more than one depending on the project's size and nature. A first-aid man is often employed as a part-time assistant to the superintendent, or he may be required to do other work intermittently, such as equipment maintenance. His costs may amount to about one-half to two-thirds the costs of a superintendent.

On unionized sites it may be required by agreement to employ a union shop steward to represent the membership of each trade union on the site. Although these persons are skilled workers, because of union responsibilities their productivity may not be high; therefore, all or part of their wages and benefits should be viewed as a job overhead cost.

If the site layout and survey work is not to be done by the superintendent, it may be necessary to employ a surveyor part-time.

It is not possible to generalize much about field supervision costs, and the only way to estimate them is to itemize and price each one, even in preliminary estimates—in which job overheads are often underestimated because of omissions.

General labor: Not all estimators provide for general labor as a job overhead item, and often this is an error of omission because there are always needs for labor not directly related to specific *items of work*[G] but needed just the same. General labor should be minimized by including for labor under such specific items as machine bulk excavation (for cleaning out in corners, around existing piping, and such like), continuous cleanup, final cleanup, and other similar *labor-only items*;[G] but the chances are (as experience shows) that you cannot do a job without some unpredictable needs for labor on the site. In a competitive market it is hard for an estimator to include such indefinite costs, but an estimator must be guided by experience and by the cost accounts. If a need is probable, it should be met. As always in lump-sum projects, a balance must be found between costs that are reasonably predictable and the constraints of a competitive market.

In 1982 the reported estimated average hourly cost of building laborers in the United States was $13.55.[3] The cited rates are described as to the nearest 5 cents, subject to adjustment as wage agreements are settled during 1982, and they do not include fringe benefits that add, depending on locality, up to about 25 percent to wages. Similarly the estimated average rate for skilled tradesmen was reported at $17.25 per hour. With rates like these (averaging about $20 per hour, including benefits, but before adding for overhead and profit) one can understand an estimator's reluctance to add labor costs not related to specific materials and work. On the other hand, you may have seen in your local service station a notice of the charge of $38 per hour (or thereabouts) for repairs and servicing, just to make sure that you know before you have to ask.

Permits, licenses, and fees: These job overhead items (some of which are really taxes) may total as much as 1 percent of total building construction costs, but they vary by place. They include the fees paid to local public utility companies for connections and for related services. Unfortunately they are often hard to estimate because some authorities will provide only an approximate estimate before the work is done, and there is virtually no appeal from their final bill, whatever it may amount to.

Insurances: Premiums for builders' risk insurance to give protection against losses due to damage to or destruction of the work in progress depend on the locality, the type of work, the duration of the project, and its total cost, among other things. Coverage may be only for fire, or it may be extended to provide coverage against loss from vandalism or earthquakes. If there will be excavation work or blasting with explosives the premium rates are higher.

The other important class of general insurance in building construction gives protection against claims by *third parties*.[G] Typically it is called *third party insurance*. Another name is *public liability insurance*. Some estimators do not include its premiums as a job overhead cost because their companies

[3] *Building Construction Cost Data 1982* (Kington, MA: Robert Snow Means Co., Inc.).

carry comprehensive policies that give continuous coverage for all the work that they do and the premiums are treated as an operating overhead.[G]

There is much confusion in construction about insurance and its coverages and costs, even among some insurance agents. Insurance brokers are often more knowledgeable and often provide more unbiased information. Given the basic information about a project, most brokers and some agents can quote accurate costs for estimating purposes.

Bonds: In construction the most important bond is the performance bond—if one is required by an owner. Its cost should always be classified as a job overhead. The bid bond that precedes a performance bond has but a nominal cost. The rates for performance bonds, like insurance premiums, depend on circumstances and requirements.[4] Rates are readily obtained from bonding (surety) companies.

Safety provisions: Usually the costs of safety are not high, even if not compared with the losses that they may prevent. Larger projects require a temporary building as a first aid station and for storage of safety equipment. Local regulations lay down minimum safety provisions. On one $25-million project the costs for safety provisions was about $12,000 for a period of 18 months, excluding the costs of the first-aid person.

Temporary buildings, fences, signs, access roads, and parkings: All these job overhead items should be costed separately in every kind of estimate because they bear no relationship to other project costs. The costs of furnishing and fixtures in temporary buildings are usually included here.

Temporary electric power and lighting: For several reasons the highly varible costs of these items are hard to pin down, but often they are considerable and they they cannot be treated lightly. They are based on costs of setting up and dismantling temporary meters, poles, and services and on actual consumption. Some utility companies have specialists in construction project requirements to provide information and service to construction companies.

Temporary water supplies: On reviewing costs for this item prior to making an estimate for a major project, it was found that these costs were 5 to 10 times greater for some projects than for others. This apparent discrepancy was in fact due to the different needs of precast and cast-in-place concrete work.

Temporary heating: In many places this is required for curing wet work such as concrete, plaster, and painting. A precast concrete building in Hawaii would probably need none; a poured concrete building in Edmonton might need much. In places such as the latter a careful estimate of these costs must be made by considering the season, the project's duration, the weeks of cold weather, the temperature differences, and the specific heating requirements. It is a minor undertaking for a project's mechanical engineer. General labor will probably tend to the heaters. Temporary enclosures must be allowed for either here or elsewhere in an estimate. Once a building is closed in and the permanent heating installation completed other provisions may be unnecessary. But there are operating costs still to be allowed for, either here or in the heating division of work.

General-use construction equipment. Probably among job overheads this item gives the greatest trouble to estimators. The technology and complexity of construction equipment are formidable and becoming more so; yet it is not possible to make a realistic estimate without careful consideration of *plant and equipment.*[G] Some technical knowledge of equipment is essential. Much equipment is rented; rental companies are good information sources, but they are not without bias. The first question is, Exactly what equipment is needed? In the answer the battle is half won.

Most multistoried buildings require a hoist for workers and materials, and a 100-foot high hoist may cost about $10,000 to erect and dismantle, with rental costs for the period in between. In 1982 at a site in the Pacific Northwest a tower crane with a 128-foot long jib and a capacity of 1,100 kip-feet cost about $40,000 to rent for a four-month period, including the operator, erection, and dismantling. These costs are less than those indicated in current building cost publications but published rental rates are useful only as a guide and for preliminary estimates. In a large city, with some idea as to equipment needs for a project, an estimator can probably get from several rental companies all the cost information that he needs in a matter of a few hours. A cost publication is particularly useful as a guide to asking the right questions. Even for a preliminary estimate it is necessary to make up a schedule of equipment needs before estimating the costs. Do not forget to add the costs of the operators, often the most expensive item.

Rubbish removal and clean-up. Some or all of the labor costs for this work may be included in general labor, but it is better to isolate as many specific overhead items and their costs as possible. Allow for rental costs of rubbish bins and for their periodic emptying. Do not overlook the need to clean up behind dump trucks, especially on streets near a site. The *final cleaning* of a newly constructed building

[4]For more on bonds and insurances in construction see the author's book, *Managing Construction Contracts.*

is a normal item usually performed by a subcontractor and not, therefore, a job overhead item.

Security against vandalism and theft. In bid estimates an estimator may explore the possibilities of hiring mature students or even someone living nearby to keep on eye on the site when work is not in progress. Various possiblities may exist to lessen costs. For preliminary estimates it may be safer to allow for costs based on regular security services. It depends on local conditions. Professional thieves have stolen entire truckloads of lumber from sites before the lumber could be unloaded. Small and expensive material items and tools provided by employers are especially vulnerable. Theft is a major cause of higher overhead costs.

Builder's yard and shop costs. Some would call these operating overhead costs. If they are allowed as a job overhead, they are often related as a percentage to total labor costs. The builder's yard is a kind of halfway house through which materials, tools, and equipment may be routed to and from sites to be recorded, overhauled, and repaired. The incurred costs arise from rental costs of land and buildings (even if owned by the contracting company) and from yard labor. The kind of cost that should appear here should not be missed from an estimate. Typically these costs may be as much as 5 percent of total labor costs, a not inconsiderable amount.

Transportation. This is for special and incidental cartage and delivery of small orders of materials or documents urgently needed at a site. It is not for the normal delivery of materials, which is part of materials' costs. Under this heading may be included the rental of a truck for the superintendent or other site staff. Transportation of head-office staff such as engineers, managers, and the general superintendent should not be included here but under a separate heading or in operating overheads.[G]

Consumable items. These may include office and first-aid supplies, light bulbs, and numerous minor items normally used on a site. Some items may be included under other headings, such as temporary buildings and temporary lighting. It is not possible to itemize everything and the total costs of consumable items may be estimated as a percentage of total job overhead costs, or of total labor costs. Why is it always *labor costs*? Why not base such items on the total costs of materials? Labor costs more accurately reflect the size and compexity of a project. Overhead costs are hardly affected by differences in the quality of fixtures and finishes in a building, for example.

Traveling costs. Whether such costs are incurred depends on the relative locations of a site and the contractor's office. Therefore they can be estimated only by detailed calculation. For so-called out-of-town jobs traveling costs may include fares and expenses for meals, room and board, and travel-time allowances. For workers, such costs are better classed as a labor cost.

Payroll charges. Some include them here, but these costs too are really part of labor costs. They include the costs of workers' compensation, unemployment insurance, Social Security, and Medicare, and as they are required by law, they are often called *statutory charges.* They vary according to place and trade, but most estimators make an average total addition to the total amount of direct labor costs,[G] which is one reason why labor costs are kept separate in an estimate. The average addition for all these charges is 20 to 25 percent of wages. But there are places in which these charges are much higher than the national average.

Sales taxes. Some add these to individual items and others to the total costs of materials in the summary. In some projects (particularly, those for government) it is useful to know the total amount of sales taxes because they may be refundable.

Contingencies. Not all estimates can afford to carry an allowance for contingencies, for the unforeseen and the unexpected. The construction market may be too competitive at times to afford bidder this kind of security. Yet in some projects the bidding documents are worded in such devious ways that any bidder who does not allow for contingencies—that the owner-designer says are solely the contractor's problem—stands to lose a lot should a major contingency arise. In some kinds of contracts and their estimates there is no doubt that contingency allowances are essential.

Contingency allowances should generally be entered and described in the appropriate section of an estimate, and they should be for a specific purpose (for example, one might make an allowance in site work for removal of rock or boulders). General contingency allowances are appropriate in a final summary only when they apply to the entire project (for example, an allowance for possible lower productivity of workers in a particular location).

Operating overhead costs. These are usually added in an estimate as a percentage of the total direct costs of the work. They represent real expenditures by a contracting company that must be paid for by its customers. The main problem is how a contracting company should allocate these overhead costs when bidding in a competitive market for work to be done within lump-sum contracts. Or, in the case of a bid for a cost-plus-fee contract,[G] how is one to assess the "plus" portion (the fee) and remain competitive? This is one of the most difficult questions for a contracting company seeking work. It is no less a problem, although a more academic one perhaps, for an estimator making a preliminary estimate. What should he allow in his estimate for a contractor's operating overhead? We may talk of the commonly mentioned percentages for such markups, but they are a mere guide. In the extremes of a market when work is either scarce or plentiful there are no valid guidelines. In a buyer's market often projects are bid for with no proper provision for operating overhead in the estimate, and in a booming seller's market the high markup may have no connection with the actual operating costs. As with real estate, so in construction; the answer lies in the marketplace.

Profit. What more can be said about this that has not already been said about operating overheads? The two go together both in fact and in an estimate. To an estimator making a bid, not much can be said; the decision as to markup should be made by a company officer, not by the estimator. An estimator making a preliminary estimate has a different problem. Does he want the estimated cost to equal the lowest bid that the owner will ultimately receive, or does he want it to equal the average of all the bids? The difference between the lowest and the average may be as much as 10 percent of the average bid or more in some instances.

If conceptual estimating is more than a science (and I believe that it is), it is also an art; some of the estimator's art lies in the assessment of the markup for operating overheads and profit. However, it may yet be demonstrated that the estimator's art is mostly intuition, that intuition is the collection of unconscious responses to perceived information, and that a good computer program can achieve the same.

Large building projects in urban and suburban areas employing union labor (or nonunion labor paid at union rates) in a relatively stable market at a time when no new wage agreements are about to be negotiated, and with government or blue-chip companies as owners—these are the projects with the least risk and with therefore the lowest markup for profit, but there are too many other variables to be sure. The private owner (as distinct from government) often selects the individual bidders for his work, while the government usually must call for bids from all who qualify as bidders. Should the markup on government projects perhaps be less? Yet bidders for government work may increase their markup to allow for time lost through form-filling and hassles with truculent inspectors. Where does the answer lie? Some contracting personnel are fanatical about the secrecy of their costs, especially about the amounts for markup. Yet any cost consultant can with reasonable ease and accuracy assess the amounts of markup by making estimates and by comparing them with bids so that the range of markup is revealed. There are no real secrets about construction costs because direct costs are a matter of fact (more or less) and indirect costs make up the difference.

Construction Management Estimates

So far in this chapter we have discussed estimates primarily as *bid estimates* leading to lump-sum contracts, with some reference to preliminary estimates—mainly, *conceptual estimates*—by way of comparison. In this section we consider those estimates made for Construction Management (CM).[G] (See the Glossary for the definition of this capitalized term, as distinct from *construction management* in a more general sense.)

The foundation of good CM is a budget estimate made by the construction manager (CM), as an agent and adviser of an owner, and I have included discussion of these CM estimates in this chapter because a CM estimate often is more like a bid estimate than a truly conceptual estimate. A CM estimate having started out as a conceptual estimate is usually revised and expanded as design information becomes available and as construction (design and production) is completed. Indeed, effective CM estimates must be more like detailed bid estimates if they are to be used as a basic tool of the construction manager.[5]

In CM projects, bid estimates are still made—by the bidders for the numerous individual construction contracts entered into by the owner and managed for the owner by the construction manager. The CM estimate by the construction manager is made earlier to anticipate these bid estimates, and if the CM estimate is well made and does in fact accurately predict the subsequent bid estimates, it becomes the means for controlling the project and keeping its costs within the budget established by the construction manager for the owner, hence its title, budget estimate. In this respect a CM estimate is similar to the general estimate of the traditional general contractor,[G] if not in all its detail at least in its purpose and usage. Let us see the ways in which these two kinds of estimate are similar but different.

[5]The procedures of Construction Management (CM) as an alternative to traditional contracting are fully discussed in, *Managing Construction Contracts.*

CM Estimate Summaries

A summary from a CM estimate is shown in Figure 1-3. It is different from a bid-estimate's summary; for one thing it has no entry at the end for operating overhead and for profit. This is because in a CM project there is more than one contractor; the traditional contractor's management function is performed by the construction manager to whom the owner pays a managment fee. The fee is not usually shown here.

Contractors' Work. The summary in Figure 1.3 shows work in 16 standard divisions of Masterformat and in a CM project there will probably be at least one contractor (and possibly one or more subcontractors) for each division of work. Within the CM estimate are separate summaries for each of these divisions of work. In some CM projects the construction manager is required to take a dual role: to be both the construction manager (as an agent of the owner) and a contractor engaged to do a portion of the work or, more usually, to provide specific job overhead items (sometimes called *general condition items*[G] in this context. In this event the CM estimate summary will probably contain a separate heading for this work by the construction manager.

Job overhead items. Some of these items that appear in CM projects are not normally found in traditional lump-sum building contracts because of the contractual arrangement of and procedures in CM projects. There are at least two kinds of CM companies: (1) those that evolved from contracting companies—sometimes as a sibling to a contracting company sharing common facilities and key personnel—and (2) those CM companies with different origins, often incorporated by engineers, architects and construction people from other backgrounds such as cost engineering, quantity surveying, and engineering design. The former tend to operate more like contracting companies, with the advantages and disadvantages that implies—stronger in site supervision, weaker in design aspects and conceptual estimating, and more inclined and able to play the dual role of CM and contractor—while the latter tend to have the opposite strengths and weaknesses.

The content of a CM contract made between owner and CM company and the kind of CM company employed will influence the content of a project's job overheads. One CM company may have several superintendents on staff; another may hire them as needed, possibly as independent contractors.[G] In one project field supervision may be covered by the CM fee, in another as a job overhead. In another CM project the CM may have to perform construction work or provide tangibles (buildings, equipment), while in another he may not.

General Labor. Usually this is essential in a CM project, and it may be provided by the CM company also acting as a contractor. Either way, the owner pays and the procedure chosen is a matter of expediency. Some CM companies try to avoid providing labor and thereby identified as a contracting company and therefore subject to workers' compensation assessments. In 1981, general labor for one major CM project amounted to over 1 percent of total construction costs, but a large part of that was for labor stripping and handling forms, work not easily contracted out in a CM project because it is sporadic and not intensive.

Insurances. The extra premium cost to protect a construction manager against third party claims is not much; otherwise insurance in a CM project is essentially the same as in a traditional project.

Drawings and documents. In a traditional project these are included in the owner and designer's agreement. In a CM project (in which many of the documents may be provided by the CM) the not insignificant costs may be treated as a job overhead. Because of the multiplicity of contractors in a CM project more copies are needed.

CM Estimates Generally

Often there are other differences in a CM company's estimate when it is compared to a traditional bid estimate. If a CM also acts as a contractor, his estimate will be more like a contractor's estimate. If a CM acts purely as a professional manager and owner's agent (the better role) his estimate will be a budget estimate made for management and control, with no additions for payroll charges or sales taxes as these costs are already included in the construction contracts, the total amounts of which are part of the CM estimate. If a CM company is not a contractor, it does not do the same kind of cost accounting as a contractor, so it does not need the same kind of estimate. Consequently the estimates of the CM company working solely as manager and agent tend to be more like preliminary estimates made by cost consultants,[G] but these CM estimates must be based on detailed trade estimates, unlike the preliminary estimates often made by others.

In making a proposal for CM to some owners, such as to GSA/PBS[G] of the United States government, for example, a potential CM may have to submit a

QUANTITY SHEET

PROJECT SHERWOOD PLACE — ESTIMATOR KC — ESTIMATE NO. 1
LOCATION NOTTINGHAM, CA. — EXTENSIONS ES — SHEET NO. 1/16
ARCHITECT ENGINEER WARD SMITH, CONSTR. MANAGER — CHECKED ✓ — DATE 05/21/83
CLASSIFICATION FINAL CM ESTIMATE SUMMARY

DESCRIPTION	NO.	AMOUNT	per SF	PERCENT (approx)
GEN. REQUIREMTS	01	582,700	3.64	5.75
SITE WORK	02	386,800	2.42	4.00
CONCRETE	03	3,527,100	22.05	35.00
MASONRY	04	14,000	.09	.25
METALS	05	201,200	1.26	2.00
WOOD & PLASTIC	06	20,500	.13	.25
PROTECTIONS	07	308,000	1.92	3.00
OPENINGS	08	471,000	2.94	4.50
FINISHES	09	1,194,000	7.46	12.00
SPECIALTIES	10	57,000	.35	.50
EQUIPMENT	11	805,000	5.03	8.00
FURNISHINGS	12	62,000	.39	.50
SPECIAL CONSTR	13	54,000	.34	.50
CONVEYING	14	333,000	2.08	3.25
MECHANICAL	15	1,322,000	8.26	13.00
ELECTRICAL	16	756,000	4.73	7.50
TOTAL COSTS		$ 10,094,300	$63.09	100.00

Gross Area (excl. balconies & parking) = 160,000 SF (Parking = 86,000 SF)

ANALYSIS: 140 UNITS @ $72,100 each = $10,094,000
160,000 SF @ $63.09 = 10,094,300

FRANK R. WALKER CO., PUBLISHERS, CHICAGO

FIGURE 1.3 Summary of a Construction Management Estimate

detailed management proposal containing a proposed preliminary estimate and a project schedule. If the CM company's management proposal is accepted, the CM's preliminary estimate, with modifications, may become the control estimate for the CM project. Such a preliminary estimate, as a proposal estimate and as a potential management control estimate, must be from its outset the best possible preliminary estimate, and it is here that we see the merging of the conceptual estimate and the proposal estimate. This kind of proposal estimate, unlike the traditional bid estimate, must be conceptual, at least in part. At the same time, it must be sufficiently detailed to be used to control the design, costs, and production of a project's work.

Other Proposal Estimates

Most construction proposals[G] refer to *contract costs* and to *contract time*, whatever kind of contract or contractual arrangement they may propose. And even in the least risky and least onerous kind of contract for a contractor—the cost-plus percentage-fee contract[G]—some kind of rudimentary proposal estimate is necessary to establish for the contracting company the required fee and, if sufficient design information is provided, the feasible contract time. In examining this kind of proposal estimate, we see again the characteristics of the preliminary estimate. It must be so, as in this kind of contract, almost by definition, there is not sufficient design information on which to base a lump-sum contract; yet, if the owner is well advised, he will seek competitive proposals, even for this kind of contract that stands as the direct opposite of the traditional lump-sum contract.

Proposal Estimates for Cost-Plus-Fee Contracts

In proposals for this kind of contract only the fee may need to be estimated, but there may be more to this than there seems. First, what must the fee provide for? It must include: the contractor's profit *and* the operating overhead costs[G] (in U.S. standard contract forms, the nonreimbursable costs[G]) as defined by the bidding documents.[G] The definition is important. We have already seen indications that job overheads and operating overheads do not always include the same things in all contracts and that an item in one project may be one kind of overhead and in another project another kind of overhead.[6] If an adequate amount of design information is provided, the owner-designer may reasonably require bidders seeking this kind of contract to relate their proposed percentage-fee to a stipulated contract time.

If a cost-plus contract is for a fixed fee it is essential to be able to estimate the contract time with some accuracy, because it is primarily to contract time that overhead costs are tied, and part of the fee is to cover part of the overhead costs. In fact, in this kind of contract it may be desirable, even necessary, to make an estimate similar in nature and content to one made for a CM proposal; in other words to make a detailed conceptual estimate as a proposal estimate.

Proposal Estimates for Maximum Cost-Plus-Fee Contracts

This kind of contract is a hybrid, part lump-sum and part cost-plus, sometimes called a target contract. Like the mule that it is it has other names, and it is a strong contender for popularity with the old traditonal lump-sum contract. The reason for its popularity is that a building's design need not be completed before work can start at the site, but unlike the ordinary cost-plus contract the costs in this contract are limited to a stipulated maximum sum. There must be enough design information available to make a proposal estimate that establishes a maximum cost (the target figure) which means in turn that this kind of proposal estimate too is at least in part conceptual and not as detailed as a traditional lump-sum bid estimate.

We see that many proposal estimates are not like those detailed bid estimates made for traditional lump-sum contracts, that many proposal estimates are in part like the preliminary estimates made by cost consultants and designers. Although the detailed estimate (with all the work measured and priced in detail) remains the exemplary estimate that we should like all other estimates to emulate, practical reasons require us to make other kinds of estimates that need to be almost as accurate as the detailed estimates, even when they are based on less design information. And that is the subject of this book.

[6]This question is examined more fully, *Managing Construction Contracts.*

Questions and Topics for Discussion

1. Define the following terms as used in the text: (a) *proposal estimate* (b) *bid estimate* (c) *preliminary estimate* and (d) *conceptual estimate.*

2. Describe in detail the relationship between *estimating* and *cost accounting.*

3. Show diagrammatically the relationships among the three main phases in the cycle of construction and their various parts; use notes and descriptions as required.

4. Is it necessary for a builder to estimate the costs of houses that he intends to design, build, and sell speculatively? Explain your answer in detail.

5. Why should it be recommended that we measure construction work *net, as installed in place*, according to contract drawings and their dimensions, when we know that invariably there will be waste of materials so that more must be used than is actually installed?

6. Explain in detail how the function of calling the shot can be applied to estimating, and explain why it should be applied.

7. Why is *total labor costs* invariably kept separate from other costs in an estimate summary? Why are some other totals of other kinds of costs sometimes mixed? Design a typical estimate summary sheet for a major building project.

8. Explain in detail why we cannot say that specific job overhead items and specific operating overhead items are always in the same category and that an item of one category can never be in the other.

9. Describe what the text calls a traditional contract and explain why it requires the kind of estimate that usually preceeds it; then describe another kind of proposal estimate and the kind of contract that it usually precedes.

10. What makes *job overhead costs* different from other direct costs of construction work? Explain how this difference affects estimating job overhead costs.

11. How should a contracting company estimate *operating overhead costs* and apply them to an estimate and a bid for (a) a stipulated-sum contract, and (b) a cost-plus fixed fee contract?

12. What should a bidding construction company consider when adding mark-up for profit in an estimate prior to bidding? Explain your answer.

13. What kind of estimate is required of a Construction manager before work starts in a CM project, and why?

14. Why should the job overhead items in a CM project be the same as or different from those in a traditional lump-sum contract?

15. What kinds of estimates may be required before starting construction in a cost-plus percentage-fee contract? Explain in detail.

2

Preliminary Estimates

In talking about preliminary estimates, we have primarily referred to *conceptual estimates*, but the topic of this chapter, *preliminary estimates*, may be *conceptual*, *abstract*, or neither, being instead practically the same in content as a detailed bid estimate. As the name indicates, *conceptual estimates* are made in part from concepts of construction work in the mind of an estimator, concepts initiated either by design sketches or by some other design information[G] in one form or another, such as a verbal description or a schedule of space requirements.

Abstract estimates consist of an abstract, or summary, of cost data applied to quantitative design information, such as the gross floor area of a building. The most common example of an abstract estimate is the commonplace area times unit cost. Alright, you may say, but why use these fancy terms for estimates; do they mean anything? Yes, they do. The main purpose for these terms is simply identification, so that we may discuss and compare different kinds of estimates, because there are indeed several different kinds of estimates that we need to be able to identify and compare. Let us review them:

1. *Preliminary estimates*

(a) Abstract—with an abstract of costs applied over to simple building measurements

(b) Conceptual—with unit costs applied to elements or to measured quantities of items of work

2. *Proposal estimates*[G]

(a) Conceptual—as the preceding

(b) Bid estimates—with unit costs applied to measured quantities of items of work.

You will have noticed that *conceptual estimates* are shown as common to both preliminary and proposal estimates, which is an indication of their importance; they do in fact link together the different kinds of estimates. Among preliminary estimates (and with approximate quantities) they may be more detailed than others. As proposal estimates they are usually less detailed than other bid estimates and usually are made for bids or negotiations for some kind of cost-plus-fee contracts. In this chapter we discuss these proposal estimates and their advantages and disadvantages.

Selection of an Estimating Method

In selecting the kind of preliminary estimate to employ, there are a number of things to consider:

1. The needs of the owner, thence the purpose of the preliminary estimate
2. The amount of resources available (time and money) from the owner to make an estimate
3. The amount of design information[G] and experiential information[G] available
4. The resources (information, data, skill) of the estimator[G]
5. The prevailing construction market.

(1) The needs of the owner are paramount, because the owner is the estimator's client (or employer). (Generally when referring to a preliminary estimate we assume that the owner is the estimator's client and that the estimator is a consultant or a consultant's employee; with a proposal estimate we generally assume that the estimator is an employee of a contracting company, but other relationships are possible, and where necessary they will be made clear.) If the owner is a layman unfamiliar with construction and estimating, the estimator-consultant must explain to the owner the exact purposes and possibilities of different kinds of estimates, and their limitations. The owner should also be told the maximum fee that he will be charged for an estimate, or other mutually satisfactory arrangements for payment for estimating services should be made at the outset according to good professional practice.

Possibly the greatest cause of disappointment and misunderstanding with owner-designers[G] over estimates is a lack of definition of the scope and limitations of an estimate by the estimator, especially for preliminary estimates, because their objectives and their substance are less definite than those of proposal estimates. As a profession it seems that appraisers are more efficient than estimators and cost consultants in defining the scope and limitations of their work, in part, no doubt, due to the existence of professional appraisal institutes that educate and regulate their membership. All estimates of all kinds should contain the following minimal information:

1. Full standard title of the project (standardized at inception to avoid the common proliferation of alternative titles).
2. A preamble containing a brief verbal description of the project, its work, the site, and the purpose of the estimate.
3. A statement summarizing the total estimated costs for the work, with specific reference to the effective date of the estimate and to the time period for which the estimate is valid.
4. Any assumptions on which the estimate is based. (In a preliminary estimate these will be numerous, and all should be listed). Also, any limitations to or omissions from the estimate.
5. A final summary sheet of the estimate showing the total estimated costs and its contents.

Whether the body of an estimate (and not only the summary) should be provided to the owner depends on the circumstances, the owner, and the agreement made between owner and estimator-consultant.

(2) The amount of resources available for an estimate means how much estimating time the owner is prepared to pay for and how much time a particular preliminary estimate is worth. Some owners cannot understand paying several hundreds of dollars for an abstract estimate reported on one sheet of paper, yet such an estimate may be worth thousands of dollars to an owner. It should be explained to an owner at the outset that an abstract estimate will cost so much and why, that the first few hours of work for the estimator consist of examining the design information (sketches, space schedule, etc.) and seeking relevant data, no matter which kind of estimate will be made. At the same time an owner should be told how much a conceptual estimate at

that stage of the design phase will cost and how much subsequent updated estimates at different stages of design will cost. Perhaps the best basis for estimating fees is hourly rates coupled with a guaranteed maximum fee.

(3) The amount of information available often determines the kind of estimate that can be made—but not necessarily. For example, given a few critical dimensions and outline specifications, an estimator shoud be able to take off (measure) fairly accurate and detailed quantities of work for a standard building, such as a warehouse or a light commercial building, whereas for a nonstandard building much more design information is necessary. For standard buildings for which much cost information is available, an abstract preliminary estimate may be sufficient.

(4) The resources of the estimator, or lack of them, is most important. In preliminary estimates experiential information often must be substituted for design information, and if an estimator cannot supply the needed experiential information, an effective estimate may not be possible. *Information*^G and *data*^G are essential but useless without an estimator's skill. Even if the needed cost data for an abstract estimate are not available, a skilled estimator can nevertheless make an effective conceptual estimate with approximate quantities of work, in other words by analyzing the work into its smallest discrete parts and pricing them. Abstract estimates, to be effective, must be based on abstracts of valid cost data from comparable buildings. Conceptual estimates (based on elements or items of work) can utilize more valid cost data because their parts are smaller and more basic.

The skill required to make effective preliminary estimates—contrary to popular belief—should embrace and surpass that required for bid estimates. Unfortunately most preliminary estimates are made by unqualified persons, hence the large number of defective preliminary estimates and the resultant general idea that effective preliminary estimates are not possible. There have been too many bad ones. The popularity of simple abstract estimates (area times unit price) comes from their brevity and deceptive simplicity. Designing buildings, writing books, and making estimates are among those things (as opposed to, say, brain surgery or extinguishing burning oil wells) that many people are convinced that they can do, just given the time. Consequently most houses are badly designed, most books are junk, and many estimates are worthless and often positively misleading.

That estimators experienced in making bid estimates are not necessarily equipped to make conceptual estimates is evidenced by some words from a federal government publication:

"Conceptual Estimating . . . is an area in which CMs [Construction Managers] should develop more expertise," and, "CM Estimates need to acquire more reliability *[sic].*"[1]

This was reported in the light of the fact that most CMs (construction managers) employed by the federal government's Public Buildings Service are or were contracting companies or were persons with a contracting background. The knowledge and skills of estimators with contracting companies are fundamental to conceptual estimates, but they are not enough. In addition, conceptual estimators must be able to visualize buildings and their components and systems in three dimensions; they must have a basic knowledge of the design principles of the work that they measure and price; and they must be able to design and measure a great variety of detailed work. But nothing is required that is foreign to or far removed from the competence and experience of estimators in contracting, only a different and more imaginative approach. Similarly many conceptual estimators^G (quantity surveyors, designers, and cost consultants) need a better knowledge of bid estimating and of construction contracting, especially in the areas of equipment, overheads, and the effects of the construction market on costs. How can estimators and others from contracting companies acquire greater skill in making conceptual estimates and other preliminary estimates? They can by analyzing construction work and by following the precepts and practices as proposed in this book, and by developing the kinds of information and data that go with these methods.

In order to make effective preliminary estimates, both abstract and conceptual, some estimators need a greater general knowledge of trade work^G and its costs. Others need to learn and to apply the precepts of estimating *and* of cost accounting, because it is only by cost accounting that we can generate the information essential to estimating. Many of the estimating methods passed down from generation to generation are cumbersome and obstruct the conceptualization of construction work *that all estimators need to apply* in making estimates of all kinds.

Conceptualization is fundamental to estimating, even in the most detailed kind of measurement and pricing in bid estimates. Think of all of the construction work that is *not* shown, even in the most detailed and complete construction drawings: most site work; all formwork, except some fancy forms for architectural concrete (again, formwork stands out

[1]General Services Administration, Public Buildings Service, *The GSA System for Construction Management* (Washington, DC.: 1977), 15.

as the best illustrative example); all scaffolding; all masonry centering and temporary supports; falsework for steel- and wood-framed structures; many critical joints and connections; most of the members in wood framing and many in steel framing; much sheetmetal work; most work in the finishing trades; and finally almost all conveying equipment work and most work by the mechanical and electrical trades as it is only shown more or less diagrammatically. In fact, *most* work is not explicitly shown in construction drawings.

Measuring construction work, therefore, always requires a lot of conceptualizing. In conceptual estimates the need is simply increased by several degrees. Conceptualizing is not something new that an estimator must learn when he starts making preliminary estimates. All that is needed is to understand the conceptualizing done in other estimating and to extend it consciously into new areas. If an estimator can measure and price formwork, he can measure any other kind of work, providing that he understands the materials and methods of construction involved. Finally, conceptual estimating requires a clear view and understanding of the different kinds of estimates and their different purposes (hence the several classifications that we use here) and the different degrees of accuracy that they entail.

(5) The prevailing construction market may be a booming seller's market in which prices are high and unpredictable or, to go to the other extreme, it may be a depressed market, a buyer's market in which the few projects available are "bought" by contracting companies that need the work to stay in business. More likely the market is somewhere between those two extremes. The price of construction is always affected, more or less, by the market, and it may be fruitless to do a detailed preliminary estimate in a booming seller's market, or in a depressed market, because the actual price is affected more by the general market than by the niceties of measurement and unit prices. This topic was originally discussed in greater detail in Chapter 3, Part 1.

Types of Preliminary Estimates

In the introductory Part 1, Chapter 4, "*Construction Estimates*," we referred to initial, intermediate, and final conceptual estimates for a project. Let us now pursue preliminary estimates further in the context of the design phase of a traditional building project to be performed with a stipulated-sum contract by one primary contractor[G] in which we might find the following preliminary estimates made for the owner by his cost consultant:[2]

1. An abstract budget estimate based on groups of elements or building areas
2. An initial conceptual estimate based on elements, approximate quantities, or both
3. An intermediate conceptual estimate based on elements, approximate quantities, or both
4. A final preliminary estimate, no longer conceptual, based on approximate quantities of work.

While these classifications are not hard and fast, and may flow together depending on circumstances and needs, they do show the general progression of preliminary estimates from inception to the end of a design phase, their types and their increasing complexity, and, we hope, their increasing accuracy, as the design develops and unfolds and as more design information becomes available to the estimator-consultant.

As already indicated, in the design phases of other than what we have called the traditional project (with a stipulated-sum contract and one contractor) and ultimately in the bidding for other kinds of projects (typically with some kind of cost-plus-fee contract), the preliminary estimates (if any) and the bid estimates may be more like the preliminary estimates than the bid estimates for a traditional project. In other words estimates for other than so-called traditional projects are often more like preliminary estimates than traditional bid estimates.

Having seen the kinds of preliminary estimates that are available as a project proceeds, let us continue by examining them in greater detail.

The Abstract Budget Estimate

This kind of preliminary estimate was first cited in Chapter 4, Part 1, and its uses, particularly in appraisals, were discussed in Chapter 5, Part 1, but so far we have not had too much good to say about its contents. Its poor reputation expressed here comes from the gross abuses to which it has been put by so many in the past, yet it has its valid uses when properly made and applied.

A building's gross plan area multiplied by a single unit price (which may be composed of several or

[2]In North America if an owner has a cost consultant at all, he may be the owner's designer who is also charged with specific responsibilities to the owner in respect of project costs, or he may be a professional quantity surveyor[G] or a cost engineer.[G] In other parts of the English-speaking world and in many other countries he will probably be a chartered quantity surveyor[G] and a member of the RICS.[G]

many unit prices all related to the area) is, as already indicated, the most common kind of abstract estimate in use today. In Chapter 5, Part 1, the common fallacies and the necessary qualifications and considerations for this kind of estimate were mentioned. There is no need to repeat them. When the cost data used in this kind of estimate are obtained from buildings similar in design, size, and location to the subject building, then an abstract estimate may be accurate. In fact it can be as accurate as any detailed bid estimate, especially if differences among the buildings are adjusted for in the abstract estimate. All kinds of estimates can be effective given good information and data and given good understanding and good work by an estimator. But we cannot stress enough the importance of these provisos, especially the proper use of good data.

The Initial Conceptual Estimate

The design information needed for this estimate may consist of preliminary sketches with overall dimensions and notes about the main construction materials and methods proposed. Additional notes may be added by the estimator as he gleans more information from the owner and designer and as he helps them to crystallize their ideas by asking questions about the project. Among this information there should be at least the following:

1. The size of the site and the gross area of buildings required
2. The areas of major site works, such as pavings and landscaping of different kinds
3. The areas of cut and fill[G] and the cut and fill lines[G]
4. Information about any special site works, such as watercourse diversion, piling, preloading, and other abnormal works
5. Information about any needed demolitions, clearing, blasting, and such
6. Information about subsurface conditions, soil conditions, underground watercourses, and the need for dewatering. The locations of all utility services
7. Requirements for hoardings, fences, and similar items
8. Requirements for hoardings, fences, and similar items.

In fact what is needed is a site survey as described in Chapter 1, Part 1, and all the information that it should provide. Also needed is information about the general requirements for the project's work that will eventually be embodied in the bidding documents, including information about:

1. The scope and general quality of the project's work;
2. The kind of contractual arrangement intended and the kind of construction contract or contracts (traditional, construction management, stipulated-sum contracts, or others);
3. The expected period and dates for construction; and
4. Any special contractual requirements (e.g., liquidated damages[G]) for late completion; nominated suppliers[G] or nominated subcontractors[G].

Ideally this initial conceptual estimate should in the owner's best interests be made *before* the owner enters an agreement with a designer for the preparation of working drawings and a project manual[G] (i.e., bidding documents[G]) in order to ascertain that the preliminary design (the subject of a separate agreement with the designer) is feasible and economic.

In addition to indicating whether a preliminary design is feasible and economic, and assuming that it is, the next purpose of an initial conceptual estimate is to implement *cost control*[G] during the design's development. At this stage a budget should be established or confirmed and related to the approved preliminary design drawings and outline specifications, so that everyone involved knows the general objective: this design and this amount of buildings and other works for this amount of money. An owner is well-advised to include both the budget and the preliminary design information in the agreement with his designer for preparation of the bidding documents and also to include in that same agreement effective wording about the cost consultant's implementation of cost controls and cooperation between the owner's agents and employees during the design phase.

The initial and subsequent preliminary estimates are to be used as guides in controlling future construction costs. What this means is that the cost consultant monitors the development of the design and if he sees what he considers a deviation from the general concept already agreed to (and embodied in the budget), he points it out to the designer and offers advice concerning the design and the costs in order to ensure that the budget is maintained. Of course, an owner is at liberty to increase his budget amount, although this may necessitate revised agreements between the owner and his agents and consultants.

Human nature seems to make us more careless with agreements and their details at the outset of projects and just when the greatest care is required to establish a firm foundation of explicit and equitable agreements among those involved concerning the scope, nature, and costs of construction work. So often one sees design and contractual principles and premises that are not suitable or desirable established in the early days of projects and carried through to the end—only because they have been embedded and unquestioned since inception. Of course, hindsight is always easier, and I am not suggesting that we can be prescient, only more careful.

Cost planning[G] should be a part of the design phase of every building project. As a building is planned, its costs should be planned and controlled at the same time. A project should begin with a budget and a clear view of the overall needs of the owner. As the needs are met by planning, the budget should be met. Similarly during the production phase the contractor (or construction manager) is required to plan the construction work. So planning is necessary throughout construction and is an integral part of construction, and cost planning is as important as any other planning. Cost planning[G] means allocating realistic costs to each part of a building. Cost control means ensuring that the costs of the parts are controlled and kept within the budget amounts so that the total costs are also controlled and kept within the budget.

Cost controls mean constant monitoring of drawings and specifications during their preparation for consistency with the preliminary design and the project budget by both the designer and the cost consultant. Understandably at times there is friction between these two agents of the owner, especially when a designer feels that another is encroaching into his area of special expertise—design. But design is more than aesthetics and more than one person's concepts of how a project should be constructed. At the same time many building projects (other than the most utilitarian) require an individual and integrated design concept, and this need and the need for economy and for a budget that is not exceeded must be balanced one with the other.

The Intermediate Conceptual Estimate

At this point let us anticipate the next two chapters by pointing out that there are two methods of making a conceptual estimate:

1. By analyzing the work into *building elements*[G]
2. By analyzing the work into *items of work*.[G]

(In addition both methods require the itemization of *the general requirements*[G] for the pricing of job overhead costs.[G]) These two methods are fundamentally different and there are arguments for both. At some point after the intitial preliminary estimate has been made, a more advanced and detailed intermediate estimate is required. It will probably still be a conceptual estimate, because the design probably will not be fully formed—if this estimate is made as early as it should. It should be made by one of the two methods listed, and in some projects two estimates should be made using each method: one with *elements* to allow quick and easy adjustments to building dimensions (if it is necessary in the design phase to change a building's size and form) but with a firm hand on the cost controls to ensure no excesses, and, following the other estimate and after the design concept and budget are confirmed, one with *items of work* to anticipate as closely as possible the subsequent bids and their analyses into the costs of trade work.[G]

The Final Preliminary Estimate

Essentially, if all has gone well, the final estimate should be nothing more than a refinement of the intermediate estimate; as indeed the intermediate estimate may in many projects be a refinement of the initial estimate. This final estimate should be made as early as possible after the design is no longer conceptual, and consequently it may not be a conceptual estimate at all but instead similar to a detailed bid estimate. That, however, depends on many things, including the types of earlier estimates made and the nature and circumstances of the work. For example, it is possible that a final preliminary estimate might be made even though some important design questions, such as the type of floor structure to be used, might not yet have been resolved. This may be possible because as, we have seen, unit costs for utilitarian elements and systems of different materials and methods of construction often do not vary greatly.

A final estimate in any case should, however, be based on trade work rather than on elements so that it can be used in making comparisons with the subsequent bid estimates and later with the schedule of values[G] subsequently submitted by the contractor. The contents of elemental estimates can be compared only with the contents of other elemental estimates, and the final preliminary estimate of a project is needed to make comparisons with the contents of a bid estimate that underlies the contract sum.

Preliminary Estimating Methods

The two main methods recommended for making preliminary estimates—(1) by elements and (2) by items of work—and their estimating precepts and practices are examined in detail in the next two chapters (Chapters 3 and 4). These are the prefered methods, but other methods have been mentioned and may be used wherever it is expedient. The only requirement is that they work. In summary all the estimates cited so far are:

1. Abstract preliminary estimates based on building space and on abstracts of comparative costs,
2. Conceptual preliminary estimates chiefly based on elements,
3. Conceptual preliminary estimates chiefly based on items of work,
4. Preliminary detailed estimates chiefly based on items of work measured more or less accurately from practically completed drawings,
5. Bid estimates for other than stipulated-sum contracts based on any of the preceeding estimating methods, and
6. Bid estimates for stipulated-sum contracts based on items of work accurately measured from fully detailed bidding documents.

The methods are listed according to the degree of detail of analysis, measurement, and pricing involved, with the least detailed first. Of course, these are only classifications for identification and comparison, and any particular estimate may fall between any two classifications. In fact we might have added another class of estimate—the mixed method estimate—which perhaps best describes most estimates, but that description would not get us far in understanding estimates. Nevertheless, many estimates, with perhaps the exception of the last in the list, are made by using more than one method among the various sections and divisions of trade work.

According to one reported survey from a reliable source (this sounds like an introduction to a commercial message) a majority of professional quantity surveyors prefer to use *approximate quantities of work*[G] even as early as the sketch design stage:[3]

Approximate quantities method	57 percent
Cost per area times unit price method	27 percent
Elemental method	13 percent
Interpolation method	3 percent
Total	100 percent

The report says that "at detailed design stage" the "approximate quantities method" is almost unanimously preferred.

The Cost per Area Method

The preceding cost per area method apparently favored over the elemental method by many quantity surveyors (contrary to my own preferences) is used with unit prices that are abstracts of a number of unit prices (all related to area) for different elements, as described earlier. In addition, to improve further the effectiveness of this area method the following procedures are used and recommended:

First separate the elements to be used into three groups: (1) those related to building area, (2) those related to building perimeter, and (3) those related to neither of the above; make abstracts of unit prices for each of the first two groups.

Next price the work by applying the abstracts of unit prices to the appropriate measurements, i.e., to the gross building area and to the total building perimeter, add the costs of the other remaining elements (arrived at by the best means available).

Any number of elements in the two groups related to area and perimeter may be used, but generally the more the better, providing valid costs for them are available. If area unit costs are not available for any part of the work, do not hesitate to use any other estimating method that is appropriate and expedient for that part: e.g, costs per unit area of element, unit costs and approximate quantities, or even unit costs related to use and function (e.g., elevator costs per floor over and above a basic cost per shaft, depending on type; escalator costs per foot of rise, added to a basic cost per installation, and so on). In estimating, flexibility in the use of different methods in the same estimate is often necessary.

The Cost per Volume Method

Although this method is out of fashion, and even though I have belittled it—and used it—this method still has its valid uses, as evidenced by the example in Chapter 7, Part 2. It is abuse that ruins a reputation. This volume method produces abstract estimates—if properly used with an abstract of unit costs, as described for the last method. Or, if used as in the example just cited, it produces an abstract estimate without an explicit abstract of unit prices.

[3]Reported in an article, "An Alternative Approach to Estimating," in the monthly periodical, *Chartered Quantity Surveyor*, February 1982.

Instead the estimate contains an implicit abstract in the single unit price originally taken from an almost identical building constructed earlier and used at the outset of the estimate. To be able to use costs from an almost identical building—as in the example in Chapter 2, Part 7—is unusual and fortuitous, but it certainly saves a lot of work.

Even if two buildings (one existing, the other proposed) are not identical but only similar, one might be able to use this volume method. Look out for similarity in both (1) quality and (2) density, as explained next.

Quality in buildings: To use costs from one building in estimating the costs of another, there must be similarity in quality; but no two buildings are exactly the same in quality. However, that need not deter the diligent estimator, who should be able to seek out the differences in quality, price the new work at a basic unit cost, and then (as in the example) make adjustments for quality by use of approximate quantities and unit prices.

Density in buildings: This means similar room sizes, a similar distribution of interior walls, and a similar number of fixtures (of whatever type the particular buildings contain)—in other words a similar amount of building mass (density) in a similar volume of building space. In most cases this probably means two (or more) buildings with the same function but not necessarily. For example, it is conceivable for one building to contain a school's administrative offices and another to be a commercial office building and for them both to have similar density and similar costs. It is primarily *enclosure*, not *space*, where the costs of buildings lie. Incidentally there is nothing to limit an estimator to using costs from more than one existing building; in fact more information usually means better data.[G]

One quantity surveyor of the old school[4] had a reputation for being able to make fast and accurate estimates using the volume method. He would study the drawings for the building project, select a particular part of the building, take off all the quantities for all the trades in that part of the work, price the quantities, and, from the total costs and the volume of the selected part of the building he had measured, calculate a unit cost per cubic foot for the entire building. It was a method suited only to large buildings, and no doubt it required judicious adjustments to the basic unit price to allow for differences in density or quality. Probably such skills were more easily developed in times of greater economic stability and less critical labor costs and when materials and methods were more traditional and less untried. No doubt, too, such skills were more readily developed after several years of practical apprenticeship than in a classroom.

[4]That is, trained as an articled pupil by an experienced member of the profession with whom the pupil has a legal and personal relationship (as with lawyers and others) instead of going to a university.

Estimating with Mixed Methods

This topic deserves special attention (and at least a subheading) because it indicates a simple technique not used often enough in making preliminary estimates. (Perhaps it could also be used more to advantage in making some bid estimates?) That is the use of more than one estimating method at a time in an estimate. Many estimators are conservative. As a result many like to stick closely to the rules—or to what they believe are the rules of estimating—in which they were once instructed.

Consequently many estimators take pride in consistency even when it is not a virtue, and they are reluctant to experiment with something as serious as an estimate. It was the opportunity and inclination to experiment that led me to the discovery that one of the ways to successful estimating is total expediency in methodology, to use whatever method seems useful and likely to succeed at the time. There is absolutely no reason why an estimate for any purpose should not contain half a dozen different methods of estimating costs—providing that they all work together.

Estimating with Alternative Methods

It is true with all estimates, but especially with preliminary estimates, that one of the most satisfactory ways to establish an estimated cost is by making two different kinds of estimates—as with appraisals. But do I hear an estimator say, That may be alright for appraisals, because they are less detailed, but who can afford to make two estimates? And who needs two estimates anyhow? Which would be correct? To respond to these questions, first let us say that length is not necessarily an indication of accuracy. With good information and data and the right method, or with the right computer system (hardware and software), an estimate can be less time-consuming and less expensive.

As with appraisals, more than one estimate is a good idea, and the only arguments against it are either economic or one that says that while appraisers have some distinctly different approaches to value (market comparison, income capitalization, replacement cost) there is only one way to estimate costs. There is some truth in that (although some argue that in fact all appraisals are really based on market comparisons), but we must not overlook the importance and potential of *information* and *data* in estimating. In fact there is always more than one way to make an estimate, but perhaps more important there are often more than one source of valid information and data to be used. And if it is not economical to make more than one estimate in the old way, then estimates should be made with a computer. As to which estimate would be the correct one, this question is dealt with every day by appraisers, and it is up to an estimator (as it is to an appraiser) to make that decision in the light of the available information.

Precepts for Preliminary Estimates

All of the precepts for bid estimates in Chapter 1, Part 3, cannot be directly applied to preliminary estimates because the two kinds of estimates have different objectives. The main objective of preliminary estimates is to anticipate subsequent bid estimates, and it is usual to expect to make preliminary estimates with less time, effort, and expense. Not only does an owner prefer it that way, it is also expected because usually there is less design information available with which to make a preliminary estimate. Therefore, it is believed, greater expenditure on a preliminary estimate is not justified. But we should consider what the answer would be if adequate cost information and data were always available for preliminary estimates.

An even more fruitful question might be: Is there anything to be gained economically by putting more effort into a preliminary estimate? The seemingly impassable obstacle to an incontrovertible answer is the fact that because every building is unique it is not possible categorically to show quantitatively that one building is more economically constructed than another. If through an effective preliminary estimate and cost controls during the design phase a more economic design and building resulted, apparently there is no way of absolutely demonstrating that fact. The most likely owner for whom that fact might be established beyond a reasonable doubt is one who buys a lot of sequential construction work in buildings designed for the same function, so that apples might be compared with apples instead of with oranges, or even passion fruit.

The following comments on precepts of estimating (examined in the last chapter) are relevant for preliminary estimates:

(1) The precept of veracity remains essentially the same for all estimates: that is, to allocate costs correctly. But it is often much harder in a preliminary estimate in which many subjective concepts may be used and in which at times one may be pricing a function rather than a system of specific construction work. Nevertheless, we should not abandon the precept; instead we may at times consciously break the rule when it appears expedient to do so. The important thing is not the rule but that we should know at all times what we are doing. The rule in this precept is, That which give rise to a cost shall bear that cost. At times, as we shall see in the following chapters, we may not follow that rule in a preliminary estimate, for example, when we substitute one minor item for another, something not acceptable in a traditional bid estimate.

(2) The precepts of measurement in preliminary estimates at times must be modified because the design information may not be precisely dimensioned, or the cost data used may be such that more detailed measurement (of labors,[G] for example) may be fruitless. Similarly the precepts of net measurement of work as done in place; of allowances for waste, shrinkage and swell, and laps, to be made in unit prices instead of in the quantities measured; and the precept of measurement according to a standard method—all of these rules are often broken in preliminary estimates. Generally this rule-breaking encourages some to abandon all rules. What is the use of rules, they say, if you deliberately break them? But that is thoughtless. The ultimate rule in estimating is that of expediency: Do what you believe is necessary, but know what you do. The enemies are delusion and confusion.

The ultimate test of any estimate is that of accuracy; as noted, accuracy is often misunderstood because it is always relative. Some cost engineers with great faith and skill in mathematics fondly believe that estimating costs is a pure science. I disagree, because I believe it is partly an art, and I do not believe that mathematics can always produce the right answer. When cost engineers use an estimating formula, it almost always contains a variable factor, and the rest of the formula often proves to be symbols for common sense. But the right factor requires art for its selection.

Summaries in Preliminary Estimates

Even though a preliminary estimate may be a mere outline of a later bid estimate, no effort should be spared in the preparation of its summary and the report that should go with it. If a detailed summary is not prepared there is a good chance that important overhead costs will be omitted, and most errors in estimates are errors of omission. Making up a summary according to the Masterformat[G] is like using a checklist. A good summary presents the whole cost picture of a project and enables us to see the costs of all the various parts of the work and to compare them with each other, with the total costs, and with the costs of other projects and their various parts. In this way omissions are often discovered, discrepancies are revealed, and brief analyses and quick comparisons and checks are made possible.

No matter the purpose of an estimate nor the method used in estimating, many estimators regularly convert all the total costs for sections and divisions of work in an estimate's summary to unit costs (per square foot or meter) and to percentages of the total. With these figures estimated, costs may appear in a new light. For example, electrical costs (no pun intended) that are only 5 percent of the total costs of a building may need reconsideration, as in most buildings the electrical costs usually are 8 to 10 percent of total costs. Similarly in ordinary buildings designed for human occupancy (not for parking or storage) the poured concrete structure usually costs 20 to 25 percent of total costs, and so on. There are many such general guidelines, but as the advertisements say, the actual mileage may vary, as may the actual proportions of costs. But they are useful guidelines when reviewing a summary.

Another reason for making a detailed summary for every estimate is the assessment of overheads and profit. This can be done much better with a good summary. In my experience one of the most important parts of making a good preliminary estimate is in the thought and reflection over the summary. This final review often results in revisions that improve an estimate because it is not until the summary is completed that an estimator can see the entire forest and not just a lot of trees—and not only the forest but the whole countryside around.

Finally do not overlook the verbal report that should go with every preliminary estimate, stating all of the qualifications and assumptions already mentioned. Even the composition of this report may result in sudden clarification so that the estimator finds it necessary to go back and ultimately make a better estimate. Estimating is a process involving intense concentration. Anything that requires an estimator to look at a project and its costs from another position is helpful, and this the summaries and reports of estimates often achieve.

Questions and Topics for Discussion

1. Define these terms used in this chapter: (a) *abstract,* (b) *abstract estimate,* (c) *design information,* and (d) *budget estimate.*

2. List and briefly describe *five* things to be considered in selecting the type of preliminary estimate to be used.

3. Assuming that a project to build a school building (apartment building, office building) will cost not less than $5 million, how much do you believe an owner should be willing to spend on preliminary estimates made for the purposes described in this chapter? Explain your answer and give valid reasons and figures.

4. Interview an experienced estimator who works with a contracting company and write a report (of not less than 1,000 words) on his views on measurement of work (preliminary estimates, the purposes of an estimate) and on your own views on the same subject.

5. Explain why an estimator with a contracting company may have some difficulties in making conceptual estimates for Construction Management purposes.

6. Why should the prevailing construction market influence one's choice of a particular method of estimating? Is this valid? Explain your answers in detail.

7. Define the following terms used in this chapter: (a) *cost control* and (b) *the precept of veracity* (in estimating). Describe their applications relative to preliminary estimates of costs.

8. Why should an initial conceptual estimate be made before an owner commits himself to a designer for the preparation of bidding documents?

9. Describe cost planning and its purposes and implementation in a project.

10. Briefly describe and compare the two best methods of making conceptual estimates, their advantages and disadvantages.

11. Prepare the summary for a detailed preliminary estimate for a high-rise concrete apartment building (without actual cost figures) showing all the information that it should contain.

12. Precisely describe the circumstances in which an abstract estimate can be accurate.

13. Describe in detail the recommended methods to be employed in making an abstract estimate based on building space.

3

Conceptual Trade Estimates

In the last chapter we observed that there are two recommended methods for making conceptual estimates: (1) using *building elements*[G] and (2) using *quantities of items of work.*[G] In this chapter we examine the latter method; in the next chapter, the former. The method examined here may be called the *quantities method*, or the *approximate quantities method*, because often in preliminary estimates based on incomplete design information[G] the quantities of work measured are approximate, as we shall see. Proper trade estimates for bids are properly based on accurate quantities measured according to the precepts cited in Chapter 1, Part 3, but in some cases approximate quantities are used, especially for common work in ordinary buildings. The standard of estimating in contracting companies varies greatly, but with the latest generation of trade contractors one often finds better estimating than in the past. For conceptual trade estimates we have as our model the fully detailed, accurately measured bid estimate[G] of the costs of trade work[G] made by bidders; as already stated, preliminary estimates generally are better estimates to the extent that they are similar to that model. But first let us examine some other kinds of trade estimates.

Unorthodox Trade Estimates

What is an *orthodox* estimate? The model just cited, the fully detailed, accurately measured bid estimate, is orthodox. Why is that? Because in its ideal form it follows the precepts of estimating, therefore it is more likely to be accurate and effective. Also it enables a contractor to perform effective cost accounting[G] as we have described it. Unorthodox trade estimates instead may use some of the methods that we have described for preliminary estimates: essentially shortcut methods used when time is lacking or when the market is booming, when prices are high, and when an estimate and bid need not be trimmed too closely.

Another reason for such estimates may exist with a contracting company that does not want to bid for a particular job but at the same time does not want to appear to refuse to bid. This company may seek a cover bid,[G] or it may decide to make up its own bid, using an unorthodox method and an inflated markup[G] so as to submit a high bid that will surely not be accepted. If by chance it is accepted, then perhaps it will make a lot of money. Typically these bids are based on what we have called here *abstract estimates*, those in which the total cost is calculated by applying an abstract of unit prices to simple quantities, such as a single gross area (of floor area, wall area, etc.). Let us look at a few typical examples of unorthodox trade estimates.

Costs for *site works* may be estimated by taking the total costs of site works for a similar project in a similar location, making some major, lump-sum adjustments, and by adding a large contingency allowance to ensure that the total is more than adequate.

Costs for a *reinforced concrete structure* may be estimated by taking off approximate quantities and pricing the entire concrete volume in the structure by a single unit price per cubic yard (cubic meter) to include all steel reinforcement (rebar) and all formwork. Serious errors may occur if the unit price is not taken from a similar building (similar loads, spans, function) and if no adjustments are made for differences in time and location and for other important conditions. This method may be surprisingly accurate if the subject building[G] is similar to the source buildings.[G] By far the largest proportion of all concrete volume in a typical multistory building is in the floor slabs (especially if the building is to have curtain walls), and the numbers of columns and other lesser concrete components have little effect on the total volume. Even foundations may vary quite a lot in size and form without significantly affecting the total volume of concrete. But be careful. Estimators should make analyses of the distribution of quantities and costs of major items (such as concrete, forms, and rebar) for the projects with which they deal. The information can be valuable in making quick estimates, which may then be used to check totals of orthodox estimates and their several parts.

Structural steelwork costs may be estimated at so much a ton, including everything: fabrication, painting, and erection. For highly functional, highly utilitarian buildings such as warehouses, for example, estimates may be based at so many pounds of steel per square foot of floor area (or kilograms per square meter). The same can often be said for economically designed high-rise office and apartment buildings. The weight of steel per unit of floor area is often surprisingly unchanging among buildings of the same class.

In addition to these more specialized methods all trades may fall back at times of great pressure and demand for estimates on the ever-popular gross area times unit price for a total cost of work. But there are usually better methods, such as those just cited, and as others become appropriate they will be mentioned. We turn next to the more orthodox estimates for various trades.

Site Works

Costs of site works are among the most dificult to estimate at any time but especially for a conceptual estimate. Costs are almost entirely dependent on physical and other conditions, and the conditions often outweigh the quantities of work in economic importance. But nevertheless, in preliminary estimates approximate quantities are desirable, at least for the major items. For lesser items of work, often costs based on equipment time at current rental rates are better than those based on quantities. Especially with this trade it is advisable to get as close to a detailed bid estimate as possible, even if design information is sparse. If necessary, design your own site works, using whatever information is available; draw your own contours on a copy of the site plan;

roughly design your own exterior drainage system; lay out your own roads, and so on. It is not hard to do, and it creates a better basis for a conceptual estimate than what might otherwise be available. This division of work may need such handling in any kind of preliminary estimate because other methods that apply to buildings themselves may not apply to their more variable sites.

Subsurface exploration costs, if not paid by the owner at the outset, should be the subject of a cash allowance[G] in a conceptual estimate. In this way attention is drawn to the item, and there is no better way of dealing with its costs at this stage than to base them on similar jobs in the locality. But do not omit this item from any estimate. If no such costs are incurred, at least there will be something for contingencies.

Clearing a site varies from doing almost nothing to felling a forest. It may even involve demolition of buildings. Get costs from local companies; they are usually able to quote accurate costs with little information. For disposal, burning may be cheapest, but do not assume that it is permitted; check first, fiddle later. Carting away debris (by truck, not cart) is usually expensive, but it depends a lot on distance to and availability of a dump site. Large trees often have value as lumber; so does topsoil, but it may be needed at the site. All work done by hand is expensive, and some hand labor is almost always needed, even with equipment. (Remember the general labor item included in job overhead costs and mentioned previously?)

Demolitions may involve salvage, either of entire frame buildings, parts of buildings, or just valuable fixtures. The last time I looked over an old building with a demolition contractor he was accompanied by his wife who was also his business partner; she could not conceal the lust in her eyes—for the solid brass fire extinguishers, door hardware, and other old-fashioned fixtures screwed to the dark-stained woodwork. Let demolition people know that you too know how much these things are worth, or warn the owner to remove them first; otherwise he will not get their true value.

Service utility lines, their moving, and their connections (temporary and permanent) are under the province of public authorities, and they are not called *authorities* for nothing. There is no appeal beyond the local light and power company when it comes to the company's service lines. After all this is power to the people, and you cannot trifle with that. Because these public bodies have no competitors, they are practically immune to cajoling and threats and can do pretty much as they see fit—for the public good, that is. As a result their estimates (when you finally get them) are not necessarily firm, as they may find that they must do other work not yet foreseen. So make adequate allowances in your estimate. Fortunately some employees of such companies are truly public-minded and offer good service. May you find such a one.

Site grading consists of (at least) two different items: initial rough grading and the final finish grading. The best way to estimate their costs is by equipment time, although some use area and a unit price. For major quantities of work, use both methods and then compare results. Any earth that must be handled more than once will cause additional costs. Topsoil is marketable, and some will pay for the privilege of removing it. If it is needed on site, allow for at least double handling. Make sure that there is enough room on site for storage, as costs for removal and return can be prohibitive.

It is pointless to try to measure an area precisely for cut or fill when the precise depths can never be ascertained; the changes in volume due to *shrinkage and swell*[G] make fun of the illusion of accuracy often found elsewhere in some earthwork calculations. Do not overlook the need for extra excavation (and backfill or imported fill) at deep cuts to form stable banks, or allow for planking and strutting (use and waste[G] of lumber plus labor costs). Similarly, large depths of fill may require compaction and the formation of banks or other means of stabilization. None of this and similar work will be shown on drawings—if you have drawings. Site work always requires a lot of conceptualization, even for bids. There are many other site works that may be needed and are rarely shown even on bidding documents, let alone in preliminary design sketches.

With site works and their costs it is important to keep a preliminary estimate simple but adequate. Do not try to make too fine a point. Measure everything gross and overall. It is important for items and their quantities to balance logically: amounts of excavation with the combined amounts of backfill, imported fill, and surplus material removed. All excavated materials must be accounted for, otherwise there will be an error of omission, the worst kind of error. But do not overlook that, in estimating, overmeasurement and overpricing are also errors. An estimator cannot simply overload an estimate just to avoid errors of omission without errors of addition. A balance is required in all things, not only between excavation and fill.

Excavating and *filling* are commonplace, and ideally they should balance and leave nothing to be removed off site — which is the most expensive of all, unless material (such as topsoil) can be sold or traded for its removal. Often removal of surplus material by truck is needed, and sometimes one must pay for the privilege of dumping it on another's property. In a conceptual estimate one should assume the reasonably worst case if firm information and costs are not available. If the native soil is heavy clay or something worse, it may be necessary to remove all excavated materials and to import all fill.

Do not omit earthworks related to underground services. In some localities they are part of a mechanical contractor's work, in others they are not. Some trenches will be within the building and may need special equipment for work in restricted areas and in compacted fill material. This work is different from other trenching and backfilling for services; it is done later, and usually it costs more.

Shoring of excavation sides for stability, or alternatively overexcavation in forming banks, pumping, and usually some hand labor are needed. These cannot be allowed for in all-inclusive unit prices for the main items if one hopes to be reasonably accurate. It takes only a moment to enter such items in an estimate and to make reasonable allowances for them; a far better way.[1]

Site drainage (and similar systems) may have to be more or less designed by the estimator making a preliminary estimate. Often there is no other way, and for most building projects an experienced estimator can do this in a few minutes. As before, the axiom is, Keep it simple, and measure the work to its farthest limits. Do not be persnickety about adjustments for minor features. If you have learned to measure building perimeters quickly and accurately,[2] you should be able to do the same with building services. The easiest way is to use tracing paper over a site plan and to draw in the services with colored crayons. Mark manholes and depths. Measure long runs as *runs*[G] and enumerate everything else, including short branches (in groups, up to 5 feet long, 5 to 10, and so on).

Both measurement and pricing are easy once you have priced the main runs. If a system contains pipes (conduits, ducts) of many different sizes, you may find it easier to estimate an average representative size and price all piping the same. Measure trenches and other work similarly. If possible combine all related work (trench, piping, bed, protection, backfill, testing) into one combined item of work and apply one unit price, even if there are minor deviations from the standards that you have created. Ingenuity goes a long way in estimating.

Do not overlook the value of measuring and pricing *extra-over items of work.*[G] They simplify measurement by separating individual characteristics of work, such as bends in pipes, for example, from the main items in which they occur, such as pipes. The pipe runs are measured as though without bends (tees and such) and the extra items are enumerated and grouped according to size. They are then priced at the extra cost over and above the costs of the main item. For example, a pipe bend may displace one foot of pipe but cost three times as much; then if the piping material costs $1 per foot, the extra cost for each bend will be $2. Do not get into the bad habit (even when making preliminary estimates) of counting a bend as 3 feet of pipe. In that mad method you end up with meaningless quantities. But, some will protest, if all the work is to be priced in dollars and added together, what does it matter how one measures providing the price is right? It does matter, for a number of reasons, and *reason* is the key word: An estimator soon loses his reason if he begins to use fictitious quantities. He can no longer visualize the work from figures if the figures do not directly relate to the work and to its real quantity. Certainly no other person will be able to understand the estimate. The estimate will be of no value as a source of information for other purposes such as ordering materials, *planning and scheduling*[G] the work, or even for estimating a project's duration and overhead costs.

In the better approach measured quantities are a translation of the work's description into another language, from graphics into a numerical and verbal language, and in a pure form it can be retranslated if needed. If its description is distorted by gimmicky methods (such as substituting one thing for another), it loses all meaning, except possibly for that of cost. Certainly cost is our first objective in estimating, but we should take the real view of what creates cost, not sacrifice it for a false view derived from a dubious method for which there is a better substitute. Thus the precepts of good estimating apply to all kinds of estimates.

But, you may want to ask, must we do all of this work for a conceptual estimate especially if in the past you have been fond of using the quick and easy *abstract estimate*[G]? Generally you reap only what you sow, but with this method of making preliminary estimates you get a far better return for your effort. Besides, the industry definitely needs a dif-

[1]Only selected, common major items are included for comment in this chapter. An estimator needs a good checklist to avoid omissions, also a standard method of measurement and a good estimating reference text (see Bibliography).

[2]See *Fundamentals of Construction Estimating and Cost Accounting,* Chapter 7.

ferent attitude and approach to preliminary estimates. Such estimates need not be expensive, but they deserve more time than is often given. Often they are given too little time because it is believed that a simple abstract estimate is the best that that can be done. This attitude arises from either ignorance or sloth.

Any division[G] of work that amounts to more than 2 or 3 percent of a project's total costs is to be counted as a major division, because few amount to more and there are 16 standard divisions of work in the Masterformat. Almost invariably site works is a major division whereas others (such as concrete and masonry) may or may not be. Site works must always be treated with respect.

Landscaping is often an important part of site works, yet many estimators know little about it and its costs. It is one of the last sections of work to be done on a building project, and often it is done by an *other contractor*[G] who comes on to the site when the building contractor is about ready to move off. Landscaping in preliminary estimates is one of those sections of work usually covered by a flexible cash allowance, reduced if money is short and needed elsewhere, and later increased if a saving is found somewhere else in the estimate. Often landscaping is precluded from a project because the owner-designer cannot see giving the contractor a possible markup on work with which it is expected he will have nothing to do. Landscaping is often viewed like the tail on the puppy: It is cute, but is it really needed, so why don't we cut it off?

In a preliminary estimate, if landscaping is to be included (often to get it into the total amount on which the financing will be based), a cash allowance should be built up from estimated amounts for the various items of work: areas of lawn, flower beds, shrubberies, graveled areas, and such. Each is priced at a different unit price; edgings, fences, individual trees, and so on, each priced separately. This is the only way to arrive at a realistic allowance, and it is the only way to be able to increase or reduce an allowance for specific items later. Anything else is probably guessing.

Generally, in a preliminary estimate,[3] an estimator must do some designing to be able to estimate realistic costs, and this is true (more or less) for all kinds of construction work. Site work is more intangible than most other work, and more variable. It therefore deserves special attention, as it is one of the divisions of greater error (of omission) in preliminary estimates. The seasonal weather likely to prevail at the time of site works must be considered.

The need for contingency allowances is high because the risks are high, because of the nature of the work involving a site and its latent conditions. Many of the items of work required are not apparent unless one studies the site and the preliminary design drawings and unless one thinks about the entire project from start to finish. It is therefore a good idea to begin a preliminary estimate with site works, to establish information and assumptions that affect much other work as well, and then to end the estimate with a review of site works to make any needed adjustments to the estimate of its costs. All of these points should be kept in mind when estimating costs of other work. Finally, check the contents of site works against the contents of Division 1, General Requirements, the division containing all of the *job overhead costs.*[G] (These were discussed at length in Chapter 1, Part 3, *Proposal Estimates*, and they are mentioned again toward the end of this chapter.)

Concrete Work

This omnipresent division of work contains at least four major sections, including.

1. Formwork for poured concrete,
2. Steel reinforcement for poured concrete
3. Poured concrete, and
4. Finishing and curing concrete.

Some prefer the term *cast-in-place concrete.* The important thing is to distinguish it from precast concrete, which is dealt with differently in estimates.

Formwork for Poured Concrete

Formwork is important in most building and engineering construction, and usually it is the most expensive and the most problematic part of concrete work. In a typical building project formwork may account for about one-third of the total costs for concrete work, yet it is never part of the permanent structure; formwork is simply a mold into which wet concrete is poured. Because of this, I have already suggested that the best way to look at formwork is as if it were a kind of *construction equipment,*[G] often made elsewhere, brought to a site, erected and used, and later dismantled and removed. Its costs include those of handling, including repairs, and the depre-

[3]*Preliminary* rather than *conceptual* only because the former is a more inclusive term.

ciation that occurs each time the formwork is used. In economic and accounting terms forms are no different from a crane or any other piece of equipment. But often this view of formwork is obscured, and confusion results, perhaps compounded because forms are often made on a particular site but later are used elsewhere, or because in frame construction much of the formwork material (lumber and plywood) is incorporated into the building, as framing members and sheathing.

In any estimate of formwork costs, if it appears that the forms must be made at the site, all the fabrication costs should be kept separate and not charged to that particular project. Instead what amounts to rental costs should be established for each use of the forms, and each project should be charged for the number of uses at the established rate for use and depreciation (often termed *use and waste*[G]) of the forms. That is the clearest procedure to follow, but there are valid exceptions. If it appears that particular forms will be used only on a particular project and that there is little possibility of their use on another project, then all the costs of fabrication should be charged to that one project, in addition to all the other direct costs of handling the forms. Similarly, if as in wood-frame shelter construction it is expected to dismantle formwork and to use the form material as framing material, only that proportion of the form material that will be actually used up on forming concrete should be charged as formwork costs. The rest of the plywood and lumber costs for material used in forming but later used for framing should be charged to framing.

It really is common sense; we try to allocate costs to the work that gives rise to the costs, not to something else. In addition to the form material costs there are costs of using special form hardware (use and waste, again, because some are always lost), costs of consumable hardware (nails and concrete ties), and, of course, the main costs, those of labor, which may amount to as much as 75 or 80 percent of the total costs of formwork (labor, materials, and equipment) in a typical concrete building structure and as much as about 25 to 30 percent of the total costs of all concrete work. The fact that the costs of formwork are often as much as or greater than the costs of the poured concrete itself means that an estimator must give it adequate attention.

Actually all of this is much easier in a bid estimate than in a preliminary estimate, because in a preliminary estimate often there is not much design information to go on. But in a preliminary estimate (conceptual or otherwise) it is usually better to estimate formwork costs in the same way—making the necessary assumptions—than to try to arrive at a realistic cost using extant unit prices applied to the measured *contact area*[G] of forms—unless those unit prices are from a recent and similar project.

The contact area of formwork is easily measured while (or just after) measuring the concrete. One need not be fussy about making deductions for minor features, especially in a conceptual estimate, as there are always many minor items that are overlooked, apart from the fact of the somewhat hypothetical nature of much formwork, especially its number of uses, both total and on any particular project. Yet at the same time (and this paradox is an essential part of all estimating) an estimator must not let hypotheses and vagueness become an excuse for carelessness and inaccuracy. There is a constant danger in estimating that an estimator will lose his grip on the tightrope stretched between the poles of accuracy and approximation.

In a conceptual estimate the only way to measure quantities of form materials is to apply established factors to the different contact areas of slab soffites, walls, footings, columns, beams, and so on. In detailed bid estimates, at least the major formwork items should be designed and measured in detail to establish new factors for each project unless, as stated before, valid information is available from recent similar work. Waste allowances should be generous, especially in a conceptual estimate. Recall the general rule for conceptual estimates: Keep it simple, measure overall, and allow for minor features in the unit prices for major items of work.

Usually the biggest problem in estimating formwork costs is to determine the probable number of uses for different forms. (Only one use should be assumed for minor formwork items.) Often a lack of information and a lack of time in which to study and analyze formwork are the main obstacles. In a preliminary estimate one can perhaps afford to be conservative and to select a low number of uses, but that is not usually so in a bid estimate. An estimator should not lose sight of which are the important items containing most of the costs and which items (usually the majority) he may price pro rata. Having established sound unit prices for the critical items, prices for other items may be based on them.

Flying forms are commonly used for multistory concrete buildings, and these forms (even more like construction equipment) require much analysis and estimating effort. These forms are a good example of the kinds of construction work for which an estimator should have information and cost data on file; otherwise when the need for it arises, there may not be time or opportunity to obtain it.

Reinforcement for Poured Concrete

Despite its apparent complexity, rebar (reinforcing bars) in concrete is not a difficult item for the estimator, even in conceptual estimating, because of the direct relationship between quantity and function in different concrete components, which perfectly expresses the principle of economic utility in terms of weight per unit of volume (pounds per cubic yard or kilograms per cubic meter). Even in detailed bid estimates accurately measured quantities of rebar should be also expressed in this way to make an easy check of quantity and to provide information for future preliminary estimates. But even in making bid estimates and while measuring from detailed drawings, one must not overlook the nonstructural carrying bars and spacer bars (to which structural bars are attached) as well as other minor but essential accessories such as tie-wire, supports, and chairs.

Rebar should be calculated separately for all major concrete components—foundations, walls, slabs, columns, beams, and so on—even in preliminary estimates. Average weights for entire structures can be used with confidence only when obtained from impeccable sources and from almost identical buildings. Weights per unit of volume can be obtained from data books, but known projects are a better source. In most concrete buildings, if one has accurate unit weights of rebar for the slabs, one has most of the rebar accurately measured. Walls come next in importance. As for items such as reinforced concrete stairs, steps, small retaining walls, and planters, often too much time is wasted in calculating their rebar quantities. Lump them all together and allow a sufficient weight per unit of volume to be safe, and the result will usually be adequate. But take care not to fall into errors of omission.

Unfortunately an estimator cannot always have confidence in the designer and always assume that structures are economically designed. I have known of reinforced structural slabs designed with twice the needed amount of steel. The last time it happened I questioned the designer and he insisted that the large amount of steel was necessary. Because I was the construction manager, and also an agent of the owner and directly concerned with the owner's budget, I tried to get another structural engineer to check the design, but no one was prepared to investigate and make a report, even though some said that they believed the amount of steel was excessive. Even the rebar bidders voiced the same opinion, and of course they had no reason to deceive. Despite overwhelming evidence to the contrary, my preliminary estimate of rebar was found wanting, and the slab was constructed with the steel shown in the designer's drawings. My apparent deficiency was unavoidable because I had made the preliminary estimate before the design drawings were made, and I had based the steel quantities on those of six similar buildings, all built during the last few years in the same city. Human deviation apart, structural design is a fairly precise science, but in making preliminary estimates the unexpected may occur.

Poured Concrete

Although formwork is installed first, followed by the rebar, it is better in estimating to measure the concrete first, followed by the formwork (some like to do it simultaneously, but I prefer to use the formwork takeoff as a chance to review and check the concrete), and then finally the rebar. With complete structural drawings, measurement of structural concrete in ordinary buildings is not difficult, except that sometimes the foundations are not clearly shown, often because the designer was not absolutely sure and preferred to keep the drawings just a little vague. But with only rough sketches for a preliminary estimate, measurement of concrete may be much more difficult.

Usually it helps to discuss the design with the designer, who often can offer additional information when pressed for a decision. Alternatively an estimator may have to calculate some of the cross-sectional dimensions of concrete members before he can measure the work. Here again a data book can be useful. One-half inch of slab thickness may amount proportionately to about 3 to 5 percent of the total costs of a reinforced concrete building structure, so one cannot be too far off. A common error in preliminary estimates of building structures is the omission by the estimator of structural shear walls needed for stiffening and stability in the structure. A column here or there does not amount to much, but structural walls soon add up to a considerable amount, especially when they occur on every floor. Assume that some such walls are needed and seek advice from the designer whenever you can.

A part of the economic design of many buildings involves the use of standard sizes for concrete structural members with reductions in cross-sections occuring every few floors in a multistory building. This reduces formwork costs, even if it causes slight oversizing of some members on some floors. Estimators should look for this feature and use it to simplify measurement. Often, even in bid estimates, it is possible to measure a few typical floors and then to make minor adjustments for variations as the dimensions of members are reduced on the higher floors. In preliminary estimates it may be possible to dispense with the minor adjustments.

Particularly when measuring highly functional building systems, such as a structure or a mechanical system, an estimator should spend some time at the outset to study whatever design information he has and to seek out and use to his advantage in the measurement of work any pattern that exists in a design and any repetition of features and dimensions made by the designer as a way to reduce design effort or to economize in construction. Standardization of cross-sectional dimensions of structural members within groups of floors in multistory buildings is a prime example. There are others.

A simple axiom of measurement states, *If it is different, keep it separate.* But in preliminary estimates it may be expedient to some degree to forego that advice. It stands, nevertheless, as a good general rule because if it *is* different, its costs are probably also different. The question is, How much different? In a multistory building the following categories usually suffice, but less may cause unacceptable inaccuracies:

- Footings below ground level
  - Continuous
  - Isolated

- Footings at ground level
  - Continuous
  - Isolated

- Walls below ground
- Walls above ground
  - Interior
  - Exterior

- Columns below ground
- Columns above ground
  - Interior
  - Exterior

- Slabs on ground
- Slabs above ground
  - Interior
  - Exterior

Other categories may be needed, including piers (attached), upstanding beams, and others, but categories should be kept to a minimum and only created for an item of significant quantity that deserves to be kept separate because its unit costs are significantly different. For example, exterior columns and beams exposed to the weather often have different concrete mixes and finishes, and often their forms are more expensive because of difficulties in access and support. But if an estimate's overall accuracy is judged incapable of justifying such fine differences, usually there is no advantage in measuring the work separately.

Costs of concrete vary greatly, and an estimator must predict which concrete will be placed one way and which another: by crane and bucket, by buggy or wheelbarrow, or by pump, for example. Therefore, even in a preliminary estimate, it is useful to measure and price separately different concrete materials and placing.

Allowances should be included for such things as special concrete mixes (for concrete fill in hollow masonry, toppings, etc.) and for admixtures. Waste should be allowed according to placing methods.

Finishing and Curing Concrete

Finishing and curing concrete and similar items are easily measured by borrowing quantities (*reflected dimensions and quantities*[G]) from other previously measured items (e.g., formwork). Fine adjustments need not be made to account for minor differences. For example, it is sufficient to take an area of forming slab soffites for finishing and curing the slab surfaces, even if the latter area should be a little less. It is this kind of acceptable *inaccuracy* in preliminary estimates that makes the use of aproximate quantities feasible while producing costs that are still much more accurate than those produced by other methods. Also, by measuring items overall, minor overmeasurement occurs where major planes (i.e., walls, slabs) intersect, and such overmeasurement helps to pay for the extra costs of additional work that so often occurs just at such locations.

Reinforced Concrete Work in General

Economy in estimating effort and suitable estimating accuracy depend a lot on a proper sense of proportion, gained mostly by experience, but that sense of proportion can in part be gained by studying the related proportions among items of work—in concrete, for example, among the labor and material and equipment costs of concrete, formwork, and rebar, as reported in cost data publications or (better) as discovered by your own analyses of work and costs:

Item	Materials	Labor	Equipment	Total
Poured concrete	24%	8%	3%	35%
Formwork	8%	26%	1%	35%
Steel rebar	19%	10%	1%	30%
Totals	51%	44%	5%	100%

This table presents some typical proportions that will, of course, vary from structure to structure, but the significance to us exists in the relative proportions, not in the precision of the percentages: They show an estimator where to apply the greatest effort and attention, namely, in estimating labor costs of formwork as they are the largest and most variable part. Material costs, apart from some variation in waste, are readily established. That popular topic on many sites—the labor time expended in placing concrete, often the subject of boasting or derision—does not even approach in importance the labor for formwork, about which we usually hear much less because it is less obvious and much less easily determined.

Masonry Work

Problems in estimating masonry costs are often associated with the minor features of masonry work: openings, piers and columns, treatment at edges of masonry and other similar *labors.*[G] Which of these costs can be assumed as included in the main items of work, and which should be measured and priced separately? This kind of question (in all trade work) is relevant in all kinds of estimates. Many companies and their estimators are unsure. A national standard method of measurement may be controversial (as all standards are), but at least it would promote discussion and resolutions if not all the answers. Not until such standards exist will it be possible to get a good handle on the economics of construction because there is no common vocabulary—but many dialects. Talk to half a dozen masonry estimators and discover their different approaches to measurement of masonry work, especially chimneys.

Even in preliminary estimates the best way is to have a basic unit price for each major item of masonry work, exclusive of variable features, and to measure and price the minor features separately; but if there is only a small amount of masonry work in a project, one comprehensive unit price may suffice. Otherwise it takes only a few minutes to measure as *runs*[G] such items as:

1. Vertical cutting abutting other work,
2. Horizontal cutting underneath abutting slab soffites, and
3. Cutting for openings, arches, and similar features.

If there is much *rough* and *fair cutting,*[G] it should be separated, as the latter is much more expensive than the former because it shows and must be neat (fair). A perennial question is, Should masonry be deducted at openings? Yes, it should if an opening is larger than about one square foot in area, but then separate labors have to be costed (forming jambs, head), or openings may be enumerated and grouped according to size (usually easier and quicker). Again, do not think that this is too much effort for a preliminary estimate. (Some think that it is too much even for a bid estimate.) Such measurement of ancillary items soon becomes almost automatic after measuring the main masonry work, and (most important) it provides a quantitative image of the work that makes pricing easier and more accurate.

The variation in labor requirements for different masonry work in different projects is great. Consider hollow concrete blockwork: In some buildings (college science buildings and hospitals, for example) almost every other block may have to be cut to fit around concealed piping, conduits, and other concealed work; the cutting must be made clean and fair on a masonry saw if the blockwork is not plastered or otherwise concealed by more than paint. If blockwork is modular and conceals no other trade work, as in warehouse construction, it is much cheaper. Therefore distinctions must be made in estimates among different classes of masonry work, and the best way to contribute to accuracy is to measure the labors separately because their effect on cost varies in each project. This precept also applies in estimating the costs of other trade work.

Concrete fill in hollow masonry work is expensive in both labor and material, and it cannot be treated as either ordinary concrete or as mortar. Building in items such as joist supports and hangers is time-consuming. But such items can neither be ignored nor discounted and one should make an allowance for them in estimates for masonry work.

Season and weather may have a significant effect on masonry costs. Wet weather is especially disastrous for concrete blockwork. Cold weather may necessitate protection and heaters or make masonry work impossible. Hot weather may require new masonry joints to be kept moist and cured.

Do not overlook the costs of scaffolding and supports at openings such as labor and the use and waste of equipment and materials. Like formwork, these items are temporary, but their costs may be significant.

Metal Work

Generally the costs of metal work are either very easy or very difficult to estimate. I remember my first brass pole for a fire station; I was distracted by finding the costs of such an unusual item, but I need not have worried; I found the right company, and to it brass fire poles were commonplace. The costs of some kinds of custom metal work always seem to be elusive. You must know the sources. With this kind of work, despite the readiness of some trade people to talk of dollars per pound for circular steel staircases, unit costs are always dubious. For a preliminary estimate, if there is only one steel staircase in the project, try to use information from a good cost data book and add a healthy contingency allowance. It will do no harm. Steel gratings and their costs are still a mystery to me. Where do they come from? How are they priced? For me an industrial plant with tons of steel gratings, catwalks, ladders, and other such ferrous esoterica would be a nightmare. But then if you had to cost much of that kind of work, you would have the information and the measurement techniques at hand and consequently, no problems. It is the fact that an item is not commonplace that keeps us ignorant and causes anxiety. And so the terrors of estimating, as of most things, are largely imagined.

Most estimators, at times, come across an elusive item and waste hours trying to track down a realistic price, when the sensible thing would be to give it the best possible guess and move on. I do not think that any estimator really guesses, as the acts of quantifying and of assessment are too deeply engrained by training and experience to allow an estimator to simply guess. Intuition often produces surprisingly accurate answers, especially if an estimator deliberately and frequently exercises and tests his judgment on all kinds of items, not necessarily only construction items.

If you have not yet done so, look around your city and seek out those obscure workshops periodically illuminated by the bluish blaze of a welding torch (scintillating like the prolonged flash of a camera), discover the unusual things that the shops produce, and seek out their prices. Firemen's brass poles are commonplace after all, and only such things as copper samovars for Cossacks or a whim-wham for a goose's bridle will raise more than an eyebrow of the metal worker.

Structural steelwork is the most common of metal work, but estimators should remember that there are at least two major classes of steelwork—framed and unframed—and that the unit costs per pound for these vary a lot. Usually, *framed steelwork* is cheaper per pound because there is usually much more of it, as in the steel frame of a building. Typically, *unframed steel* is the isolated beam or lintol built into other work. You cannot use the same unit price for that because in small isolated quantities almost any work costs more. In fact the best way to cost unframed steel is to build up a lump-sum cost, pricing only the basic material at so much a pound, and the cutting, drilling, and handling costs will be proportionately high and not much related to the total weight.

Framed steelwork must be measured from framing drawings, and if they do not exist, a conceptual estimator must make his own or get a structural engineer to do it for him. Alternatively the gross weight of steelwork may be estimated from the gross area of a building if good, valid information from similar buildings is available. As with reinforced concrete, most designers have attained a high level of design economy with steelwork, and therefore there are fairly constant relationships among building spans, loads, and areas for similar building functions. In a preliminary estimate it may be accurate enough to measure in detail one typical bay of framing or the framing of one floor and to obtain a unit weight (per unit of floor area) to be applied to the whole building. But do not overlook the necessary adjustments for atypical end bays and any atypical framing elsewhere. As always the biggest hazard is the error of omission; in this trade, for bracing, connections, splices, extra framing around openings, and similar minor items that add extra steelwork and extra weight. (An example of a sophisticated approach to estimating the costs of structural steelwork is presented in Chapter 8, Part 2.)

A few items to look out for when estimating steelwork costs follow. Hollow sections (round and square) are much more expensive per pound than solid sections, and unusual sections always cost more. Steel is generally more expensive in the West. Do not overlook the costs and the problems (and extra, hidden costs) of erection. A good cost book provides a lot of information about basic and universal trades such as steelwork. Standard, predesigned steel buildings cannot be priced by weight, and square-foot (square-meter) costs are more effective for these, but the unit costs vary according to spans as well as area, perimeter and quality.

Wood Work

There are many classes of work in what we formerly called *Carpentry*, but we shall look at only the two of most concern to the conceptual estimator: (1) framing and (2) finish carpentry. These are the most important and the most problematic.

Wood Framing

Wood framing in bidding documents requires of estimators a great deal of conceptualization because it is never fully shown or described; only the outlines are indicated. In the case of a preliminary estimate the estimator simply must go even further with his conceptualizing. In one sense it is easier for a conceptual estimator, because to do a proper job he must get into the basic design requirements for framing and sketch much of it for his own purposes before he can begin to measure it, so that he understands it better than if faced with a designer's limited information that must be translated. Once again the greatest cause of errors is omissions—of not measuring enough—and this appears commonplace.

Part of the reason for this unintentioned parsimony rests with teachers and technical writers who tell students and novice estimators to allow, for example, for *two* studs at corners in framed walls when everyone knows that at corners *three* studs are installed. But, they say, we should allow for only two *extra* studs at corners because the third stud has already been measured when the extended length of the wall is divided by the stud-spacing to give the basic number of studs. But this is a fallacy, as there is about a 10 to 1 chance that the third hypothetical stud will coincide with a corner; or to put it another way, the specified stud-spacing (typically 16 or 24 inches) is the *maximum* spacing allowed. If a wall length is not an exact multiple of the spacing, there will be an extra stud. Thus the popular methods of measurement are often deficient.

In fact even in bid estimates[4] it is often better and more accurate to use information about framing from past projects and to estimate framing quantities by using data related to gross areas, because in this way we avoid such errors of omission, especially if the data has first been tested and proven by cost accounting. To measure framing in detail (and it must be so measured at times) an estimator must have a good knowledge of framing details and practices in the locality. Many framing members are installed to support other work (drywall, fixtures). Often it is quicker and cheaper for framers to use more material and to thereby minimize their effort; as a result, in practice several studs may be laminated together where theory says studs should be spaced a few inches apart, and solid boards may be installed instead of the dainty 2 x 2 framed furrings—like that of a shoji screen—so popular with some draftsmen. Many designer-draftsmen (and draftspersons, too) undersize framing members because they rely on published tables for sizing them (especially joists and rafters) and they make the common mistake of taking *minimum requirements* of a code as *optimum requirements*, a widespread mistake in all sectors of construction made not only by designers but also by some officials, inspectors, and estimators.

Waste[G] is a big problem with framing, both in estimating and in practice (and the two are supposed to be related). In practice on a construction site much yard lumber and plywood go for uses that are unrecognized by many estimators:

1. Temporary blockings and supports for site trailers, sheds, and plant
2. Temporary fences, railings, hoardings, barriers, walkways, and enclosures
3. Temporary poles for power lines, posts for signs, and protection of lines
4. Site fixtures, including work tables, benches, shelves, and racks for use by workers and supervisory staff
5. Boardwalks and handrails to site buildings
6. Batterboards, leveling poles, pegs, markers, and such
7. Edgings for pavings, graveled areas, and planted areas
8. Shoring and supports for sides of excavations and work in excavations
9. Concrete forms and their supports
10. Temporary supports and braces for framing, prefabricated components, and for other work

[4]I frequently make such a statement to contrast *bid estimates*—the ultimate model that we have recognized—with *preliminary estimates*, which we expect to be less exact.

11. Permanent supports and blockings for sheetmetal work, fixtures, and finishes

12. Minor scaffolding and an almost infinite variety of other items incorporated or not incorporated into a project's work but nevertheless needed.

Undoubtedly in all estimates there should be, in Division 6, an item described more or less as follows, "Unspecified uses and waste of yard lumber material," that should have an amount set against it based on experience and formalized by cost accounting so as to control and estimate better the costs involved. Alternatively it might appear in Division 1, General Requirements, where it would be more easily assessed along with similar general items.

Finish Carpentry

In building construction the amount of finish carpentry work is often an indication of a project's quality: the more woodwork, the higher the general quality. This is because woodwork has become so expensive and good woods and skilled labor are harder to obtain. Often in apartment buildings finish carpenters install little or no actual woodwork (kitchen cabinets are now factory-made and installed by specialists), only minor fixtures such as shelves, clothes hangers, and bathroom fixtures, usually made of anything other than real wood. Even kitchen cabinets rarely contain any wholly natural wood; instead they are made from veneers, laminated plastics and composition boards of various qualities. If this usual condition is encountered, an estimator may simply allow so much money per dwelling unit to cover labor costs for finish carpenters. If there is real woodwork (typically, trim around openings and at floor perimeters, also, possibly at cabinets and closets, and, in extraordinary instances, in wall paneling), it is too expensive not to measure and price it with care. Because finish carpentry work is mostly embellishment and therefore one of the first items to be deleted if a budget is tight, that is often reason enough to make it the subject of cash allowances in a preliminary estimate, because these are easier to adjust.

There is no use in detailing the work until one can be reasonably sure that it will be included. Of course, certain kinds of institutional and commercial buildings usually have a considerable amount of finish woodwork (one thing that declares *quality*) and for such buildings as these, even in preliminary estimates, it is necessary to measure and price finish carpentry work in some detail. Shortcuts in preliminary estimates can be used, for example, by detailed measurement and pricing of selected areas or items of work to arrive at inclusive unit prices per room, item, square foot (meter) of paneling, or whatever other unit is suitable.

Moisture Protection

Generally, large areas and relatively low unit prices make this class of work one of the easier for preliminary estimators, who often can use fairly accurate quantities and accurate unit prices from trade companies for dampproofing, waterproofing, and roofing. For simple, overall measurement it is best to imagine a moisture protection system as an umbrella or as a waterproof tank (as the case may be) without apertures or breaks. Assume that flashings occur at all edges and at all steps and changes in plane; assume that extra costs are needed for extra work at all corners and other features in the otherwise blank planes. Allow for control joints in large areas and for inspection and testing of important work.

Even if a roofer will do the sheetmetal work, keep it separate; it makes it easier to price and to adjust prices later. There can be no constant relationship among moisture protection areas, total lengths of flashings, and the quantities of other sheetmetal work; usually each needs separate measurement. Do not overlook the possible need for sheetmetal pans under water tanks and certain plumbing fixtures, for flashings in conjunction with exterior claddings and veneers and at all abutments between horizontal and vertical exterior planes, and for ductwork from kitchens and bathrooms. A good checklist for different kinds of buildings is helpful in identifying such esoteric and expensive items that often are not shown even in final drawings.

Do not overlook waterproof membranes needed to protect occupiable spaces underground or beneath traffic areas and also on balconies and terraces. Some or all of these membranes may need flashings and cement mortar toppings, pavings, or other work over them for protection.

Insulation is often needed in conjunction with membranes as well as elsewhere, and the similar approach to measurement of insulation is to assume contiguous and uninterupted areas enclosing the interior spaces. Do not be concerned about deductions from gross quantities for minor displacements and interruptions. Often areas for work in this division are obtainable as reflected quantities from other work.

Openings in Buildings

This division of work includes windows and doors of all kinds with their hardware as well as curtain-wall systems, window walls, entrances, and all related glass and glazing and other similar and related work.

Windows and Doors

Standard items in this division are readily priced by manufacturers and suppliers, and there is no reason, even in a preliminary estimate, not to have fairly accurate unit prices, even if it means assuming certain standard sizes and selecting the items from a catalog. (*Standard* does not necessarily mean from stock, as most large orders are specially produced, which means that some minor variations from standard specifications may be available at little or no extra cost. If an order is large enough, major variations are possible at little or no increase in standard prices.) Custom items[G] are a different story, with often very different prices, something some designers have not yet grasped. It is hard to get quotations for custom items for preliminary estimates, but some manufacturers are more helpful than others. Cost data books are often not much help in this division of work, possibly because of the great variety. A friend in the glass and metal business can be invaluable to conceptual estimator.

In this division, measurement of items is often not much more than counting, but pricing can be complex ($M + P = E_c$). Sizes of glass panes (lights) affect prices greatly. So does the number of opening vents—which are frames within frames—with finish hardware; they are therefore best priced separately as extra-over items.[G] Overall quality and the quality of parts (e.g., hardware) may vary greatly; in a preliminary estimate often it is wise to get prices for at least two different standards of quality in order to provide for adjustment of costs if necessary. Some products are installed by their makers[5] while others are delivered to site for installation by the contractor or subcontractor; thus it is better to keep installation costs separate.

Site-installed finish hardware (particularly for doors) is often covered, even in bid estimates, by a stipulated cash allowance,[G] and on major projects most makers provide for potential customers the services of a hardware consultant (employed by the company) who can give expert advice on the appropriate amounts allowed for finish hardware. Window hardware is usually installed at the factory.

For specialty items, such as overhead doors, in a preliminary estimate it is better to make an allowance for supply and installation. At least two prices for doors of different quality are desirable.

[5]A shorter and therefore better word than *manufacturer* not acceptable to some because of theological connotations.

Curtain Walls

Because of the specialized nature of this class of work, one needs the help of a manufacturer or supplier to estimate the costs. Differences in quality, quantity, and installation conditions affect costs greatly. The most expensive systems may cost four or five times the costs of the cheapest systems, and as large quantities are often involved, this may become a significant item in an estimate and in a budget. Therefore costs of systems of different types and qualities will probably be needed for a preliminary estimate. The same general facts apply also (but to a lesser degree) to window walls and glass and metal entrances.

Weights and other qualities of metal members and of glass, as well as glazing methods and materials, greatly affect costs, as do metal finishes (such as anodizing and enamel) and glass properties (such as light and heat absorbtivity and reflectivity). As before, sizes of panes (panels) affect costs, with larger panes costing more because of larger sheets of glass and heavier members. Glazing materials and methods may be cheap and simple—such as the ordinary puttylike glazing compound and nothing more—or expensive and complex—such as three of four different calking compounds and sealants, all used together in a single glazing system. Extensive and high curtain walls usually require the latter, while interior glass and metal walls may need only the former. These specific differences are pointed out simply to indicate that costs in this highly variable division of work may range greatly and that one cannot use costs from one project in another without some research into the specifications.

Interior Finishes

Cost experts generally believe that while a building's exterior cladding costs can have considerable effect on a building's total costs, the interior finishes are not nearly as significant. Presumably this is because while the cost variation for an exterior is potentially great, for building interiors (within specific classes) wide variations are uncommon. Of course, one might, for example, specify an expensive wool carpet instead of a cheap artificial fiber shag (so popular in speculative housing), but that is unlikely in rental housing, for example, while costs in use may show that a better-quality and more expensive exterior cladding (clay brickwork, for example) for rental housing buildings may prove more economical in the long run.

At least for preliminary estimates (and the technique can be used elsewhere) the easiest way to estimate the costs of interior finishes is first to establish basic finishes throughout a building for all the floors, walls, and ceilings. This is not difficult because usually it has already been determined. (For an apartment building, for example, the basic finishes might be the following: floors—nylon carpet; walls—latex-painted gypsum wallboard with a plaster skim coat finish; ceilings—sprayed textured finish on wallboard). Any deviation from these basic finishes is then measured and priced as *extra-over items* (for such items as vinyl sheet goods on bathroom and kitchen floors, ceramic wall-tiling in bathrooms, and acoustic-tile ceilings in lobbies and hallways.) This simple technique minimizes measurement effort and the chances of errors of omission; errors are quantitatively reduced because the erroneous omission of a higher quality finish is largely offset by the basic finish already included.

Plaster and Gypsum Drywall

In common classes of buildings (particularly in the different subclasses of dwelling construction) it is possible to discover fairly constant relationships between a building's gross floor area and the gross total area of wall and ceilings (often 3.5–4.0 to 1 in cheaper apartment buildings). This relationship is often a fairly reliable constant because design characteristics are fairly constant (including the average sizes of rooms, the ratios of areas to perimeters for both rooms and entire buildings, and ceiling heights). Knowledge of such factors is useful in estimating, either as a quick method of measurement or as a method of checking detailed quantities.

Openings (windows and doors) make a significant difference to wall finish areas and costs because of deductions of openings from areas (a minus) and because of the creation of narrow widths of finishes at (and between) openings requiring proportionately more labor and also metal beads (a plus). Should one deduct openings? That is the question. The case for not deducting says that pricing as though an opening were not there allows for the extra labor needed for cutting in and at the reveals. But does it? If it does, it is only by chance, because there is no relationship between area and perimeter. Therefore, it is better to deduct always and to measure separately the extra work at reveals and narrow widths, even in a preliminary estimate, because the additional effort is so small. Similar considerations arise from other similar features, including returns, beam sides and soffites, small areas, control joints, and also trim at edges, corners, and abutments up to other work. If a project, however, contains but a few of such items they may be overlooked or dealt with summarily.

Suspended Ceiling Systems

As with all finishes made of standard units of products (particularly tiling of all kinds) there is always the cost of cutting to fit to work dimensions, typically with part-units of not less than half size at perimeters and with equal-sized part-units on opposite sides of an area. In other words costs are to some extent affected by perimeters as well as by areas. It is not only a question of more or less cutting and waste. More perimeter-length per unit of area also implies more separate areas and therefore a greater number of times work must be laid out, started, and stopped—with scaffolding dismantled, moved, and re-erected—thereby reducing productivity.

An expensive feature of suspended ceiling systems is vertical drops at changes in plane, and these drops often do not show up on preliminary drawings. Similar items include access panels to get at services in ceiling spaces; high ceilings, requiring extra-high and more expensive scaffolding; fire-resistant panels; and integrated mechanical services and lighting. Qualities and therefore costs of ceiling installations also vary greatly.

Floor Finishes

There are some general precepts for estimating costs of floor finishes: (1) Use a basic finish throughout where possible, and deal with exceptions by use of extra-over items; (2) ignore minor deductions of flooring, such as displacements by appliances and fixtures; (3) on stairs, measure an 18-inch width for each normal tread and riser (times the stair width); and (4) for base trim (otherwise requiring tedious measurement or guesses), measure total length by using the total length of interior partitions times 2, plus the building perimeter-length at each floor (for the base attached to inside faces of exterior walls). It is not accurate enough to assume that by measuring gross areas of flooring and *not* deducting the flooring displaced by interior walls, the extra flooring measured will compensate for the cost of base trim: it will not.

I believe in measuring and pricing straight cutting and waste at perimeters of floor finishes (basically the same total length as bases), but some argue that the costs of this item cannot be isolated (by cost accounting) and therefore it cannot be priced separately.[6] I prefer to try, so that my estimate then takes some account of room size and of layout costs, as cited. In a preliminary estimate, if in doubt as to whether a flooring is to be laid beneath certain fixtures, assume that it is. (As a general rule for preliminary estimates we assume the more expensive case.) Obtaining unit prices for this work is not usually difficult.

Specialties

In estimates for large projects the need for quick and easy measurement of many specialties is a necessity. Often it can be done by measuring groups or sets of items instead of trying to count items separately. For example, bathroom fixtures and kitchen appliances are better counted by first counting the numbers of bathrooms and kitchens and then pricing by groups at so much for all the fixtures in each bathroom or kitchen. Often it helps to make up a matrix or schedule of items and rooms, or units, especially if there are different numbers of items and different numbers of different units (e.g., one-, two-, and three-bedroom apartments, each with different numbers of closets). Methods of measuring items and ways of recording them should reflect the logic of their distribution so that the figures in an estimate become a numerical picture of the items of work and of their distribution. For example, for 27 three-bedroom apartments, an entry in an estimate of 27 times 3 (closets) makes much more sense than an entry in an estimate of a total of 81. The former says something about the arrangement; the latter, only that the estimator can multiply 27 by 3 and get the right answer.

It is only a matter of time before we develop a written language that will enable us to represent three-dimensional work other than by drawings. Theoretically it can be done, by using a system of reference points—something like the present two-dimensional grid references used in construction plans, but much more detailed and precise and, of course, three-dimensional. But we need a simple system that is easily learned, remembered, and used. Undoubtedly computers will help. Such a system may require the use of imaginary axes located through the center of gravity of a building's plan and sections, or similarly through parts (blocks, wings) of a building, so that the reference system will not be constrained by a building's outer perimeter. This concept is illustrated in Figure 3–1.

[6]The easiest way to price it is to include the cutting with the base trim.

Equipment in Buildings

This division (No. 11 in Masterformat[G)]) embraces special equipment of all kinds installed in buildings, apart from what is included in the building systems in subsequent divisions (12 through 16); for example, a dishwasher, a vault door, and a cooler. In preliminary estimates especially, this work is usually covered by cash allowances,[G] as it may also be in bidding documents and in bid estimates. The reason for this is that few designers and estimators know a lot about equipment, because it is often highly specialized, so they must rely on specialists to tell them how much to allow. This is a good place, therefore, to say something more about cash allowances in estimates.

When describing equipment and its costs, it is a good idea to use precise terminology to ensure that all relevant costs are covered by a cash allowance (or elsewhere) in the estimate. Such descriptions typically include:

Delivered to the site, unloaded, moved inside to a designated storage location, and later moved to the place of installation....

Installed complete with all ancillary items and including all necessary jointing materials and fittings, tested for use, and left in working order and ready for use by the Owner....

Be specific when seeking prices (and bids) by referring to any necessary related work to be included, or excluded, such as:

Venting for kitchen equipment
Electrical and other service connections (water, gas) to many items of equipment
Testing installations
Provision of spare parts and consumable items for equipment.

In the same way that cash allowances are used in preliminary estimates because a design decision has

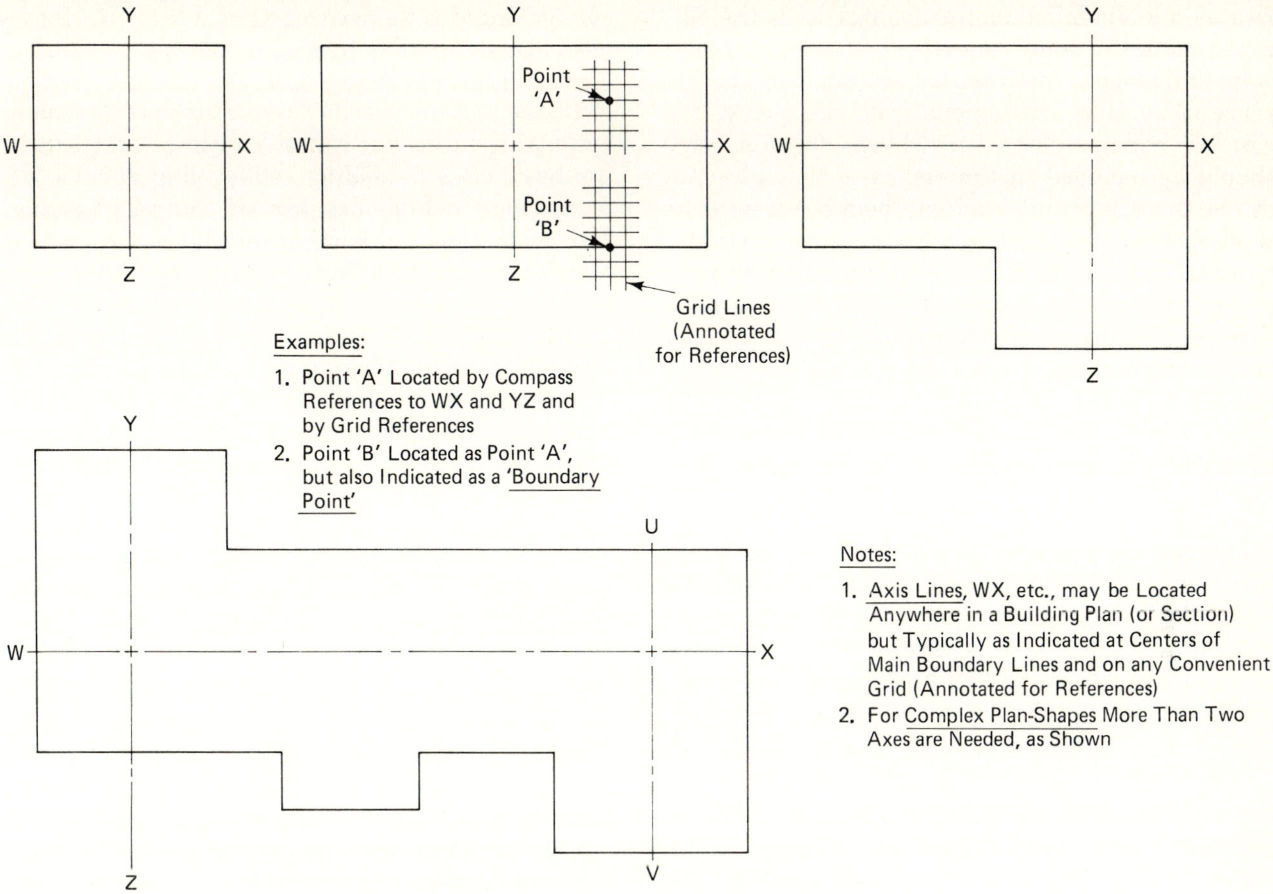

FIGURE 3.1 **Axes in a Reference System**

yet to be made—although a cost estimate is needed and cannot be left until all the design decisions are made—cash allowances are used for essentially the same reason and the same purposes in bidding documents (specifications) and in bid estimates. With much of the equipment installed in buildings, ancillary work is needed that is usually performed by the contractor to serve the needs of the specialist subcontractor installing the equipment, and for the equipment itself. This work should be so identified and measured in the normal manner. Unit costs for this work may be higher than for similar work elsewhere because of the often small quantities and isolated locations involved. Builder's work related to equipment may include:

- Floor drains;
- Concrete beds;
- Plumbing and other connections;
- Venting, hoods, and fans; and
- Electrical supply, switches, and motors.

The list is unlimited and depends only on the kinds of equipment to be installed in a building. It may even include special walls, floors, and such, but if these are not within the normal scope of a general contractor's work they may be more appropriately found in Division 13, Special Construction. Again, an estimator must judge how much time to give to such items; the important thing is to at least con sider them and their costs.

Furnishings

In preliminary estimates, especially in early conceptual estimates, it may be difficult to be specific about items of furnishings to be included, yet an early decision may be needed in fixing a budget amount. Again, cash allowances are the best means, but the earliest of estimates should state if specific furnishings normally found in a certain class of building are to be included or excluded from a project and its budget (for example, blinds and drapes, in apartments and offices). As the trend to greater specialization continues, in which kitchen cabinets have become a specialty item often found in Division 10 rather than in Division 6, we may find eventually that cabinets and similar items are totally prefabricated and classified as furnishings, as is the current practice in some countries.

In preparing a preliminary estimate in the absence of specific instructions from the owner, the cost of commonplace furnishings (drapes, etc.) should be included in the estimate. This gives an owner the opportunity to take them out later if he wishes, after discovering that the total costs are more than he expected. In some construction projects it is required to have one or more dwelling units furnished for display, and an allowance for this may be included here.

Special Construction

The subject of this division is work of such a nature that it is not normally performed by a general contractor, or by one or more of the common building trades. It is work not integrated with the general construction but performed by a specialist who requires only the minimal of attendance from the general contractor and who requires perhaps only to use the general site facilties (telephone, power, toilets); consequently a primary contractor's [G] involvement with such a specialist may not involve any actual supervision.

Typically, special construction may include the installation of prefabricated freezers, special partitions (for offices, X-ray rooms, and such), and other prefabricated components and units. (Refer to the Masterformat for others.) As in previous divisions, cash allowances are normal.

Conveying Systems

Elevators are the most common conveying systems. Builder's work associated with elevators is fairly extensive and often some of it is done by a so-called general contractor and not kept separate (e.g, the elevator shafts). Hydraulic elevators require a hole drilled in the ground for the elevator's hydraulic piston (as deep as the travel of the elevator is high). This must be done before any concrete work in the elevator shaft and may necessitate some preliminary bulk excavation; often therefore it is one of the first items of work scheduled. Other elevators require the provision of steel beams and a housing at the top or bottom of the shaft.

Costs of elevators are not difficult to obtain from specialist companies that are usually keen to get early notification of a project's need for elevators and a chance later to quote a price. The cost of elevators depends on whether they are for freight or passengers, whether hydraulic (slower) or electric operation, geared or gearless, the number of stops at floors, and the speed. There are, however, many optional features with most elevators that add to the basic costs, including self-leveling devices, express floors, cab phones, special controls, lighting, and decoration. As with automobiles the makers search for features different from those of their competitors' products so that they may have something on which to hang a selling point. For a conceptual estimate it may suffice to use a good cost data book or information from recent projects if one is in a hurry and does not have time for the intricacies of the published brochures and specifications of elevator companies.

Mechanical Systems in Buildings

This division of work (Division 15 of the Masterformat) includes a wide range of mechanical service systems, but in buildings the most common systems are air treatment and distribution systems, heating systems, water supplies (hot and cold), and plumbing. (See Masterformat for details.) What we have to say about mechanical systems in buildings applies to them all, but we shall refer to the more commonplace systems for examples and for illustrations.

Generally there has always been a deficiency in the capabilities of most designers, quantity surveyors, and other building cost consultants in dealing with mechanical and electrical (M & E) services for buildings, and all too often they relied too much on the advice and information obtained from contracting companies and engineering consultants in matters of M & E costs. In British Commonwealth construction practice the bills of quantities[G] often contain cash allowances (called instead *prime cost sums*[G]) for all M & E work in building projects, while the remainder of the work, including even the builder's work in connection with M & E work, is meticulously measured in detail according to the numerous and complex rules of a national standard method of measurement.

The contrast between the methods employed—P.C. sums and detailed measurement—only draws attention to the architects' and the quantity surveyors' inability to deal with M & E work as they deal with all other building work. In preliminary estimates, too, the quantity surveyor often must depend on others for cost information about M & E. I know, because I did the same; but I deplored it. On arriving in North America in the middle 1950s, I found generally an equal ignorance of M & E work in the New World. The only difference here was that it was not made so obvious because nobody used bills of quantities for building works. (The eminently practical engineering profession, however, did use contracts based on *schedules of quantities*,[G] something almost the same as the more traditional bills of quantities used elsewhere.)

The general state has since changed to the extent that there are now more quantity surveyors who specialize in M & E work and its costs and who consult to their colleagues, and I was privileged to work with one of the best.[7] Together we devised the tree concept of M & E layouts, with the branches and fruits and flowers (the terminal units), while teaching estimating mechanical costs at night school. His successors have since developed and refined the techniques of estimating M & E costs.

As in all good conceptual estimating the prerequisite is a sound knowledge of the construction work involved, in this instance, the mechanical services in buildings. To illustrate the procedure of making a conceptual estimate, let us go through a conceptual estimate for mechanical work in a building and review its methodology.

Conceptual Estimate for Mechanical Work in an Office Building

A conceptual estimate begins with a *facility program*, a schedule of the required mechanical facilities obtained by the cost quantity surveyor[8] from the owner-designer[G] that specifies in outline the kinds of services and systems required.

Air conditioning may be provided by one of several systems, typically in an office building (with about 50,000 square feet of gross area) either a dual-duct system or a constant-volume system. The design criteria for our example require 0.8 to 1.2 cubic feet per minute (cfm) per square foot of area, say, 50,000 cfm (using the average rate). This requires a system with one, perhaps two, air-handling units with heating and cooling coils. The subclasses of air conditioning work in such a system include:

1. Air-conditioning equipment, including coils;
2. Piping;
3. Ductwork, including dampers;
4. Piping terminal units (radiators and unit heaters);
5. Ducting terminal units (diffusers, grilles, and louvers);
6. Piping insulation;
7. Duct and plenum insulation;
8. Controls; and
9. Balancing of the system and terminal units.

A similar breakdown of work applies to all such systems. The system that we are reviewing is for an office building. Let us review these subclasses of work in detail, as an estimator making a conceptual estimate would have to review and estimate them.

Measurement of Mechanical Work

(1) Air-conditioning Equipment. Initially the owner-designer specifies the type of system required; here it is a multizone system with perimeter radiation (hot water) and air cooling. The capacity of the equipment is determined by the estimator from the building's area, its volume, the air flow, and the required cooling capacity of 250 square feet per ton. The air-handling unit requires a capacity of 50,000 cfm. (50,000 square feet x 1 cfm per sf.) The capacity of the air chiller is 200 tons (50,000 square feet of total area divided by 250 square feet per ton = 200 tons). Also, the system requires two water pumps (one for standby), each with a capacity of 50 percent of the equipment's cooling capacity, and a third pump for the cooling tower's condensate.

The heating plant's capacity is determined by the estimator from the transmission heat loss (dependent on the rate of heat loss from the building's exposed surfaces) and the heat loss from the expelled air through the ventilating system (50,000 cfm of mixed inside and outside air). This total heat loss, or heating requirement, determines the needed capacity of the hot water boilers. The two primary pumps are required for the hot water system, with other pumps for the secondary circulating systems to the perimeter-radiation heating units and to the air handling units. The washrooms and storage on each floor of the office building will require exhaust fans to provide air handling at a rate of 2 to 2 1/2 cfm per square foot of area instead of the general rate of 1 cfm per square foot elsewhere in the building. In addition the system will require expansion tanks and other minor items of equipment.

(2) Piping. For an estimator this part of a me-

[7]Julien Hardy, of Vancouver, British Columbia, founder of Julien Hardy & Associates Limited, Vancouver, B.C., Mechanical Cost Consultants and Quantity Surveyors, whose present principal is J. K. Kubicek from whom I received much assistance in writing the contents of this section on estimating M & E costs.

[8]Both deal in construction economics. A quantity surveyor, however, may be an agent of an owner, as in a contract with quantities.

chanical system is both important and tricky. The designer has only to indicate piping diagrammatically, even in completed bidding documents, while estimators must actually measure the piping. A conceptual estimator must measure the piping with incomplete drawings of the building, without any mechanical drawings. Yet 20 to 30 percent of the total costs of a heating and ventilating system are in the piping. The piping in our example consists of:

1. Chilled water piping,
2. Condensate piping, and
3. Hot water piping.

Let us look at each of these piping systems.

(a) Chilled water piping circulates from a chiller in the building's basement up to the air-handling units, usually high up in the building. The size of the chilled water mains in the mechanical room is determined from the total cooling capacity required; its total length is found by adding the length required in the mechanical room to the length of the mains (flow and return) to and from the air handling units. The hookups (connections) to equipment (air handlers, etc.) are measured separately by enumeration. Each hookup includes a short length of piping with valve, flanges, and a strainer.

(b) Condensate water piping size depends on capacity (20 tons) and is measured between chiller and condenser in the cooling tower and from chiller to the cooling coils.

(c) Hot water piping is sized and measured in the same way as the chilled water piping. The primary circuit supplies the heating coils; the secondary circuit, the radiant heating units at the perimeters. Hookups to these units are enumerated and described by averaging the sizes to simplify costing (e.g., if branches are 1/2 inch to 2 inches diameter, and the sizes are about equally distributed, then 1 1/2 inch-diameter hookups may be taken as the standard average).

Generally, supply piping and return piping are the same sizes and the lengths of each are about the same, so approximate measurement is fairly easy. Only main runs are measured by lengths; branches (hookups) to equipment and heating units are enumerated and described by average diameter and average length, and this further simplifies measurement.

(3)Ductwork. Factors obtained by the analysis of previous work are used to calculate the quantities of sheet metal in ducts in the several duct systems for:

1. Supply air,
2. Return air,
3. Outside air, and
4. Exhaust air.

The weight of ductwork is related to the system's capacity (i.e., 50,000 cfm x 0.7 pound = 35,000 pound) for the supply air system. The weight factors are obtained from previously completed projects.[9] Return air volume is taken as 80 percent of the supply air volume, and the weight factor here is much less because with return air systems there is no zoning required and so the return air system is much simpler than that of the supply air (i.e., 40,000 cfm x 0.3 pound = 12,000 pound). The outside air system also is simple, and it includes an intake louver in the penthouse designed for 100 percent capacity (50,000 cfm x 0.08 pound = 4,000 pound). Exhaust air systems from the washrooms and storage are also straightforward (10,000 cfm x 0.4 pound = 4,000 pound). Thus the total weight of sheet metal in ductwork required is 35,000 + 12,000 + 4,000 + 4,000 pound = 55,000 pound. In addition, there are flexible duct hookups to air supply terminal units and also the fire and balancing dampers to include.

(4) Piping terminal units. These include baseboard radiation units at the exterior wall perimeters on all floors and two forced-flow heaters in the entrance lobby. At the exterior walls there will be continuous (from wall to wall) baseboard cabinets. The radiation units will occupy (according to similar past projects) about 60 percent of these cabinets. The length of the heating elements (of the radiation units) is calculated from the transmission heat loss from the building.

(5) Ducting terminal units. These include supply air diffusers, return air grilles, exhaust air grilles, and louvers for fresh air (intake) and for the exhaust air.

Supply air diffusers are estimated from the total supply air rate (i.e., 50,000 cfm divided by the rating for each diffuser, say, 250 cfm per outlet, = 200 diffusers required).

Return air grilles are similarly estimated (40,000 cfm divided by the rating of 400 cfm per grille = 100 grilles). Exhaust air grilles (in washrooms and storage) are similarly estimated (10,000 cfm divided by 200 cfm = 50 grilles); fresh air louvers (50,000 cfm of outside air divided by 300 fpm [area] = 166 sf of louvers); and exhaust air louvers (40,000 cfm of return air divided by 300 fpm = 133 sf of louvers).

(6) Piping insulation. The costs may be estimated as a percentage of the total piping costs, usually 20 to 28 percent in the particular locality of the ex-

[9]Conceptual estimating thus vividly demonstrates that all estimating is dependent on the collection and analysis of historical cost information and on the data created therefrom.

ample. As an alternative method, the insulation quantities may be taken from the piping quantities and priced by unit prices.

(7) Duct and plenum insulation. The costs can be estimated in the same way as for the piping insulation. Alternatively from the total weight of supply air ductwork, an approximate area of sheet metal may be calculated by estimating an average gauge and weight per square foot of the metal (e.g., 3,5000 pound total weight, and assuming average 24-gauge metal at 0.865 square feet per pound, = 30,000 square feet of metal), thus giving the approximate area of 1-inch flexible insulation. Similarly for the outside ductwork (assuming average 18-gauge sheet metal at 0.464 square feet per pound, 4,000 pound divided by 0.464 = 1,800 square feet), the area of 2-inch internal (in-the-duct) rigid insulation.

(8) Controls. The first need is to establish the type of control system required—either pneumatic or electric—also any requirements for automation and computers. There are many control circuits and the costs depend on the extent and the number, including controls to:

1. Air-handling units (with varying air flow capacities containing heating and cooling coils and mixing air to the required temperatures),
2. Chilled water circuits,
3. Radiation circuits, and
4. Compressor circuits.

In addition, the extent and costs of controls depend on the numbers of:

1. Air zones,
2. Controlled radiators, and
3. Forced flow unit heaters.

(9) Balancing. The costs of balancing depend chiefly on the numbers of units. Balancing the air systems includes the total number of outlets and of fans. Balancing the water systems includes the total numbers of coils, pumps, radiators, heaters, and the chiller. In addition, there are the reports by the balancing company.

Pricing of Mechanical Work

All work is priced with labor and materials separated, at the bare costs.[G] Taxes and job overhead costs[G] are added to the total bare costs to give the total direct costs.[G] The indirect costs[G] for operating overhead[G] and profit[G] (the markup) are added as percentages of direct costs, as described. In addition to labor and materials there are also the costs of subcontractors' work, such as for insulation, that are kept separate because of the additional markup on this work.

Building equipment[G](boilers, etc.) are priced for material and labor, with subcontractors' work (materials and labor) separated. This is the easiest part of pricing.

Piping costs must include all the pipe fittings, and the fittings' costs (materials and labor separated) are estimated from formulas established from historical cost information. Piping hookups are priced at so much each, as described.

Ductwork material costs are estimated at unit prices per pound of sheet metal, and the labor, by labor rates[G] per pound with different labor rates for the different systems, according to their complexity.

Terminal units are priced with material and labor costs separated (as for all items). Radiation elements and cabinets by linear foot (meter), louvers and grilles by square foot (meter), and other units are enumerated.

Insulation, by a subcontractor, is priced either at a percentage of piping and ductwork costs or by quantities and unit prices.

Controls are costed at so much a circuit, and balancing, at so much a unit (expressed in man-hours). The costs of the balancing reports are based on so many man-hours plus the costs of manuals for the owner.

The estimate summary tabulates bare direct costs for the items of work (items 1 through 9) in constant dollars (labor, materials, subcontractors) with the percentages to be added to each for job overhead, mechanical contractor's operating overhead and profit, and (where applicable) subcontractors' operating overhead and profit.

Any other mechanical system may be estimated conceptually by this method, and, as we have seen, this method emulates the method used by effective estimators making bid estimates. The essential difference in this kind of conceptual estimate is the use of experiential information (based on past design information) as a substitute for the design information lacking at the early stages of a project's design phase. Notice, however, that this and most other conceptual estimates use certain essential and fundamental design information namely, in the preceding estimate, that information obtained from the owner-designer in the facility program.

Estimating Costs of Plumbing

Having just reviewed heating and ventilation in a conceptual estimate and having seen the general precepts in use, we can say a few particular things about plumbing. Usually in a building there are at least three systems related to plumbing: (1) cold water distribution, (2) hot water distribution, and (3) the drainage and venting system that removes the soil, waste, and bad air. In one sense they all together constitute a flow and return system of "good water" going into and "bad water" coming out of a building. Each of these three systems may be viewed as a tree, each with its branches, fruits, and flowers (except that a drainage and venting system contains venting loops). Each system contains main pipes (trunks) with branches (both major and minor, the latter as hookups) and with terminal units (fruits and flowers on the ends of the branches). Measurement requires lengths of main pipes (risers, stacks) and of major branches; everything else is enumerated, including mechanical equipment (boilers, tanks), fixtures, and their hookups. Pipe fittings are not usually measured in preliminary estimates and may be dealt with as described, using data from past systems in the form of coefficients and percentages. Cost accounting and the analysis and use of historical cost data have been proclaimed here as the essential counterpart of all estimating, but it is a fair question to ask, How does one start? Which came first, the estimate or the cost account? Few can start with a complete library of cost information and data sufficient to make all the estimates and to price all of the items encountered. This incisive question is answered in detail in Part 4. For the present let us take note that the answer lies in part in the difference between information[G] and data.[G]

Electrical Work

The preceding general approach to conceptually estimating mechanical work is equally applicable to estimating electrical work in buildings. The same tree concept applies, even though with electrical work the tree may be less obvious in form. A cast-iron plumbing stack with branches to sets of plumbing fixtures looks like a tree, at least like a tree sculpted by a Picasso; but the spaghettilike tangle of electrical wires that one sees exuding from the open ends of conduits appears as something totally different. It is not, however, the difference between pipes and wires that we are concerned with here but the similarity of their arrangements: mains, major and minor branches, and terminal units (panels, outlets, and fixtures) at the ends. It is essentially the same form of distribution system as before, and it can be measured in the same way: lengths of the main runs and major branches and enumeration of the rest—the minor branches (hookups), panels, outlets, and fixtures. In electrical work there are even more enumerated items, including feeder distribution and branch circuit panels, circuit breakers, safety switches, wall switches, and receptacles, also motors, fans, and controls associated with other trade work (especially with mechanical).

In addition, we should mention other wiring systems that are estimated in the same way, including low-voltage wiring (becoming more popular as house builders become aware of future uses of electronic equipment and personal computers in houses), telephone and intercom systems, and, of course, the almost omnipresent television cable system.

Summation of Techniques

Having looked at each of the Masterformat's standard 16 divisions of building construction work in a rather long chapter, it would be as well to summarize the general points made about conceptual trade estimates:

1. Conceptual estimators must substitute *experiential information* for *design information*, and the experiential information is obtained only from past projects.

2. A detailed summary and trade checklists are useful aids for the conceptual estimator.

3. Even in conceptual estimates, a standard method of measurement helps through consistency and in the comparison of information. (A checklist, cost code, and a standard method of measurement may exist as one document.)

4. Conceptual trade estimates are best if based on bid estimates, both in measurement and pricing, using approximate quantities and unit prices but only major items are measured in conceptual estimates as they are measured in bid estimates, while minor items are either measured in enumerated groups (e.g., piping hookups) or allowed for in prices (e.g., pipe fittings).

5. Measurement in conceptual estimates should be simple and comprehensive and ignore minor deductions. Extra-over items can be measured to simplify measurement. All items should be measured using the least possible number of dimensions,[10] but verbal descriptions must be adequate.

[10]For example, measure bulk excavation by area (not volume), stating the average depth. In this way a better concept of the work is retained, whereas in measurement by volume some useful information is lost.

6. In the absence of design information in making a conceptual estimate, the estimator should substitute valid experiential information. In the absence of that, an estimator should make the best possible assumptions, which should then be clearly stated in the conceptual estimate or in an accompanying report.

7. It is essential to declare exactly what an estimate includes and what it does not include, also the effective date or period at which the estimate is valid and any qualifications affecting the estimate (e.g., subject to completion by a specified date).

8. All work should be adequately described, including that covered in an estimate by a cash allowance.

9. All cash allowances should be clearly defined. Allowances for other than specific work may be included in an estimate but their purpose and scope must be stated, for example, (a) allowances for *design changes* in specified sections of work where design information is lacking and assumptions have been made, (b) allowances for *increases in costs* of labor or materials where they are expected or imminent within a project's expected duration, and (c) allowances for *inflation* (or even for deflation). Many cost consultants oppose making such predictions, claiming that estimates should be made only in constant dollars (or other currency), that because inflation affects all costs, an estimate in constant currency best represents relative costs, and that estimators are no better qualified than economists to make such predictions (which is saying much). Nevertheless, some owners insist on an allowance for inflation in their estimates. In such a case an estimator might report only on those aspects of inflation for which he has some expertise (materials and labor, the construction market); for other aspects he might make his best shot and add a disclaimer.

Questions and Topics for Discussion

1. Describe briefly (in not less than one page) the main differences between the two main kinds of conceptual estimates: (1) elemental and (2) trade.

2. Define these terms: (a) *trade work*, (b) *unorthodox trade estimates*, (c) *orthodox trade estimates*, (d) *preliminary estimates* and (e) *conceptual estimates.*

3. Explain (in not less than 500 words) why site work is distinctly different from most other work, in all kinds of estimates, and what an estimator should do about it.

4. Define these terms: (a) *fine grading*, (b) *swell*, (c) *cut and fill*, (d) *extra-over items*, and (e) *fill.*

5. Relative to estimating, describe concrete formwork in terms of its costs and economics (in not less than 1,000 words).

6. Describe the best way to measure and price steel reinforcement in concrete for a conceptual estimate of the costs of a major concrete structure, and explain why it is a good method.

7. Describe several instances of "pattern" in a major reinforced concrete apartment building and how it can be used by an estimator.

8. Explain how in an estimate an estimator can take into account the major causes of differences among unit prices for the same masonry work in a building (e.g., 8-inch hollow exterior quality concrete blockwork or 6-inch hollow interior concrete block partitions) by first identifying those causes.

9. Define the two main classes of structural steelwork in buildings and explain why and how each should be measured and priced in estimates.

10. Describe in detail the best ways of including in a conceptual estimate the costs of (a) metal flashings in conjunction with roofing and (b) opening vents in windows.

11. What is the best way of measuring and pricing an aluminum and float glass window wall? Answer in detail.

12. Describe in detail the best concept of a mechanical system for measurement purposes. Describe also how each item of work in such a system should be measured and priced.

13. In the absence of specific design information about a building, how can certain economic concepts in the mind of an estimator be justified as a basis of costing? What principles are involved?

14. Briefly describe six precepts of conceptual estimating that do *not* apply to bid estimating.

15. Describe four important kinds of building construction work that have to be conceptualized (more or less) even to make a detailed bid estimate. Explain in detail why this is so.

4

Conceptual Elemental Estimates

As we have discussed in Chapters 2 and 5, Part One, elements can be complex and contain scores of *items of work*.[G] A building may be divided into relatively few such elements, with *three* elements as the practical minimum (of area, of perimeter and others), or elements can be simple, containing only a few items, and thereby a building may be divided into many elements. But buildings generally vary a lot, and in order to compare buildings and their costs we usually need to use more elements rather than less so that each building can be more thoroughly analyzed and represented by its *basic elements*[G]. In so doing we facilitate comparisons by dealing in simple and common elements.

In this chapter we examine and discuss recommended standard building elements as described in a published list.[1] Later in this chapter

[1]Published by the Canadian Institute of Quantity Surveyors and reproduced with permission in the Appendix.

we shall see how best to summarize an elemental estimate, and we see how elemental estimates—unlike other kinds of estimates—can be used to see quickly the effects on costs of major changes in a building's design.

Elements in Substructures

The standard list of elements (in the Appendix) refers to three elements in substructures: (1) normal foundations, (2) special foundations, and (3) basements. It is not fortuitous that these come first in the list because they are among the most problematic, simply because they are part of substructures, which are themselves always problematic, and because, while all standard elements are measured by building area, area is the least suitable basis of measurement for many substructures.

In making an elemental estimate, an estimator might be well advised to make an alternative estimate of the substructure using approximate quantities to check and to modify if necessary the estimate of the substructure's costs based on area. In fact there is no need to include a substructure element in an elemental estimate at all as the substructure of a building can always be dealt with separately by whatever estimating means appears to be most effective, usually approximate quantities.

For many buildings the substructure is chiefly related to the building's perimeter and primarily linear in form. For other buildings, such as high-rise buildings with curtain-wall cladding, their substructures are chiefly in mass foundations under a central core or columns and are directly related to neither area nor perimeter. A few substructures are directly related to building area because they are built like a raft on which an entire building floats, but these are the least common.

An estimator should always choose the estimating method likely to be most effective. This is one of the golden rules of estimating, one that is often ignored because so many seem to think that once they use one method of estimating in an estimate, they must stick with it.

If one uses a substructure element based on area (usually the building's gross area), then the cost data used to price that element must be valid; that is, it must (among other things) derive from similar substructure elements, that is, from substructure elements similar in density, form, and function.

Special foundations of uncommon design should always be estimated with approximate quantities. If you wish, convert the results into costs related to area, but do not begin by trying to estimate the costs that way. Except by luck it cannot succeed, and the odds are against it.

Basements may be dealt with by using an area element if valid cost information is available, which means costs of basements similar both in form and function. Otherwise it may be easier, and it will certainly be better, to use approximate quantities. *Basement walls* are not included in this element but in the external cladding group of elements described next. I am not sure that this is the best way, as I prefer always to separate substructure from superstructure entirely. But we are reviewing, not endorsing the methods that flow from the standard list of elements. It is important to designate a definite, logical, and practical plane of separation between a substructure and a superstructure. To ensure valid economic comparisons among buildings and their costs separation planes should always be in the same relative position. Discussion of the next element—the lowest floor—should clear up this point.

Elements in Structures

Once out of the ground, as it were, the elements become more complex but more recognizable, and we shall deal with each in a separate subsection. The order followed is that of the standard list of elements.

Element: Lowest Floor

This element consists of a strata of several items: typically a floor slab on imported fill and a waterproof membrane and possibly other items, especially if it is a slab on the ground. If it is a suspended slab (perhaps over a crawlspace), it may include columns and possibly beams to support the slab. It may also include minor features such as control joints and even small pits and sumps. Finishes (in the standard list) are kept separate, as are items related to other elements such as concrete bases for mechanical and other equipment. There should be no problem in using building area for an element of this kind as it is the logical unit of measurement; it is measured overall, from outside to outside, and no deductions are made for such features as openings for ducts and elevator shafts or for stairs.

Visualize a suspended slab as a table with legs (columns; probably more than four) and perhaps, like a table, with horizontal supports (beams) between the legs and below the top. A multistoried building structure then becomes a stack of tables.

We can accept minor features that vary from building to building as part of an element because they are just that—minor—and therefore not consequential in a preliminary estimate.

Everything below and separate from the *lowest floor element* is part of the substructure.

Element: Upper Floors

If the upper floors are all the same (minor features aside), they may be added together into one upper floor element, in which case their total area plus that of the lowest floor should equal the total gross floor area of the building. (This gross area is measured overall, to outside wall faces, and is therefore greater than the rental area and even greater than the walkable floor area.) As before, the description of this element excludes finishes of all kinds, according to the standard list. Like the lowest floor element it is esentially a plain table or a stack of such tables.

Element: Roof

In many flat-roofed buildings this element is identical to the upper-floor element, but in other buildings it may be larger, sloped, or of different construction; thus it is kept separate. It excludes all roof coverings and related work but includes integrated overhangs or projecting eaves. Any additional items such as parapets not related to other elements are best handled as separately measured quantities of additional work. (Some contiguous items are often related to other elements; e.g., an elevator machine room on a roof maybe part of the elevator element.)

That is the sum of the elements of structures. Note that they are all horizontal elements, like a stack of tables. The next group of elements enclose a building.

Elements in Exterior Cladding

This is a large group of elements because of the variety available. In some estimates it may be convenient to combine some of the elements separated in the standard list cited here. They all are all measured by area (according to the standard list), although for some, area may be rather meaningless. For example, the area of a door has only a relatively minor effect on its costs.[2] Sometimes, especially for standard and common buildings, it may suffice to have only one or two elements of exterior cladding and to include in their descriptions a percentage of total area for windows, doors, and such.

How these elements are described and used depends on the form and function of a specific building and how its design compares to others of the same class. It has been said that this group of elements, external cladding, potentially has more effect on a building's costs than any other. Certainly it has a great potential for variety. This may be the cause of a shortage of useful and valid cost information, because few buildings of a specific class may have similar external cladding. Adjustments for differences among elements may be made by deducting and adding the costs of the differing items of work. Such adjustments are theoretical (when compared to the reality of the actual costs of elements obtained by analysis of previous projects' costs), but theoretical adjustments may be the only way to obtain useful cost information.

In describing these vertical elements, it is useful to include the following information:

1. The percentages of the total area of an element represented by *openings* of different kinds (windows and doors) and by different *finishes* (e.g., brick veneer, stucco plaster).

2. The ratio of exterior-cladding area to building-plan area. (This ratio is an indication of plan design economy. Some say, "design efficiency," which is inaccurate. Mathematically the most economical plan comprised of straight lines is a square. Ratios can be used to compare the relative economy of plan shapes.)

Element: Roof Finish

This element has the same area as the roof element in the structure. Some estimators may combine them, but separation may make two more useful elements as a basis for cost information and data, especially in the light of widespread variations among roof insulation and roof coverings. (This question is pursued further under interior finishes.) In this element the standard list of elements includes in this element items such as fascias and parapets. Presumably this is in an attempt to achieve uniformity by having all elements measured by area, but as with foundations this ignores or sets aside an important relationship with perimeter and imposes one with area. An estimator must be aware of these relationships and, where necessary for accuracy in estimating, depart from standards and use his common sense—and make any necessary adjustments with measured quantities of work.

[2]All costs can be seen as containing two essential parts: (1) the fixed lump-sum part of the total costs independent of size or quantity, and (2) the variable part that depends on an item's size or quantity. Total costs therefore are rarely directly related only to size or quantity because of the more or less unchanging lump-sum portion of the total costs.

Element: Walls below Ground Floor

In this element the standard list includes perimeter basement walls, which is suitable so far as it goes; but does it go far enough? The crux of the question is, again, that there is no constant relationship between the area and the perimeter of a plan. If there is a fairly constant relationship between area and perimeter for a specific class of buildings, then within that class one may minimize the problems that otherwise arise when perimeter is ignored and when only area is measured.

Otherwise it seems that perimeter must be taken into account. How should that be done? One way is to depart from the standard list of elements and its definitions and to separate perimeter walls from other interior walls (which the standard does, but only those perimeter walls included in external cladding), and to divide foundation walls into (1) perimeter foundation walls and (2) other foundation walls. There is, however, a penalty for departing from any standard, and that is a potential reduction in effectiveness. (These vertical elements are measured according to their own area, not the area of the building; their costs in an elemental estimate are related to gross floor area, as in the example reproduced in Chapter 9, Part Two.) The standard list includes waterproofing and insulation but excludes interior finishes.

Element: Walls above Ground Floor

As with the last element, measurement here is by wall area (perimeter times wall height), ignoring minor plan-projections and recesses for items like columns, windows, and decorative features. Describe the proportions of different exterior materials. Exclude all interior finishes (to be taken as separate elements); however, the standard list includes insulation and exterior finishes.

If there are extensive features, decorative or otherwise, it may be advisable to separate them so as to get down to a basic element[G] and to cost the additional features separately. This is a general precept arising from the concept of basic elements so that whenever possible historical cost information is collected for basic elements, thereby making it possible to compare dissimilar buildings.

Element: Windows

Following another estimating precept—if work is different, keep it separate—each kind of window should be kept separate, and the proportion of its aggregated area to the entire exterior cladding area noted. But take notice that we are primarily concerned with *costs* and that if two kinds of windows cost about the same, they may be combined. However, should some areas of windows contain many opening vents, and other areas contain few or none, then these different window areas should be kept separate as their costs will differ considerably. (You will, of course, realize that the precepts related to examples here are general precepts applicable to all elements, whenever appropriate.) It is useful to describe the area of opening vents as a percentage of total window element area, as opening vents so much affect costs.

Element: Exterior Doors

Often exterior doors are so few that they may be ignored and included with another element. In other buildings they may be treated as enumerated extra-over items. Use common sense in deciding how to deal with doors; they can be expensive, some running into thousands of dollars per opening when the finish hardware is included.

Element: Balconies and Projections

Experience shows that in discussing and comparing costs of apartment buildings, balconies can be a source of serious misunderstandings. The question that always needs to be settled at the outset is, Are balconies included in the building's gross floor areas? If balconies are affixed to and projecting from a building's face, the answer is simple: It is best to *exclude* such balconies from a building's gross area and simply to treat them as fixtures.

In some designs, however, balconies are created by setting in parts (panels) of the exterior wall (often of glass and metal) from the building's exterior perimeter (and that of the structure) to create what is in effect an outside room on a common floor deck. That kind of balcony is essentially different from the projecting balcony as a fixture. In the case of a building with set-in balconies, if the balconies were eliminated by moving the exterior wall at the balconies out to the perimeter line, as illustrated in Figure 4-1, the only additional costs would be for the heating and ventilation of the newly created and additional inner space, and this would be offset in part by elimination of expensive balcony railings and possibly some insulation. Other costs would be hardly changed. In such a case, balconies should be taken as part of the building proper. Errors must be avoided by ensuring that historical costs are valid and comparable. Balconies are a case in point. (Refer to the standard list of elements on balconies and

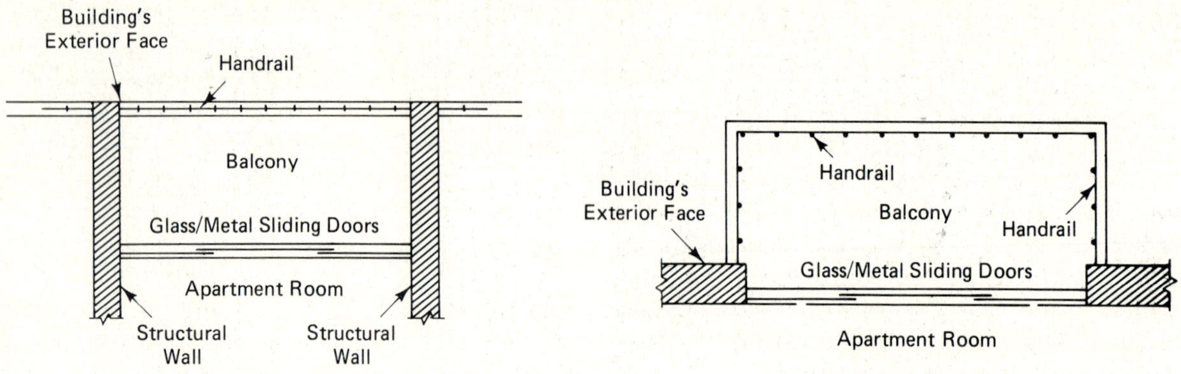

Typical Integrated Balcony (Plan) Typical Projecting Balcony (Plan)

FIGURE 4.1 Balconies

projections. It is better to keep them separate if only to keep the other exterior cladding elements as basic as possible.)

Element: Interior Partitions and Doors

These too are measured by area—their own area, length times height—and the standard list says to measure doors also by their area, but I cannot agree with that. It is better to enumerate them as extra-over items than to measure them by area. Most of the costs of a doorway are unrelated to the doorway's area. Each type of partition and each type of door should be kept separate, but again the criterion is costs; if two kinds have almost the same unit cost, then an estimator might reduce the number of items in an estimate by combining items of similar costs, especially in a preliminary conceptual estimate.

A useful ratio exists between the total length of partitions and the floor area, because in some kinds of buildings (for example, apartment buildings) the ratio is fairly constant, although it varies slightly according to the mix of dwelling units (classified by numbers of bedrooms) and according to building quality. The relative quantity of partitions in an apartment building containing bachelor and one-bedroom suites for rental is probably at a maximum, while the minimum might be found in a building containing luxury-quality two- and three-bedroom apartments. These ratios of partitions and floor area are useful because for many preliminary estimates planning has not gone far enough to show actual layouts and the extent of partitions. Historical data expressed as ratios can provide reasonable design information about the probable quantity of partitions in a building, as well as other features. (See the example in Chapter 9, Part 2.)

Similarly in some buildings (such as apartments) the numbers of interior doors relative to lengths of partitions may be reasonably constant but varying according to similar criteria to that affecting quantities of partitions, i.e., numbers of rooms in a given area. The estimate for an apartment building shown in Chapter 3, Part 2, demonstrates the value and use of relating some construction costs to the number of dwelling units (suites) in a building.

Elements of Vertical Movement

Stairs include everything in the stairwell but not the enclosing walls, which are part of the structure walls. Stairs should be fully described and measured according to the combined total height of risers. If stairs are of a typical width for a particular class of building, it is as well simply to measure the numbers of risers, stating their width across the staircase. Usually staircases are not numerous and while they are expensive, they usually do not have a great effect on a building's total costs. Consequently, if one is dealing with a common class of building, it may suffice simply to enumerate staircases and describe them according to the number of stories.

What has been said of stairs applies also to *elevators and escalators*, except that with the latter an estimator is better off obtaining costs from manufacturers. For a hurried estimate one can often get fairly reliable cost information on elevators from a good cost data book.[3] But do not overlook the fact that there is builder's work[G] in connection with elevators and escalators other than just the shaft-enclosure; for example, forming or cutting holes and

[3]For example, *Building Construction Cost Data* (Kington, MA: Robert Snow Means) listed in the Bibliography.

openings, providing and installing steel supports, and also electrical services.

Elements of Interior Finishes

Some estimators question the need to separate interior finishes from their bases as separate elements, and certainly in hurried and early estimates there are reasons for combining finishes with the other elements on to which they are applied, especially when one is reasonably confident about the intended specifications of the finishes because of familiarity with the class of building, with the owner-designer, or with both. The other method is to obtain the total areas of interior finishes from the areas of the other elements, using only the total areas for calculations. The measurement of items such as base trim should be similar: At each floor, measure twice times the total length of partitions plus once times the perimeter, with no deductions for minor omissions. Common sense will show the way. In order to simplify the measurement of interior finishes, estimators should always calculate the relevant proportions of the areas of:

1. *Finished* floors, walls and ceilings to *total areas* of floors, walls, and ceilings;
2. Different *kinds* of finishes and areas with no finishes to *total areas* of finishes,
3. Total *floor* finishes to total *wall* finishes;
4. Total *floor* finishes to total *wall and ceiling* finishes.

As this type of data is collected and used, measurement in future preliminary estimates becomes more effective. Reflection on and research into the quantities of different buildings' designs and layouts will produce other useful quantitative design factors for use in preliminary estimates as a substitute for more specific design information, as an easy way to reasonably accurate and rapid measurement and, in bid estimates, as a means of checking accuracy.

Elements of Fittings and Equipment

The costs of such items as fittings, fixtures, and equipment installed in buildings should always be itemized, no matter how slight the design information at the time of making the estimate. Itemizing at least presents a rational basis for costs, even if the items and their numbers are assumed. To try to include costs for such things on the basis of floor areas is foolish and futile—unless there is some direct relationship between area and the items (for example, in some instances, regulations or other requirements may stipulate that there be one item for every so many square feet of floor area). By itemizing such things in a preliminary estimate an estimator creates an understandable basis for future discussion and for decisions by the owner-designer, whereas an allocation of so many dollars per square foot of floor area for gymnasium equipment, for example, simply leads to the obvious question, What will that amount of money provide?

Also, to repeat a precept, all preliminary estimates should be specific about what they include and exclude. This precept is particularly relevant to equipment, fittings, fixtures, and other things that are often nebulous in the early stages of design and budgeting and that may not be required to be included in a construction contract.

Elements of Services

Often mechanical and electrical (M & E) services are among the most nebulous parts of a building's design in the early stages, and by their nature they remain, in their details at least, among the most nebulous even in bidding documents.[G] Plumbing, for example, is so much taken for granted that often bidding documents only show graphically the general type and location of plumbing fixtures, and all else is left to the trade contractor and to the plumbing code—except for the colors of the fixtures. Often much the same can be said about electrical work, and it is more or less true of all M & E work. As a result, and because of the general ignorance of M & E with many who are not specialists, typically these services are often poorly handled in preliminary estimates.

Element: Electrical

It has been observed that the costs of electrical work in most buildings *as a proportion of total costs* normally do not vary greatly from 8 to 10 percent. (A cursory survey of a cost data book[4] indicates that over 70 percent of the classes of buildings listed with their median square foot costs show their electrical costs close to that range. Look for yourself.) The costs of electrical services in hospitals, as we might expect, may be 50 percent higher than those of other classes

[4] *Building Construction Cost Data 1982.*

of buildings, while in warehouses electrical costs may be only half that of the general median amount. Nevertheless, because electrical services generally are utilitarian (except for some fancy fixtures), often their costs as a proportion of total costs do not vary greatly, especially within a particular building class.

Ideally through research it is better to establish (for different classes of buildings) a basic lump-sum for the costs of the electrical mains service going into a building and a cash allowance for the lighting fixtures and then to price the rest of the electrical work on the basis of gross floor area. In this way we can achieve more accurate estimates.[5] In large buildings with more complex electrical services this separation of costs might with advantage be taken further, especially if there are subsidiary switchboards or major branch circuits, in which case the costs for each major circuit might similarly be divided. Other service systems for such things as telephones, television, computers, and special equipment and lighting may be dealt with in a similar manner.

Alternatively electrical services (and similar systems) may be estimated on the basis of the number of mains and branches, service outlets, receptacles, switches, motors, and such. This requires some rudimentary design and some knowledge of the work, but this method may prove more accurate, especially if valid historical data suitable for other estimating methods are not available.

Element: Plumbing and Drainage

The costs for these systems are most useful if they are separated into the following groups:

1. Fixtures
2. Stacks (drainage and vent)
3. Branches (drainage)
4. Cold water service
5. Hotwater service.

If this separation is not possible or practical, costing should at least be done on a per fixture basis and not by building area, for the reasons given in Part 1. There is little rational connection between building area and plumbing and drainage costs. If you think of a series of fixtures, say, eight water closets all in a row, and imagine the piping that connects the fixtures, you will quickly understand that even relating costs to the numbers of fixtures can be misleading. A ninth water closet at the end of the row will cost much less than the first one. But in most conceptual estimates we must sacrifice something for speed, and fixtures are much more reliable as a basis for plumbing costs than building area.

Of course, if the owner-designer is prepared to spend for the fees and the time, it is always possible to hire a cost consultant who is expert in M & E work to make an estimate of the costs of plumbing based on approximate quantities, as described in the last chapter, even for a conceptual estimate in which the rest of the work is based on elements. With many buildings this is recommended. The consultant's fees may well be exceeded by savings made in building costs. In estimates for plumbing and drainage, the only logical element is the fixture or, following the axiom about the divided nature of costs, the bank of fixtures when appropriate. Some simple designing may be necessary, but it does not take much to make rudimentary layouts of plumbing for a conceptual estimate. The same approach may be used for other services' systems containing piping and terminal units, or fixtures.

One system that can be estimated accurately on the basis of area is a *fire sprinkler system*, because the sprinkler heads are laid out in a regular pattern to provide fire protection to specific areas. In other words in the case of sprinklers there is a direct relationship between area and costs based on the number of sprinkler heads and the lengths and arrangement of piping. The case for sprinklers actually illustrates the poverty of the argument for using area as a basis for estimating the costs of other plumbing systems.

Element: Heating, Ventilation, and Air Conditioning

Similar to those of electrical services, the costs of heating, ventilation, and air-conditioning (HV & AC) services are better divided into costs of:

1. Central plants (boiler, refrigeration)
2. Distribution systems (hot water, condensate, cold water, and air) and
3. Terminal units.

A typical breakdown was cited in detail in the last chapter. Of course, such detail may not be possible in some elemental estimates (although it usually is desirable), and it may be necessary to have a simpler

[5] As stated, total costs for all kinds of items consist of these two basic kinds of costs: (1) those costs related only to the item, no matter its size, extent, or quantity and (2) those costs that directly relate to an item's size, extent, or quantity.

basis for estimating HV & AC costs. In that case, costs should be related to building area *and* volume and also to the number of tons of refrigeration required. In addition, cost analysts might consider introducing another factor, *the total area of external cladding*, as this obviously affects HV & AC costs. Do not overlook the costs of fixtures, controls, and nonmechanical items such as chimneys and other similar and related items that must be included in this comprehensive element of HV & AC. Fixtures and controls especially vary in quality and costs; trim to fixtures (the parts first seen and the most used) varies even more.

Element: Site Development

This title is taken from the standard list of elements (see the Appendix) and it refers to what we have previously called site works, all the work on a site that is outside the buildings. The extent and variety of this work are great, and the methods of measurement and pricing also vary according to the nature of the various works, but most are obvious and similar to those of similar work contained in other elements already described. If in doubt or difficulties with an important item, measure the approximate quantities, even if it means designing the work first. That is far better than applying some highly theoretical unit cost that may be meaningless or involves guessing.

Elemental Estimate Summaries

Summaries of preliminary estimates are as critical as those of bid estimates but for some different reasons. For preliminary estimates (whether made with elements or with quantities of trade work) a summary is a place for doing some of the most important estimating work, including consideration of the costs of the several parts of the work in the light of the projected total cost and in the light of the costs of all the other parts, the proportions among the costs. These are essential to an estimator's overview of a project's costs: the percentages for each of the parts related to the whole, the costs of each part related to the building area, and the design functions of parts related to their costs (such as electrical lighting costs per lumen, structural steel costs per ton).

In elemental estimates, summaries consist, of course, of the costs of each major element (described in the sections above), but subsidiary summaries should show the costs of the minor elements that comprise the major elements. All of the results of the cost analysis just mentioned should also appear in these summaries, and the layout of a summary should be such that it will facilitate comparisons within the summary as well as comparisons with other summaries. This clearly calls for standard summary sheets (at least within each company) that will also serve as checklists for elemental estimates.

The standard list of elements to which we have referred has final elements of overhead and profit and contingencies. Enough has been said about these topics in the last chapter that no more need be added, except to refer to the common contingencies allowed for in preliminary estimates for design, for escalation, and for post-contract contingencies.

Design contingency amounts are intended to fill out the estimated costs to cover anything that has been missed or anything that is reasonably added by the owner-designer. It is intended to fill in the inevitable gaps that occur in all preliminary estimates, but one must be sparing, otherwise an estimate becomes meaningless. This contingency should be allowed for in each separate element (or in each trade in trade estimates) because the need for design contingencies varies with each part of the work depending on its nature and on the amount of available design information. It is pointless to allow for an overall design contingency of, say, 10 percent (or any other amount) added to the total costs when certain parts of the work are inevitable and unlikely to be changed. It is far better to allow nothing for contingencies related to the firm parts of the work and to allow such amounts as deemed necessary for design contingencies for the less definite parts.

Escalation contingency amounts are allowed for increases in costs occurring after the estimate is made and before the submission of bids—but not for foreseeable increases during the actual performance of the work, such as increases in labor costs due to an expected new wage agreement. Foreseeable increases should be calculated and included in the body of the estimate, each in its appropriate place. As before, these contingency amounts are best included with the appropriate elements.

Post-contract contingency amounts are intended to allow for changes in the work during construction. This is a valid purpose, but again such amounts cannot reasonably be based on the total estimated costs; they must be related to specific parts of the work and assessed according to probable need.

It is interesting to note the ways in which the standard list of elements recommends dealing with these contingency allowances. It proposes allowances without discussing any of the possible causes that may give rise to contingencies and make allowances in estimates necessary. Of course, the range of causes is unlimited, but some guidance might be useful as some estimators are too ready to insert an allowance rather than attempt an analysis and a rationalization of costs. In good conceptual estimating, contingency allowances should be used sparingly, and an estimate's summary should indicate the total of all contingencies allowed, expressed as a percentage of the total estimated costs, as well as the individual amounts.

Manipulation of Elemental Estimates

The great advantage provided by elemental estimates that is not found with other estimates is the ease with which the elemental estimate can be adjusted according to changes made in the form, dimensions, and capacity of a building in the early stages of its design. In order to develop a parcel of land to its highest and best use (that is, with reference only to investment economics and not to social or aesthetic standards), an owner often needs to know the construction costs for a number of design alternatives. For example, should a warehouse have only 10 20-foot bays in its length, or more than 10, and what will additional bays cost? Should an apartment building contain more or less apartments? Or, in another building, more or less floors?

All of these questions and their answers depend on a number of things, but almost always a prime consideration for an owner is, How much building can I afford? Elemental estimates provide the answers most readily. By making an elemental estimate and by manipulating it—by shrinking and expanding the building's design and by adjusting the estimate accordingly—an owner can quickly be given the answers that he needs. To do this accurately with abstract estimates[G] is practically impossible. To do this with trade estimates based on approximate quantities usually takes much more time and effort. In fact, elemental estimates were first devised just for this purpose.

Figure 4-2 shows the bays for a simple warehouse-type building. If an owner-designer is not quite sure at the beginning of the design phase of construction how large a building he wants to construct, an estimator might find it prudent to measure the contents of one typical end bay and one typical intermediate bay. He can then quickly estimate the costs of buildings of different lengths by adding the cost of two end bays to the costs of any number of intermediate bays. Often it is more accurate to measure *end bays* rather than *end walls* because often the other (side) walls in an end bay contain features and work (e.g., bracing) not found in the intermediate bays. In fact, whether an owner-designer will need the alternative estimated costs or not (and an estimator never knows for sure), it is always more efficient to measure a building by typical bays, or floors, or other multiple part as indicated by the design and type of building.

Figure 4-3 shows a similar application, but this example involves floors (stories) of a building. The top floor, the substructure, and an intermediate floor might be measured, and the latter may then be multiplied for buildings of different numbers of stories. This application, however, is not quite as simple as the last because more stories require more support from the foundations and from the lower

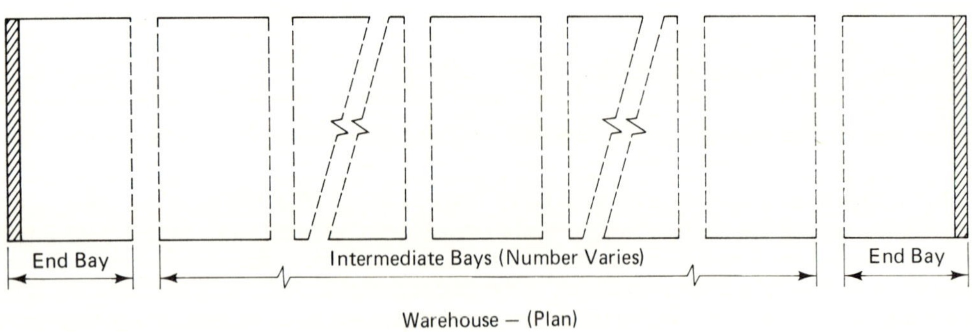

Note:

There are Always (in a Rectangular Plan) Two "End Bays" and a Variable Number of "Intermediate Bays" to be Measured – by Measuring a Typical One of Each (X Number) and Making Minor Adjustments to Measurements as Needed to Account for Design Differences, if any.

FIGURE 4.2 Warehouse Bays

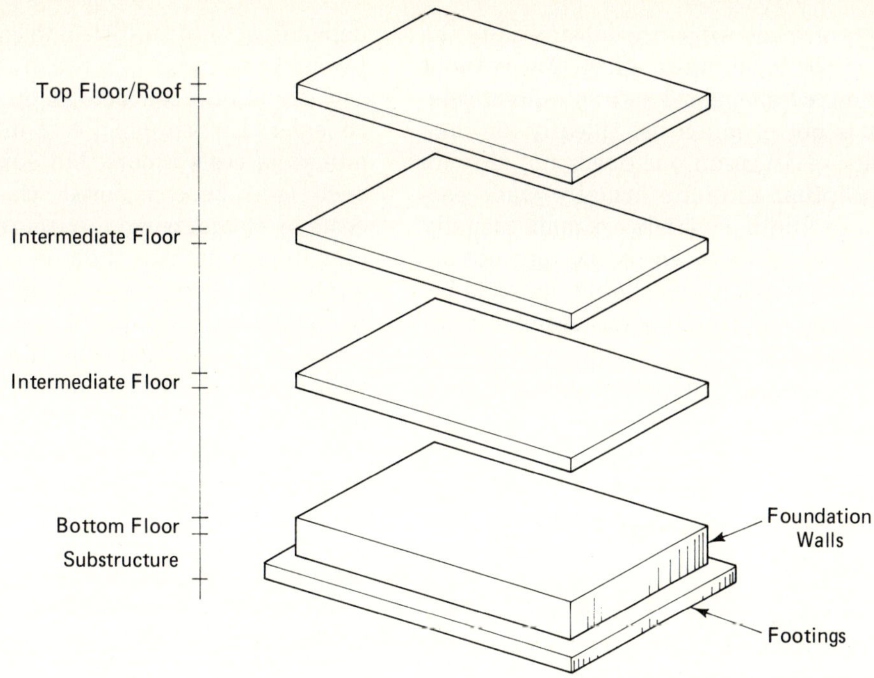

Note:

There is Always a Bottom (Lowest) Floor (and a Substructure) and a Top Floor/Roof Deck, with Intermediate Floors Between to be Measured Separately, for example: Substructure and Bottom Floor; Top Floor (Roof); and a Typical Intermediate Floor (X Number); with Minor Adjustments as Needed to Account for Design Differences Among Floors. (E.g., in Columns' Cross-Sections; Beams' Depths; Steel Reinforcement.)

FIGURE 4.3 Building Floors

stories. In a reinforced concrete building this may only mean increases in the sizes of some steel rebar and a slight increase in its total weight in the structure, but only up to a point beyond which sizes of concrete members must be increased, with commensurate increases in concrete and formwork costs. This is not difficult to allow for as an adjustment to an elemental estimate, but the estimator may need advice from a structural engineer.

These techniques can be applied in the same general way in preliminary trade estimates as well as in elemental estimates, but unless the need has been foreseen, and unless a trade estimate has been measured so as to facilitate such adjustments (by measuring typical parts of buildings separately), it takes much more effort to make such changes in a trade estimate.

Another way by which elemental estimates can be easily manipulated to provide alternative design costs is by the simple substitution of elements. For example, an estimator may assume one external cladding element (made up of numerous different trade items), and later he may reveal the effects on total costs of substituting other external cladding elements, simply by applying a different inclusive unit price for a different element. There is no limit to the variations that can be made in this way. But caution and care are advised, because there are hidden traps for the careless who overlook the less obvious changes that are required, such as increased weights of steel rebar, increased sizes of concrete members, or both, required by increasing a building's height.[6]

Pricing Building Elements

Previous elemental estimates are the obvious source of elemental unit prices, but they are a good source only if the estimates are verified by cost accounting[G] on completion of the work; otherwise by using them one may be only perpetuating error. This means not only verifying that the total actual costs at completion are the same as the total estimated costs (or, otherwise making the needed adjustments in costs) but also verifying that the costs

[6]A heavier structure is not the only additional requirement for a higher building, it is only the most obvious. In addition, increased resistance to wind and earthquake stresses, additional safety requirements, and better weatherproofing, are only some of the other things that may have to be considered.

for each section of trade work are substantially the same; that requires much more effort. But without such verification an elemental estimate for a completed project is not of much use. Ideally for complete and easier verification one needs not only an elemental conceptual estimate (usually made earlier) but also a preliminary trade estimate (usually made later). At the time of verifying, any needed corrections to elemental prices should be made before filing the cost information for use in future elemental estimates.

The use of elemental unit prices obtained from historical sources should be discriminatory: One cannot simply use a unit price for an element in one building (for example, a three-story light commercial building) to price a similar element in another building (say, for example, a multi-story apartment building) without making several significant adjustments to the unit price. All such adjustments are more or less theoretical and subject to errors of judgment. But in the absence of more valid information or data, adjusted unit prices from other kinds of buildings are probably better than unit prices that are entirely theoretical (see the following section). Reinforced concrete walls and slabs and their costs are typical examples of elements that are easily abused in this way because the things that chiefly affect their costs—formwork and steel rebar content—are the least obvious of the constituent items. In addition to the differences in direct costs of labor, materials, and equipment for similar elements in different kinds of buildings, numerous but less obvious costs are less directly affected by a building's design and function and more by the ways and means of constructing the building.

For example, a light commercial building (such as a small retail store) may be built by a contractor using nonunion workers and with a minimum of overhead costs. A high-rise apartment building on the other hand may be built using union labor, with proportionately greater job overhead costs, and with an even larger operating overhead. Furthermore, a government office building may be built under similar conditions but with much more supervisory time spent in meetings, form-filling, and dealing with officials and inspectors, apart from the other increases in costs arising from the complexities of institutional buildings and compliance with government regulations. Each of these three buildings may share a common item of work—reinforced concrete walls, 8 inches thick, carrying similar loads and performing essentially the same function—yet the costs attributable to the wall in each building may vary as much as 20 percent between the highest and the lowest, depending on the nature and conditions of the building project.

Such descriptive facts as those just cited must be considered when using cost information and when adjusting costs among buildings. Even more must such facts be considered when synthesizing unit costs of elements from basic information. The best cost information in the absence of truly valid and applicable elemental unit prices is that obtained from trade contractors,[G] for example, such as is provided in the verified schedules of values[G] of lump-sum contracts.[G] From such schedules an estimator may synthesize accurate elemental unit prices. The best way to do this is with a minimum of analysis, as shown in the following example:

> *Data obtained from contract schedule of values (project X):*
> Total value of masonry subcontract; exterior masonry walls (6,000 square feet) = *$48,000*
> Total value of rigid insulation subcontract; insulation applied to interior faces of exterior masonry walls (5,400 square feet) = *$1,500*
> Total value of window subcontract including glass and metal entrances (2,000 square feet) = *$13,000*
> *Synthesis of unit price for external cladding element (project Y):*
> Total value of work in external cladding element (project Y) (8,000 square feet) = $62,500
> Elemental unit price = $7.81 per square foot.
> Proportion of window area in element in this Project Y = 20%
> Proportion of window area in element in other Project X = 25%
> *Adjusted Elemental Unit Price for Project Y = $7.93* per square foot.

Is this adjusted elemental unit price valid? Mathematically, yes, but is mathematics enough? In estimating, generally it is not, despite the effort of some to make it appear so. Mathematics is useful for dealing with facts, but in estimating we are trying to predict facts (cost facts), and there are always other factors to be included that can be obtained only through subjective judgment, and most cost engineers recognize that fact. In this hypothetical case one might observe that a reduction in the amount of windows would not reduce their costs in direct proportion.[7] The same can be said of the masonry work, but in a reverse sense.

So, in part, the actual differences would be self-

[7]The dual nature of all costs of an item arises again: those costs related and those unrelated to quantity.

canceling, but not completely. In such a case I might accept and use the mathematical answer, but are we aware of the thinking needed that goes beyond the purely mathematical? In another more complicated case with larger quantities involved, one might have to make a subjective judgment or run it through a computer using a program designed to take into account as many economic factors as one can reasonably include. Historical costs of trade work must always be viewed in the light of the actual circumstances and conditions (discussed in Part 1) that prevailed when the work was estimated, then bid, and later performed. Not all of them can be known, but usually some of the more important can be considered and taken into account.

But what does an estimator do in an elemental estimate when he is unable to obtain valid information from past projects that he can identify? Then he must turn to less reliable information and data from past projects of which he has no direct knowledge, that which is published in the cost data books.[8] Cost data from publications are often third-rate because they come from projects that we cannot identify, and we should use superior data whenever possible. This requires the classification of information and data. All of this is explained in Part Four.

Elemental Estimates of Engineering Works

We have discussed in this chapter elemental estimates for buildings, and it would be misleading to give the impression that unlike trade estimates they are suitable only for building works.[G] One of the difficulties, however, in describing elemental estimates for engineering works[G] is the tremendous diversity of those works and, therefore, proportionately fewer basic elements.[G] Some engineering projects lack advantage in the use of elements because in many engineering works the identifiable elements are made up of only a few items of work. Therefore, one might as well stay with approximate quantities because they are almost as easily handled as simple elements. There is not the same number of common elements in engineering work that there is in building works. In almost all buildings there are elements similar to elements in other buildings. In engineering work, common elements are limited to smaller and more specialized fields. Further, many engineering works are much more subjected to site conditions than most building works. This does not mean, however, that elemental estimates are of no use in engineering work.

To discover the basic elements of work (of any kind) that may be used in making conceptual estimates, an estimator must look to the design fundamentals of the work. In buildings the fundamentals are shelter and support. The enclosing planes (walls and roof) provide shelter and the floor decks provide support. An interior environment is created within which the conditions (of air purity, temperature, humidity, and lighting, etc.) are controlled by building services. What are the equivalent fundamentals in an engineering project? Let us examine some examples.

One simple example is a pipeline built for the conveyance of oil or another product. Essentially it consists of a number of pumping and valve stations interconnected by piping. The piping itself may consist of a number of items of work: layout, excavation, preparation of pipe bed, stringing of pipe, welding of pipe, wrapping of pipe, installing pipe, testing, backfilling trench, and so on. The pumping and valve stations can be analyzed into building elements, with some special elements peculiar to pipelines and their related equipment. The piping might consist of three different elements: (1) piping below ground, (2) piping above ground (on supports), and (3) piping under water (at water crossings). The pumping stations and other buildings are additional complex elements but they present no new problems.

Highways might be dealt with in a similar manner, as might many other industrial engineering works that consist of conglomerations of enumerated equipment connected by a system of piping, ductwork, or other linear systems, many of which could be successfully handled (by those estimators who fully understand them) in the same way as M & E services in buildings.

Many engineering works are much more specialized than buildings, all of which exist to provide shelter and support and therefore have a large number of common and similar items. But although extreme specialization is a disadvantage in obtaining historical cost data from widely different sources, it is also an advantage in that it demands a depth of perception of the design of the work that can be utilized in making conceptual estimates. Otherwise all the precepts observed in the making of elemental estimates of building costs are generally applicable to elemental estimates of engineering works.

[8]Unfortunately there is no guarantee that published data are valid data and we must rely on the reputation of the source and on testing.

Estimating Costs of Major and Complex Projects Quickly

Some multimillion dollar projects are neither pure building work nor pure engineering work but a mixture of both, and their formidable size and complexity often require an attack on many fronts. Allow me to risk the hazards of direct, personal reminiscence in order to bring from my own experience an actual case as an example.

One project involved the construction of a military cantonment, or permanent headquarters, for an armored division of a foreign army on a broad Asian plain hedged by high mountains to the north. It consisted of some 20 to 30 barrack blocks, a small number of married quarters, more elaborate quarters and houses for officers, several very large maintenance sheds and garages (more like warehouses) for tanks and other vehicles, and dozens of other miscellaneous buildings, together with all the usual services required by what amounted to a small town situated in a remote location that otherwise provided no facilities or resources except for small amounts of dairy and agricultural produce, clay bricks (after we had helped to rebuild and set up the kilns), and lots of cheap labor. Everything else had to be imported and then transported for several hundred miles along one major road from the nearest port. For all the buildings in this project I was required to make an estimate of costs and to submit a bid within less than two weeks. The site services were to be the subject of a separate contract. (The incompetence of certain colleagues which precipitated this critically brief period for bidding is another and more colorful story.)

As I was the only estimator with the company in the country at the time, it was necessary to utilize all the local help that I could obtain in ways that would leave me to deal with those parts of the estimate and bid with which only I could deal. With about six clerks of varying skills and with no mechanical office equipment, other than a typewriter and an adding machine, we prepared a multimillion dollar estimate and submitted a bid within about 10 working days. The bid was successful—in an unexpected way—and I stress that point to show that one never knows what kind of return one's efforts may bring.

For the first three days of the brief bidding period I did nothing but plan how to proceed. I studied the drawings and other bidding documents and made list after list of notes. Eventually I concluded that the only way to get the bid completed in time was to make elemental estimates of the most important buildings, those that were most typical of the work as a whole. This included the barrack buildings, the maintenance sheds, and a few other miscellaneous buildings that were typical of many of the other smaller buildings.

While I defined and measured the elements of the most important buildings, several of the more adept clerks measured the floor areas of other buildings and the areas of other work that I had designated on the drawings. (There were hundreds of drawings, but many were repetitious.) This was the first work of this kind that they had ever done. One of the clerks had some construction knowledge, and this proved most useful as he was able to measure the elements of the less important buildings and to monitor the others.

My next task was to price the elements of the critical buildings as accurately and carefully as possible and then to arrive at the total estimated direct costs of those buildings. With some adjustments these elemental unit prices were used for the other buildings. Those buildings of little importance in the estimate, of which there were only a few, and the small and simple ones, were quickly priced using each building's area and a single unit price derived from the typical buildings that had been priced by elements. To facilitate and control the pricing by the clerks, I devised standard estimate and summary sheets containing lists of items of work for two or three of the more common larger buildings (to obtain the most complete and representative lists of items possible), and as each building was measured and priced, the clerks were required to make estimates and summaries of costs on the standard sheets, to delete any items from the standard sheets not required for a particular building, and to insert any new items. With this format I was easily able to check the completed sheets.

Having priced the elemental estimates for the important buildings and, from those estimates, having established unit prices for elements or for floor areas for the rest of the buildings, I was then able to oversee the completion of the measurement and pricing, to answer questions, and to check the summaries of estimates for individual buildings as they were completed. We were working 12 to 14 hours a day in rather sultry weather.

About the eighth day, and two days before bids were due, I laid out a general summary sheet and began to enter the estimated costs of individual buildings. Before completion of all the building estimates by my staff I started to enter approximate totals for the remaining buildings on which they were still working by making quick square foot estimates. With these approximate estimates I was able to get in advance a tentative grand total of the project's direct costs and start to think about the markup amounts for the complex indirect costs, for overhead and profit, and for such exotic items as allowances for foreign customs duties, local taxes with strange-sounding names, the establishment of a complete branch office near the site (several hundreds of miles from our national branch office), and cash allowances for numerous common and unusual items, such as a local theocratic consultant (to ensure proper compliance with all local laws and religious customs), armed guards (local retired veterans), a blacksmith's shop on the site (to fabricate a thousand-and-one miscellaneous items from steel rebar and standard steel sections), and water carriers, each with his own regulation-size goatskin water bag.

To be brief, only two bids were submitted: ours and another. The other was lower. I vaguely knew the other company's chief estimator, so I called to congratulate him and was met with a surprising wail of woe. He had grossly underestimated his reinforcing steel in the building structures, he said, so I commiserated with him. I was not surprised therefore when that firm received the contract for the buildings. Apparently our bid was not much higher, and the outcome was that our company was asked to negotiate with the owner's agents for the services' contract (everything else on site but the buildings), which amounted to several million dollars. Subsequently our company also performed a series of other similar contracts for the same owner in various other places. I was rewarded with the job of starting up the services' contract for the cantonment.

As for the estimating aspects of this story, our bid and its ultimate outcome indicated that in some instances even major projects can be successfully estimated and bid using the *elemental method* of estimating, but there is a larger risk with that method than with more detailed estimating methods. In the case just cited, the conditions that made the elemental method particularly suitable included an unusual market in which there were few contracting companies able to handle such a large project and unusual and exotic conditions that made any kind of estimating difficult. Important things, such as labor productivity, availability of materials, even firm prices for materials and many other things could not be precisely determined. Consequently, precise and detailed measurement and pricing were less critical than normal. Measured elements in which minor items of work were allowed for in the unit prices were as accurate as anything else could be under the circumstances, and similar circumstances can at times prevail in domestic situations.

Only a traditional lump-sum contract (still the most common kind) requires completely detailed measurement and pricing because the bidder knows that should his bid be accepted he is bound to perform the work for the price that he has stipulated. In any other kind of contract, some of which are more beneficial to an owner under certain conditions, a contract sum need not be quite as precisely stipulated at the outset. A common alternative to the estimation of a stipulated-sum, the establishment of the maximum cost (or target figure[G]), can be well done by elemental estimates. Yes, there is a place for elemental estimates beyond the design phase of construction, in bidding for and making construction contracts, and even beyond that to the eventual appraisal of the value of existing buildings.

Exercises, Questions, and Topics for Discussion

1. Obtain a set of building construction drawings for a recent project and in a report illustrate and describe the *major elements* under one (or more) of the following titles: substructure, structure, exterior cladding, interior partitions, vertical movement, interior finishes, fittings and equipment, services, site development, overhead and profit, and contingencies.

2. Analyze the elements described in 1 into their constituent *items of work* (include and describe minor items with the appropriate major items) and, using a cost data handbook, price the elements according to the costs in your area. Create fictitious quantities for each element (or measure them from the drawings) in order to compare your results with those of others.

3. From actual drawings of a building's substructure, measure all of the work in the following divisions: site work, concrete, and masonry. Then price the work according to the preceding directions. After obtaining a total estimated cost for the substructure, calculate an elemental unit price showing its constituent parts by costs and by percentages of total costs. Divide the items into two groups: (1) those directly related to the building's perimeter and (2)

the remainder. Calculate two elemental prices based on those two groups of items. Examine and briefly report on the significance of your findings on the costs of building substructures generally for that class of building.

4. Write brief but adequate descriptions for the following building elements (according to the standard list): lowest floor, upper floor, walls below ground, walls above ground, and balconies and projections.

5. Write brief but adequate descriptions for the following building elements (according to the standard list): interior partitions, stairs, interior finishes, fittings and equipment, and site development.

6. Write brief but adequate descriptions for the following building elements (according to the standard list): electrical; plumbing and drains; and heating, ventilation, and air conditioning.

7. Explain the relevance and use of contingencies allowances in elemental estimates in an essay of not less than 1,000 words.

8. List and discuss the primary considerations for an estimator when using elemental unit costs from past projects to make an elemental estimate.

9. Select a building project and describe and illustrate the application of elemental estimates during the design phase to assist an owner in obtaining the most economically desirable building development. Use both written explanations and sketches.

10. Examine and discuss in detail, estimating the costs of building mechanical services in elemental estimates. (Use not less than 1,000 words and simple diagrams.)

11. Enlarge on the analogy in which *items of work* are described as atoms and *elements* as molecules.

12. Following the analogy in the last question, describe the atoms and molecules of a selected *engineering project* with a view to making elemental or other kinds of preliminary estimates of the project's costs.

13. Devise and describe in detail a simple, streamlined method of estimating building costs for a selected common class of buildings (schools, offices, supermarkets) using building elements.

14. List four basic types of construction contracts, precisely describe the primary features of each, and describe also the most economic and effective type of estimate that might be used to arrive at a competitive bid for such a contract. Explain your answers in detail.

15. If items of work are atoms, and elements are molecules of buildings, what are the equivalent of polymers in buildings and how might these be useful in estimating costs?

5

Estimating Applications in Appraisals

In Chapter 5, Part One, I broached the topic of appraisal of real property values using the classical methods, including that of the depreciated reproduction or replacement cost,[1] or simply the *cost approach to value.* My purpose here is to convince appraisers that there are better methods in the cost approach, to persuade them to use them, and to use the cost approach more often and more effectively. In this chapter I continue to move toward that goal and to do so I must be critical.

All generalizations about races, populations, denominations, sects, sexes, professions, and other groups of people and their actions are fallacious, and so if I refer critically to some appraisal practices, you

[1]The difference between reproduction cost and replacement cost of a building is discussed in Chapter 25, Part *1*, and summarized by the definitions of these two terms in the Glossary.

must understand that I refer only to the practices of some appraisers and that I intend only to be constructive.

As for the nonappraisers who have come to this chapter, let me assure you that your interests are not disregarded. In Part Two we read of an approach to building costs for estimators that goes beyond simple capital costs into *costs in use* and examines construction costs as an investment; aspects of estimating that are similar to aspects of appraisal practice of direct concern to appraisers. It is here that we reach a common ground for construction estimators and cost consultants on the one hand and for appraisers on the other.

An Appraisal: A Critical Review

Before me, I have a copy of an appraisal of value for what is popularly known as a high-rise condominium building. The document is thick; it contains about 90 pages in the appraisal report proper and in addition more than 100 pages in the several exhibits and miscellaneous supporting material. It is weighty and impressive. The appraisal is obviously detailed and thorough. It concludes that the building under scrutiny is in 1983 worth over $30 million. To arrive at that final evaluation the appraisers used all three of the classical approaches to value: (1) the cost approach, (2) the market approach (by comparison of sales), and (3) the income approach (by using gross rent multipliers). The results of these three approaches in the appraisal were each multiplied by a weighting factor, according to the appraiser's subjective view of the effectiveness and importance of each of the three approaches, and the aggregate amount was divided by the aggregated weighting to produce the final amount of value.

So far, so good—I certainly agree with and advocate the weighting of the three values obtained, and not all appraisers do. But let us look more closely at the details of the *cost approach* used in this appraisal. Of the almost 200 pages in the appraisal report, only *5 pages* are given over to the cost approach.

Direct costs of the building project[2] are simply calculated from total building areas (for the apartments, their balconies, and the covered parking) and from other areas (the utility areas, the landscaped areas, and so on), each multiplied by a single unit price per square foot, to the total of which are added certain lump-sum costs for design fees, testing, and miscellaneous items. As an estimate of its kind (building area times unit price), it is as good as you can get. It contains about a dozen different unit prices for the different kinds of construction—although more than two-thirds of the total direct costs are produced by the one unit price for the apartments themselves.

The weakness of this estimate is the lack of detail and support for the critical unit price that produces almost $14.5 million of value, that is, about 50 percent of the total value of the property, and the same goes for the other less important unit prices. The cost data represented by the unit prices are too slight, too arbitrary, too unsubstantial, and hardly supportable as presented. The appraisal report declares that the construction costs used are derived from an analysis of an estimate submitted by the developer and from cost manuals. That is all.

I have been involved as a witness in enough court cases over real property to know that if this appraisal report were exhibit A in a court case in which one party were trying to discredit the appraisal, this is where an attack would be made: on the validity of the report's unit prices. In an appraisal report of about 200 pages containing almost every bit of supporting evidence that the appraisers could find (including photographs, maps, and demographic and economic studies) there is hardly anything to support those unit prices—other than the statement that I have summarized and cited. The fact that most of the support for the unit prices comes, according to the appraisal report itself, from an estimate presented by the developer could be construed as another weakness. At the least, in support of the unit prices that it uses, the appraisal report should contain:

1. A copy of at least the summary of the cost estimate submitted by the developer, if not the entire estimate (which itself was probably also made using simple building areas and unit prices);

2. The titles of the cost manuals referred to, and quotations therefrom itemizing the critical unit prices, with the necessary adjustments indicated for time and location differences;

3. Details of the direct construction costs of the other projects reviewed by the appraiser and used in making judgments about construction costs used in the appraisal report; and

4. The specific methods used by the appraiser to arrive at the actual unit prices used in the appraisal, based on information similar to that outlined above

[2]In development, as distinct from actual construction, *direct costs* are the *hard costs*, the total construction costs that include what we call in estimating both direct and indirect construction costs. The indirect, or *soft*, costs in development include the costs of financing, advertising and sales, fees, insurance, ground rent and taxes during construction, and the costs of land rights.

(e.g., application of factors, subjective judgment, expert opinions).

With such supporting material included in the appraisal report, the appraiser might be able to make a good defense of the unit prices that he used, but perhaps not a good enough defense of his methodology, for it can be demonstrated that the use of building area times unit price without extensive and detailed adjustments for such things as area-perimeter ratio, building height, gross volume, mix of apartments of different sizes and content (e.g., numbers of rooms, fixtures, appliances), type of structure, quality of finishes, and numerous other facts, is liable to serious error.

Rather than building area as a basis of construction costs, I would urge the use of a more accurate and therefore more defensible method that employs either building elements, approximate quantities of the items of work, or both. Reasons for this choice are indicated throughout this text.

The use of simple and inclusive unit prices applied to the gross areas of buildings as a means to estimating costs has become so embedded in the everyday practice of so many professions that it is widely accepted as the normal way to obtain accurate costs. This is wrong. If in these United States there were a profession that specialized in building construction costs, its members would say that it is wrong. But there is not. In North America, apart from a relatively small number of professional persons not within the contracting business, the whole subject of construction costs is generally only understood by contracting companies, and they have obvious reasons not to encourage a deep interest by others.

There are some good cost engineers and quantity surveyors—and designers—whose business it is to know about construction costs, but they are few and far between. Every day owners, developers, designers, finance companies, bankers, accountants, insurers, real estate agents, assessors, appraisers, and many others talk authoritively about construction costs in terms of *dollars per square foot.* Why is that? They do not know better, and presumably they are satisfied not to know. The subject is complicated, and if everybody is willing to go along with the illusion that they are talking sense when they talk of the costs per square foot, why should they change?

A Better Cost Approach to Value

In 1978 three members of a graduating class from the British Columbia Institute of Technology's department of building technology received diplomas that included construction economics as a major elective they had selected, and then almost immediately they went to work with an appraisal company on replacement-cost appraisals. I met one of the three graduates some weeks later, and he told me that he was "flying all over" doing appraisals and using not only the same techniques that he had learned at BCIT but also the same cost manual that we had used in BCIT's construction economics' program.[3]

The techniques that he and his classmates had learned were essentially those explained in this book, and particularly those described in the previous chapter. Because they were not qualified appraisers, I knew that the new graduates were not doing appraisals on their own, but I also knew that they were going to many places to inspect and measure buildings and to collect data and that they were producing elemental estimates of costs on which replacement-cost appraisals were based. These young men had studied building construction, and they knew how to measure buildings, to analyze them, and to cost them. They understood the principles of depreciation in its several theoretical forms, the varying relative rates of depreciation for the different parts of buildings, and they knew how to produce a detailed estimate of depreciation based on a study of a building's elements. They had done several practice appraisals using several different appraisal manuals and cost data books, and they knew that the method they were using was effective. They also knew that the best methods to make replacement-cost appraisals are those described in this book for preliminary estimates, using either quantities of trade work or building elements, and a good cost data manual as a guide to current prices.

Why do I tell you this? It is not to make claim to some dubious credit as a teacher (although I had some feelings of pride when I met that graduate) but simply to dramatize an important fact: There is a strong link between construction estimating and the valuation of existing buildings and they have certain techniques and the need for certain kinds of information and data in common.

If you are an appraiser, I hope that you have read the previous chapters before you read this one because the content of the previous chapters is more or less relevant to replacement-cost appraisals (RCA); in this chapter we shall in outline apply it to RCAs using data from two specific sources of cost information for appraisers: (1) the Robert Snow Means, *Building Construction Cost Data* annually published manual and other related manuals by the

[3]Robert Snow Means, *Building Construction Cost Data* (issued annually). See Bibliography.

same publisher (here generally and collectively identified as the RSM) and (2) the Boeckh building valuation manuals and related publications (here generally and collectively identified as the BVM). (For further information on these publications, see the Bibliography.) *However, you do not need these manuals to gain something from reading this chapter.* My objective is to convert you, if necessary, to the use of such manuals as these—with certain provisos—in making RCAs, ideally in conjunction with a computer and a data base. Primarily it is my intention to convince you that appraisers can and should make better RCAs by using superior estimating techniques.

Use of the RSM Manuals in Appraisals

Briefly the basic RSM manual (*Building Construction Cost Data*) is organized primarily for estimators. It uses the standard 16-division format of the Masterformat[G], which is easily followed, but use of this manual also requires some general knowledge of construction, if only to understand the language. There is no escaping the fact that to make good RCAs you must know construction, and that is one reason why so many poor RCAs are made. At the same time it does not require an engineering degree to make a good RCA. In less than a year any intelligent adult can acquire sufficient construction knowledge to make good RCAs, assuming that all the other necessary knowledge is present. Incidentally, study of the RSM is an excellent aid to learning about construction.

If you use the RSM, do not overlook the Foreword, as it contains essential information, specifically: the economic basis for the cost information in the manual, including notes on the variations in local prices with reference to the indexes provided in the manual to adjust the manual's average costs for scores of cities in North America; the primary factors affecting costs; notes on material and equipment costs; labor costs; notes on how the manual is arranged; overhead, profit, and contingencies; information on doing construction work; and on training seminars offered by the publishers. You will not be able to use the cost data in the book properly if you do not read this Foreword. (I never read a book's foreword until I started to write them and then realized how important they can be.) Even if you are not yet sure if it is worthwhile to get a copy of the RSM, or even if you are sure that you will not, read on. You will find something of interest and value if you are concerned at all with the costs and values of real property.

In the RSM manual, after the foreword, the next section on standard crews may not be of immediate use to an appraiser. It is useful information to refer to when one is ready to make refinements in a technique already learned. The next part,"*Division No. 1, General Requirements*", is another part that the appraiser need not delve into right away. (You will remember that in our examples of estimates this Division, 1 containing job overhead costs[G] is usually the last division to be dealt with.) The manual contains data on all the standard divisions of work.[G]

Let us turn to a division containing one of the most common and tangible kinds of building work: Concrete, in Division 3. Do not be put off by all the detailed data in the RSM; instead go almost to the end of Division 3, where the columns of figures begin to break up and are separated by paragraphs of wording. The order of work for all the constituent parts of concrete is formwork, reinforcing steel, and finally concrete itself. Find the small table numbered (in a circle) 45 in the 1982 edition,[4] entitled "Concrete For Entire Building," and study its information. Look at the total average cost per cubic yard for concrete for an entire building ($215.60 in the 1982 edition and $223.70 per cubic yard in the 1983 edition, an increase of just under 4 percent) and work out the approximate percentages for the constituent parts. The costs of concrete, forms, and steel are about the same in proportion. All "materials" represent about 50 percent of total costs. You can use this average total figure to price all the concrete in a building structure, but do not forget to apply a multiplier for differences in time and place. (Look for the city cost indexes in the book.)

If you must appraise a multistoried building with a reinforced concrete structural frame, the costs of the frame will represent about 20 to 25 percent of the building's total costs. It can be measured and priced as an element in an hour or two, or a few hours depending on a building's size and complexity, and the results are so much more reliable than those obtained from unit prices related to building area.

In a few hours! you say. But if you expect to complete an entire RCA and not just a building's structure in a few hours, you must use another method (which we shall examine shortly), and as with any other product your client should get what he pays for. Only you, the appraiser, can judge what is required. Does your client need an approximate RCA, which in court you would have to describe as taken directly from an appraisal manual and based on the

[4]In the 1983 edition, in which the publisher has returned to the earlier format with the analyses grouped together, it is numbered 44.

closest published example that you could find. Or does he need a RCA that takes account of the unique features (and not just the area) of a building? Of course, your client must expect to pay a fee accordingly. It is up to you to convince him of the worth of a fuller and better appraisal—if it is needed—and you should be able to make such a case for such an appraisal. Alternatively you might employ a construction consultant, but you are obliged to give your client the opportunity to choose what he kind of service he will obtain. Many who require appraisals do not know exactly what is available.

For building elements[G] used in appraisals (as in some estimates) you can synthesize unit prices from the cost data provided in the basic RSM manual, you may use elemental prices from other sources, or you may use another cost data manual of the Means series, the *Building Systems Cost Guide*, also published annually. This guide contains a similar foreword to its companion manual, but its cost data are arranged differently, not in trade divisions but for building elements. The elements in this second manual are not exactly the same as in the standard list to which we have referred, but they are close enough that you will have no problem relating to them. They are as follows:

Division 1,	Foundations
Division 2,	Substructures
Division 3,	Superstructure
Division 4,	Exterior Closure
Division 5,	Roofing
Division 6,	Interior Construction
Division 7,	Conveying
Division 8,	Mechanical
Division 9,	Electrical
Division 10,	General Conditions
Division 11,	Special [Items]
Division 12,	Sitework
Division 13,	City Adjustments
Division 14,	Square Foot Costs
Division 15,	Reference/Aids
Division 16,	Index
Division 17,	Estimating Forms.

Divisions 13 through 17 are not related to divisions of work in Masterformat but to other contents of the manual. Division 1 is what we have previously called *substructures*, and Division 2 includes slab on ground. Each division (1 through 9) contains selected and alternative specifications for each kind of element, and typical alternatives are illustrated by clear diagrams. All the estimator or appraiser must do is to look for the specification that he needs to price each element. Of course, as with all manuals, the choice is limited, but alternatives can be made by additions and deletions to and from the elemental unit prices provided, using data from the *Building Cost Data* manual.

Looking at two of the more important elements in the RSM manual for systems' costs, namely, superstructure and mechanical, we see that the information provided is sufficient for use in the most detailed elemental estimates:

Cast-in-place concrete suspended floor systems—a one-way slab with beams. Given are bay sizes from (in feet) 15 x 15 to 35 x 40; superimposed (live) loads (psf); minimum column sizes; slab thickness; total load on slab; and finally, the costs per square foot. For that example the unit costs range from (in the 1982 edition) $6.20 to $10.20 psf. That wide range itself is a signal indication of the need for *elemental estimates.* Without them, such wide cost differences simply cannot be accounted for. In the example cited the difference in unit costs is one of over 60 percent. Even among slabs of the same bay size but with different loads, the difference is more than 30 percent in some instances. Need more be said to justify the need for and the use of such information—or to condemn less effective methods?

Turning to mechanical (plumbing) for an example, we find one (on page 209 of the 1982 edition of the *Systems' Cost Guide*) of plumbing costs for a 20-unit apartment building. Essentially the pricing method used is that described in this text in Chapter 3, Part 3: Fixtures are enumerated, and piping is measured and priced per linear foot or as a percentage of the total costs of fixtures (based on historical data). Only sprinklers are priced according to *building area* (and this is acceptable because of the regular layout of sprinklers), and there are different unit prices according to the numbers of sprinkler heads required. The manual provides information for quick estimates and also for more detailed elemental estimates.

For more detailed information and data on M & E costs, Means also publishes annually another manual, *Mechanical and Electrical Cost Data*, that is more detailed than any of the M & E information in the other manuals. It is of more use to estimators of building construction costs than to appraisers doing RCAs for buildings, although it would be invaluable to industrial appraisers specializing in M & E work.

The RSM group of manuals just cited provides the best collection of detailed cost data available in

North America. I have seen other cost manuals that at first sight appeared to be more detailed, but on closer examination of these bulkier manuals I found that most of their extra bulk consisted of tables that were simply "ready reckoners," something no longer needed in these days of electronic calculators and personal computers. There may be other cost manuals of equal quality but I have not seen them. Finally, Means has just (in 1983) advertised its General Estimating Program and Unit Prices' Cost Files, among several computer-based cost services (more on that in Part 4).

The BVM Manuals in Appraisals

Reviewing the long-established and respected appraisal publications of Boeckh,[5] we find that there are three basic volumes entitled: (1) *Residential* and *Agricultural*, (2) *Commercial*, and (3) *Industrial* and *Institutional*. In addition, periodic supplements are issued for the basic volumes containing new cost studies of new building models (examples) together with the latest "time-location modifiers" for major cities in North America and brief reports on building economics and markets. This material is specifically published for appraisers, and I am sure that all appraisers are acquainted with it. My purpose here is to comment on the material and to show how it may be enhanced. As I said earlier, material by this publisher was among the first that I used in making conceptual estimates and appraisals, and I found it very useful. Later I came to see its shortcomings (primarily in lack of detailed information for making conceptual estimates, which is not its declared purpose) and later to see how the Boeckh data might be enhanced for making both conceptual estimates and replacement-cost appraisals.

Looking at the second volume, *Commercial*, we find three estimating methods explained:

1. The general estimate method
2. The model method
3. The component method

The general estimate method is based on:

1. A "ratio graph," from which one finds a "ratio factor" according to the proportions of a building's length, width, and area. (We have referred to this factor as the area-perimeter ratio.)

2. A model building selected from among a limited number of illustrated models (although the number of models is constantly being updated by the periodic modifier publications, for which one pays an annual fee).

3. A "base cost" per square foot of *superstructure* from a table (A).

4. A "base cost" per square foot of *basement* (if required) from a table (B).

5. A cost adjustment for *exterior wall finish* (if needed) from a table (C).

6. Additional costs (if needed) for an *air-conditioning system* and for *elevators* from a table (F).

7. Additional costs (if needed) for other physical features from another table (E) for several different trades, (F) for mechanical work and elevators, and (G) for special fixtures and construction.

8. Time-location modifiers.

With this "quick method," an estimate-appraisal (or RCA) might take less than an hour to one or two hours depending on a building's complexity. You can see from the preceding list that several important factors, particularly the ratio of area and perimeter, are more or less considered. The weakness of this method is its reliance on a base price from a model selected from the manual that may resemble a subject building to a greater or lesser degree. If one is appraising older buildings, there are more models from which to choose. Some models for other, more modern buildings are cited in the periodic supplements published by Boeckh.

The model method is based on:

1. Essentially the same basic data and information used in the first method but with some refinement induced by using additional, more detailed data as described next.

2. Costs of building components from a table (D) that analyzes and presents the details of the cost information given in tables A and B. Essentially these are *building elements* and their costs, including building frame, slab on ground, finishes, fixtures, M & E services, and miscellaneous items—all priced per square foot (psf) of building floor area.

This method makes it possible to make further adjustments to bring the standard models closer to a unique subject building[G] that one intends to value. If one is dealing with fairly typical and commonplace buildings, the method is quite accurate in providing the relative costs of a building. If one is dealing with an uncommon building, more adjustments are

[5]The Boeckh Division of the American Appraisal Company, Milwaukee. See Bibliography.

needed according to the degree of difference from the selected model.

The Component Method is intended to deal with the *atypical* building that cannot be represented by a model building selected from the manual to begin with. It requires a description of a building's elements (comparable to those we have already discussed) and their costing from Table E. This method approaches that of an estimate based on approximate quantities, but I have never been able to use it successfully because of the limited contents of Table E. Instead (as even the Boeckh manual shows in its example of this method), one is practically limited to using the model method with adjustments from Table E.

The enhancement that I propose to apply to the BVM is derived from my own experience and from the elemental information and data provided in the RSM, as described—in effect, a combination of the two, because from the typical appraiser's point of view, the one complements the other. The RSM may be too technically detailed for the majority of appraisers doing RCAs, and most appraisers may not feel inclined to go down to the actual foundations, as it were, and to measure (even in the most simple ways) such obscure components as steel rebar. But that same appraiser will at times undoubtedly recognize the inappropriateness of the models that he finds in the BVM and also the inadequacy of the cost information provided in the tables for adjustments to base unit costs. The answer is to use both the BVM and the RSM and get more than their contents provide when used separately.[6]

Another fact adds extra worth to this suggestion to integrate two data sources: The Boeckh data are related to the base year, 1967, and the RSM were in publication at that time. One can therefore compare data and indexes and establish mathematical relationships between them. The subject of data and data integration is examined further in Part 4.

The BVM refer to an analytical approach to the calculation of *depreciation*—that bugbear of all RCAs—and they give an example, but it uses insufficient components as a basis of calculation. The better way is at least to consider the various kinds of depreciation (economic, functional, physical) and their effect on each and every major element. It is easy (especially if a computer is used) to calculate the depreciation along with the replacement cost, because replacement cost is 100 percent of "new value" in a RCA, and the depreciation is deducted. It is only with depreciation that one finds something that is not common to both estimating and appraisal; otherwise everything else that we have examined and discussed here is relevant to both. In life-cycle costing[G] depreciation is considered, and therefore we may say that estimating and appraisal concatenate in life-cycle costing.

Many of the valid objections of the past to more sophisticated methods of estimating and appraisal centered on the prohibitively large amounts of information and calculations needed, but those objections are invalidated by the availability and cheapness of personal computers and of software that can be used by both estimator and appraiser alike with relatively little time needed for training. This topic is examined further in Part 4.

[6]Users of cost data should not, however, overlook the existence of *Means' Square Foot Costs*, the publication mentioned earlier. But more models and data are always desirable. This is a good example of a general fact about information and data: their value and effectiveness increase at a rate greater than that of their increase in quantity when they are collected and combined.

Exercises, Questions, and Topics for Discussion

1. Use an apt and illustrative analogy to compare (in not less than 500 words) estimating and appraisal using the cost approach to value.

2. Obtain a copy of an appraisal manual not mentioned in the text and published for use in making replacement cost apraisals (RCAs) and write a constructive criticism of that manual (in not less than 1,000 words).

3. Obtain a copy of a cost data manual published annually for estimators but not mentioned in the text and write a constructive criticism of that manual (using not less than 1,000 words).

4. List, describe, and criticize at least four methods of costing a building's mechanical services for appraisal purposes.

5. List, describe, and criticize at least four methods of costing a building's reinforced concrete frame for appraisal purposes.

6. List and describe (in order of importance; most important first) the 10 most significant physical characteristics of buildings that should be taken into account in any RCA. (Do not discuss specific materials and methods, only general characteristics.) Discuss briefly the best ways of dealing in appraisals with each characteristic.

7. Define the two main kinds of depreciation and describe systematic procedures in RCAs for analyzing and calculating their amounts as accrued in a building.

8. Consider and then explain in detail why elemental estimates are the best kind of cost estimate for use in RCAs. Alternatively, argue that they are not.

9. List all typical major building elements in order of importance. In making RCAs, which building elements should receive the most attention? Generally explain your answer by stating the fundamentals involved.

10. List the three major elements in a typical building that, in your opinion, are likely to give an appraiser making an RCA the most difficulty in achieving accuracy. Explain your answer.

11. Describe in general terms and ordinary language (not in computer language) a computer program for the appraisal of buildings using the replacement cost approach. List and describe systematically the input and the output.

12. Write an introductory essay designed to inform those not familiar with computers about computer software suitable for use in making replacement cost appraisals.

PART FOUR

1

The Nature of Information and Data

In Part Four we take a small step into the vast and largely unknown area of information and data; an area that in our typical hubris we believe we finally may have begun to conquer because of our newfound technological ability to handle immense amounts of the materials found there. It seems however that we are looking at a beginning, not at an end, and that the future will bring before us things about information and data that are not as yet even conceived. In the relatively small sector of construction (not an area in which one commonly finds the latest developments in the sciences of information and communications) we see developments in *computer-aided design* (CAD) that foretell further advances in new directions that may leave behind those things, techniques, and procedures that we now assume will be with us forever because they are so much a fundamental part of construction.

For instance, construction drawings, as we have known them for decades—centuries even, for they have not radically changed in centuries, not even to the extent that books have changed—appear to be about ready for radicalization, and that will mean profound changes in the ways that graphic information is communicated and translated into other formats, including numerical quantities of work and verbal contract documents (and from there into costs). And as the monks continued to copy their precious books with quill pen and ink and embellished them with elaborate illuminations for years after the first book came off a printing press, so no doubt there will be draftsmen using their favorite grade of pencil to shade and embellish the paper surface of working drawings after construction graphics and other information are produced and conveyed by electronics to the building site.

But prescience and prediction are only slightly more expensive than hindsight. I only want to urge where necessary (and especially in places of education) that we look forward, not backward, and that we start to move into the future of construction, not cling to the past. Computers, especially the smaller and lighter portable computers, are a symbol of the future in construction, and that is why I, experienced in construction but a computer novice, presume to say something about them here.

At the outset it is important that we understand the terms as they are used in the title of this chapter and in the previous parts of this book. *Data* is a popular word and has been ever since the terms *electronic data processing* (EDP) and *automatic data processing* (ADP) were first coined, in fact, ever since the Information Age began, probably sometime during the 1950s. I do not claim to be an expert in the science of information; I simply want to make some points about getting and using information and data in estimating and appraisal because I have worked in this field for over 30 years and because I have seen the results of misunderstanding and of misusing information and data.

The two terms, *information* and *data,* each have several meanings, and hundreds of books have been written about them. Some meanings you can find in a dictionary; others are still new coinage and have not yet been banked.

In computer terminology, *data* is defined by one dictionary[1] as "a general term used to describe numbers, facts, letters, symbols, names, etc." It is followed in that dictionary by about 150 other terms, each beginning with *data.* In the same dictionary *information* is defined as "(1) knowledge not previously known, (2) an aggregation of data, or (3) the meaning assigned to data." Right there you get today's views of the meanings of those two words. The words are many-faceted and often confusing. For our purposes I shall use the two terms in their classical senses. Webster's Dictionary (third edition) says of *data* simply, "*pl[ural]* of Datum," and that is how I shall use it, from my days as a surveyor when a datum was (and still is) a reference point, level, or surface; which is substantially the second meaning given in Webster's, wherein the first given meaning of datum is "something that is given, either from being experientially encountered or from being admitted or assumed for specific purposes: a fact or principle granted or presented...."

In standard dictionaries, given meanings of *information* are, however, quite different from the given meanings of *data* or *datum,* and that is why I choose to use the two terms differently—despite the popular trend toward making their meanings virtually the same—in order to make a clear distinction between two related things. Following the general dictionary, we shall use (and already have used) *data* in reference to something devised as *a reference level* (of cost, of productivity), and *information* in a more general sense as knowledge not previously known and as knowledge communicated by others or obtained from investigation, study, or instruction. Incidentally thereby strike a light but deliberate blow at a formidable foe: the dissolution of the distinct meanings of many English words in modern usage.

Information for Estimating

In Part One the point was made several times that effective estimating is not possible without cost accounting[G] or more specifically without the *information* obtained by cost accounting. Cost accounting is the construction industry's way of keeping account of what it is doing and of doing routine research into the economics and costs of construction. It is the only systematic way of getting the information needed for management and estimating. Yet in most companies cost accounting is still rudimentary if it exists at all in any systemic form.

In my first book[2] I advocated the pooling of cost information, and in one early review this proposal was condemned as being socialistic, thus making the text unsuitable for teaching. I had hoped that the

[1]Charles J. Sippl, with JoAnne Coffman Mayer, *The Essential Computer Dictionary and Speller* (Englewood Cliffs, NJ: Spectrum Books, Prentice-Hall Inc., 1980).

[2]*Fundamentals of Construction Estimating and Cost Accounting.*

book might have been banned for political reasons, but it is still in use and I am not yet in exile. But, although the technological means exist and are patently economical, no great advancement has been made in acquiring and disseminating information and knowledge of construction economics, particularly at the grass-roots, on-site level of construction—in the field, as it were. It seems that practically the only dissemination of this kind of information is by commercial publications such as those mentioned in Chapter 5, Part Three, and earlier. I am in favor of private enterprise, but some things need a profession for their development. Only a minority of contracting companies have effective cost accounting systems, and most of these systems shrink and practically disappear when the volume of work is small. Therefore, two things are needed: (1) an effective professional institute of construction economics and (2) education in construction economics and research by its graduates and teachers. In the meantime construction companies and estimators need to improve their information handling practices. Next, I offer some proposals for the codifying and collection of cost information.

An Existing Cost Code and Data

The problems and expenses connected with setting up most business management systems are well known, yet nothing can be done effectively without a system. For building costs such a system already exists both in form and substance: the cost manuals published by Robert Snow Means Company, Inc. (RSM) and referred to in Chapter 5, Part Three, and earlier.[3] As I have said, I believe that these manuals present the best of cost information in North America, and this opinion is shared by acquaintances in the field. The manuals are well organized and coded for processing by computer. The primary manual, *Building Construction Cost Data*, is organized with the standard 16 divisions of work, and each item is individually coded with a five- or six- digit numerical code number, for example:

1. Code # 3.1.450—forms in place, footings, continuous wall, 1 use
2. Code # 3.2.050—rebar in place, footings, #4 to #7
3. Code # 3.3.020—concrete, ready mix, 3500 psi (material only)
4. Code # 3.3.190—placing concrete, labor and equipment, footings, continuous, shallow, direct chute.

[3]See the Bibliography.

The advantages of using this manual and its ready-made coding are several:

1. It exists, therefore it is economical. You have only to buy the manual.[4] Starting in 1983, Means is offering its data and system for sale in *computerized form* as software and at a reasonable price.[5]
2. The RSM is based on the Masterformat's standard divisions: the nearest thing in the North American construction industries to a universal standard format.
3. The original RSM is widely used throughout North America and it is published with several other related cost manuals specializing in various trade works that have developed from it.
4. The RSM are updated and expanded every year with increasing amounts of basic information and data.
5. The RSM company provides for users of their manuals to provide information for the manuals. Potentially it already has a common pool of cost data.
6. The cost data provided in the RSM are a ready-made foundation and framework for your own cost information and data. (This aspect is enlarged on in the next section.)

Collection and Development of Cost Information

The content that you get in a copy of the RSM is *cost data*, but really you need *cost information.* (Note well the difference.) Different persons (individual and corporate) in construction have access to different sources of information. With data we can create and enhance information, or, more precisely, we can create more information. Data is the compost that causes information to grow. But first we need information as seeds with which to start.

Designers, construction managers, and owners involved in construction contracts have access to one of the best sources of information: the *schedules of values*[G] and the *applications for payment*[G] submitted by contractors according to the conditions of their stipulated sum contracts. Often this source of cost information is not fully used, neither for the im-

[4]I have no financial or other commercial interest in promoting this publisher's material. My interest is solely professional and academic.

[5]After writing this chapter I received material from the Robert Snow Means Company advertising "Cost Data on Computer—ESTEK™" saying: "Now you can have access to the R. S. Means Data Base of construction costs through MCAUTO Computer Services." The 1983 R. S. Means catalog, "New Ideas for Building," contains a "Microcomputer Information Request."

mediate contractual purposes and duties nor for any other purpose. Of course, there are legal and ethical constraints on using certain kinds of information, such as that provided by one contracting party to another, but this constraint can be overcome without any breach of trust or breach of contract. I am not proposing anything that even flirts with the unethical. The necessary ethics for this kind of situation have already been established where contracts with quantities[G] are normal.

Once again the distinction between *information* and *data* is important, and I hope that anyone reading this who initially disagrees with my proposal and intends to criticize it will first fully understand what I am saying. Often information is required to be private and protected, but data are different. (Data may be kept private and protected by those who have created them, but they have no obligation to do so.) Once information from one source is combined with information from another source, it loses its individuality and immediately is on its way to becoming data. The exposure of such data is not the same as disclosure of information. If I state publicly that my neighbour, Old So-and-So, is a drunkard and a practicing thespian, I may be presenting information, but I may also be guilty of slander or defamation. If I quote statistics by saying that 1 neighbor in 10 probably is a drunkard, that is data.

Certainly the distinction between information and data is not always hard and fast, but it can be a clear and definite distinction if we make it so. It is not, therefore, difficult for anyone privy to information to convert it into embryonic data (not yet fully formed or ready for use); it simply needs to be mixed with other, similar information to be divested of its individuality. Typically, professional people do it all the time. If a doctor looks at a patient's sore and says, "That's herpes; I know because I just correctly diagnosed the same disease on Mr. X," he is disclosing confidential information. But if he simply uses the knowledge of a disease that he has gained through treating Mr. X in his treatment of Mr. Y, that is normal professional practice. Inevitably the professional person combines in his mind the information from different cases and thereby increases his expertise, and likewise the cost consultant uses cost information from one project on another.

Contractors have the greatest access to the best sources of construction cost information, namely, the projects in which they perform and supervise the work. Much of that information is needed for their routine purposes: for payroll, bookkeeping, and general accounting.[G] Therefore, the collection of much of this information is also routine and unavoidable. Yet so often no further use is made of the information beyond the routine uses just cited. If it costs a contracting company $1,000 per month to collect this information routinely, it might cost $1,300 per month to collect it *and* the additional information needed to make it useful for other purposes, including estimating and construction management. But can the additional expenditure of $300 a month be justified?

Unhappily information cannot be weighed and priced at so much a pound, so how do we evaluate it? Perhaps, only through its use in practice, and that never is conclusive, which is perhaps why cost accounting is not popular. Yet can anyone deny this common-sense statement: To estimate properly it is necessary to have historical information recorded from previous projects? If not, then all that remains is to decide how the information shall be collected and processed—haphazardly or systematically? This is not the time or place to describe a cost accounting system, only to insist on the need for such systems and to propose that we may begin by utilizing the data, format, and the system found in the RSM and similar and equal sources.

A hypothetical estimator, whether working with a contracting company or with a construction manager or working for a cost consultant, needs information and data. Where does he begin to collect it? He is obliged to use information and data obtained by another before he can generate his own. The R. S. Means Company and other publishers provide an immediately available body of organized cost data, but it is not valid for immediate use in a particular construction project. What is the next step?

Uses of Cost Data in Estimating

How can the cost data provided by sources such as the RSM (and there are others) be converted into useful information for the estimator? The RSM shows how, but the estimator eventually must make his own way, having once been given the direction. His main deficiency is in *information*, and what he has is usually incomplete and inadequate. The RSM tell him to use their data, to use the published costs, and then apply an appropriate city cost index[6]. This is all right for preliminary conceptual estimates in the earliest stages of construction, but for later estimates I suggest that estimators in effect create their own index (or indexes) for each project—in some estimates, for each major trade division of

[6]It is imperative that users of the RSM and any other manuals first read and understand and then apply the requirements for use set down therein.

work. As an example, let us suppose an estimator is estimating the costs of a multistoried building with a reinforced concrete frame; let us suppose that he has the following information from recent projects in the locality of the subject building:

Average Costs from Three Projects

Concrete in place in structural frames (12 to 18 stories): $78.50 per cubic yard
Steel rebar in place: 53 cents per pound
5/8-inch formply: $638 per thousand square feet
Yard lumber, standard and better: $350 per thousand board feet
Average weight of steel rebar per cubic yard: 160 pound
Rodmen (per hour, including fringe benefits): $20.25
Laborers (ditto): $14.90
Carpenters (ditto): $18.75

From this information the estimator calculates that a typical flat reinforced concrete slab will cost $274 per cubic yard for all costs other than *operating overhead and profit*.[G] He then checks the unit price for the same item in the cost manual and finds it to be $250 per cubic yard;[7] a difference of about 10 percent over the manual's price. But he notes that the manual's unit price assumes only *4 uses* of forms while he knows that on local projects the average is much higher, in fact, about 12 uses. He then estimates an adjustment for 12 uses and calculates a revised unit price of $279 with a difference of about 11.6 percent over the manual's price. He then compares the preceding material costs and labor costs with those in the manual and finds a difference of about 10 percent.

Eventually he is able to find that the average cost of the concrete structure for two of the previous projects was $238 per cubic yard. Neither had any extraordinary foundations or other unusual features. The manual shows an equivalent figure of, say, $215 and a difference of plus 11 percent. With this information the estimator begins to feel confident that he can use the unit prices for concrete in the manual if he adds 10 to 12 percent. (This fictitious account assumes that our estimator has considered a number of other things, such as any price fluctuations since the earlier projects were completed, the conditions in the construction market, and differences and similarities in design and construction.)

Our objective is to establish from limited information a factor so that we can then apply that factor to the data and thereby increase the potential of limited information, but it must be done judiciously. It also needs to be done at two levels: at the microlevel and at the macrolevel. It is not sufficient to compare only one or two unit prices from actual projects with unit prices from a body of data; too many things affect costs to make this reliable. Neither is it sufficient to go to the other extreme and to compare the total costs for major elements, such as a reinforced concrete structural frame.

There may exist self-canceling costs for different items whereby a project's overall costs are made misleading. For example, a project may have formwork costs that are lower than those in published data, but that particular project may contain features that have increased other costs of the structure. Ideally, therefore, an estimator[8] should compare unit costs for important *items of work*,[G] such as formwork to suspended slabs, and basic material costs and labor rates; he should also compare average unit costs for entire elements. If both kinds of comparison show similar differences between data and information (published unit costs and costs from known projects), then he can establish a factor to apply to the data.

The efficacy of factors created to supplement information by comparing information and data may depend on the kinds of comparisons made. Here we should remember the 80/20 rule, which says that a large proportion of total costs (figuratively 80 percent) is accounted for by a small but critical proportion (figuratively 20 percent) of the total number of items of work involved. Those items chosen for comparison and calculation of factors for data should be from the smaller, critical group. Clearly not all items, information, or data, are equal.

Relative Values of Information and Data

Within any building project some items of work constitute most of the project's work and costs, and the other items, while necessary, are relatively insignificant. Think of a multistoried office building's structure before the exterior cladding is applied. Typically it consists of a central core-tower containing the elevators and a series of floor slabs separated and supported by a few columns. If one is able to pin down the costs of the floor slabs, one has already perhaps 80 percent of the structure's costs accounted for.

[7]All figures are assumed and for illustration only. See current editions for actual figures.

[8]As elsewhere in the text, the term "estimator" is used in the most general way to include all who estimate costs.

It is instructive to divide the total volume of all the concrete in a building's structure by the total slab area. The resultant equivalent gross slab thickness may then be compared with the actual slab thickness to see the difference that comes from the other concrete members. Within certain building classes it is convenient to classify concrete structures and their costs according to their equivalent gross slab thicknesses for future reference.

Similarly we may look at a mechanical system as though it consisted of only three elements (or subelements): (1) the central plant, (2) the piping, and (3) the fixtures. (In plumbing systems there may be only the last two elements.) Analyses such as these give a good idea of relative values among elements and their major subelements in projects, but what about *the relative values of information and data* among projects? How do we establish such values quantitatively? We begin by rating them qualitatively, by creating classifications within which to work.

Not all information and data are of equal quality. We may classify them according to their sources and thus according to their qualities, as follows:[9]

(1) First-class information and data obtained from sources well known to the collector by using systematic means (i.e., a cost accounting system) so that the collector is able to relate them to the actual economic conditions and other conditions under which the work that is the subject of the information or data was performed. Typically the collector might be a cost accountant working on a project in progress or just after its completion. This class of information and data is supported by an estimate and a cost account.

(2) Second-class information and data obtained in a similar manner to first-class material but after processing passed on in some way (e.g., as the contents of a publication) so that they become for the ultimate user at least second-hand. Consequently the user's knowledge of the data's origin is limited and he relies more or less on the credibility of the originator from whom he has received the material. Typically this class includes material from the RSM and similar publications, although not all such material is equal in quality, as evidenced by the quality of descriptions and supporting analyses provided in some publications and not in others.

(3) Third-class information and data—are the remainder, those that cannot be placed into one of the superior classes. It includes some published data that typically are inadequately described and supported.

This simple attempt at classification is based on common sense, but it embodies an important possibility that requires much research and development, namely, a much more precise and extensive classification of information and data, so that from such classifications quality factors might be derived in order to be applied to data; a kind of weighting of different data in the calculation of probable costs from historical cost information, starting with unity for an absolute fact and decreasing as the reliability of the information decreases.

Some such factor should be applied to all information and data, and third-class material alone should never be used. It must be understood that while first-class information can be used to create information from second-class data, the result is always second-class information, and that data always must to be converted into information before it can be used in an estimate.

There is in the program evaluation and review technique (PERT)[G] a formula for the calculation of probability of the time for an event:

$$T_e = \frac{(a + 4m + b)}{6}$$

in which T_e = expected time; a = the most optimistic time; m = the most likely time; and b = the most pessimistic time. This has the kind of configuration that one would expect to apply also to the probability of the costs of an item, particularly because *time* is the most variable and the least predictable element of most construction costs. (The same configuration appears in the formula for the calculation of the volume of a prismoidal shape.) Similarly one might look for a similar configuration in a formula related to the probability of information and data, but much research is needed.

It does seem strange that at a time when information is such a common commodity that it gives its name to the epoch in which we live that we do not have practical means for rating or weighing the value of information. Anyone can see that there is so much information around and that even in everyday life we need to be able to classify and grade it. In the meantime in construction estimating we can be selective by rejecting all third-class material and by using only what is better. We might also consider experimenting with such a formula as that cited so as to give relative weightings to data.

Ask me to predict a labor rate[G] for an item of work, and I will offer an opinion as to the most likely

[9]This was first done in my earlier book, *Managing Construction Contracts.*

rate. Ask me for the most optimistic rate or for the most pessimistic, and I can express a reasonable opinion. It seems likely that if we could plot actual observed rates, they would fit into the classical bell-shaped curve of probability and that the preceding formula or something similar would apply. By extension cannot we similarly rate all information? Certainly with cost information it is desirable and it should be possible.

Questions and Topics for Discussion

1. Write an essay (of not less than 800 words) on the differences between information and data.

2. In what practical way can published cost data be used to provide a datum in estimating the costs of a group of items of trade work?

3. Examine the primary RSM publication (or other similar publication) referred to and write a criticism of its use as the source of a cost code for information that is to be integrated with the data in that publication.

4. Compare the cost code in the RSM with that published in the Masterformat. What are the advantages and disadvantages of each?

5. List and describe five different generally accessible sources of construction cost information and rate them according to their reliability (in your opinion). Explain your ratings.

6. Devise a method that has relatively simple means of application for rating the significance of any item of cost information.

7. Define these terms: (a) *data,* (b) *information,* (c) *cost datum,* (d) *first-class information,* and (e) *second-class information.*

8. Apply in a written discourse the primary principles expounded in Azimov's *Foundation* novels to construction economics.

9. Discuss the pros and cons of a national organization to collect and disseminate information about construction economics, especially in connection with competitive bidding.

10. Discuss the proposition that in building construction competitive bidding should extend only to matters related to efficiency and profit and not to matters involving the factual accuracy of measurement and pricing and the interpretation of bidding documents.

11. Propose an alternative source of coded cost data to that proposed in the text and compare the two in detail.

2

Information and Computers for Estimating and Appraisal

When I wrote my first book on estimating and cost accounting,[1] I included a section entitled "An Example of Estimating and Cost Accounting by Computer," and I included information about a Swedish system[2] that seemed more advanced than anything I could find elsewhere. The hardware included a quantity calculator (the QC-95) with

[1] *The Fundamentals of Construction Estimating and Cost Accounting.*

[2] Developed in the 1960s by a Swedish company, Samdata AB, Huvudkontor: Vretenvagen 8, Fack, 171 20 Solna 1, Sweden.

an electronic pencil for measurement of work and the software was specially designed for construction estimating and management, including planning and scheduling.[G]

The advances in computer technology since the QC-95 was first developed have completely changed the size, efficiency, and the costs of computers. Then it cost several tens of thousands of dollars (hundreds of thousands in equivalent dollars today) to set up a computerized estimating and cost accounting system. Now the necessary (but somewhat different) hardware and software can be had for about $10,000. The setup time might conceivably cost a few thousand more. Systems and applications similar to the Swedish system but using microcomputers are now advertised in North America and Europe.

A Personal Computer for Business Applications

I purchased a personal computer designed for business applications halfway through writing this book, and I agonized for a month trying to decide whether to complete the writing with my IBM Selectric typewriter or whether to switch to my new Kaypro II computer and use the Perfect Writer™ word-processing software that came with it—along with other software—all for less than $2,000, including state sales tax.[3] I switched, and I am glad I did—because it has added a new dimension to my writing; it has made it so much more pleasurable.[4] The IBM Selectric is one of the best of typewriters, and it was a dream to use after poking at the manual model I used for my first book, but no typewriter can compare with a desktop computer for ease and flexibility in writing—now called *word processing* when done by computer, a term that I abhor.

I am no touch-typist, and it does not help much to be a touch-typist for writing because, for most of us, writing is rewriting and rewriting. Corrections and retyping are one of the biggest problems in producing a typed manuscript. With a computer and a word-processing program, writing and corrections (the computer term for both is *editing,* which conveys the truth about writing) are so much easier. You can read reviews of both the Kaypro and the Perfect brand of software that comes with it in several computer books and magazines.[5] My purpose is not to promote hardware or software per se but to promote better estimating, cost accounting, and better usage of construction cost data; these are more easily achieved *through the use of personal computers.*

The construction industry can overcome the obstacles to better information-handling by using computers on construction sites as well as in the office. One of the primary obstacles, that of cost, hardly exists any longer. The remaining major obstacles are refractory attitudes of potential users and the lack of awareness with many in the computer business that in the long run it is service that counts. The computer industry has not yet matured and it is still like a whizkid who thinks that brilliance is all that matters.

It is frequently said, and I have no reason to disbelieve it, that today's microcomputers (such as the one I am using) are as powerful as the mammoth Univac computer of the 1950s that occupied not half a desktop but hundreds of cubic feet of space and weighed not 20-odd but hundreds of pounds. Microcomputers can be a great aid to more efficient construction and they are already are in use in many fields: in graphic design; in structural, heating, and other complex calculations; in accounting and other office routines; in planning and scheduling; and to some extent in estimating and cost acounting. But, at least in estimating and cost accounting we have yet to get to the root of the subject.

Computer Systems in Construction

One company publishes a monthly newsletter on computer applications in construction[6] and is therefore a useful source of news about the development of computers, especially of micro- and personal com-

[3]The Kaypro computer is made by Non-Linear Systems Inc., 533 Stevens Ave., Solana Beach, California, 92075. The *Perfect*® brand of software, including *Perfect Writer, Perfect Speller, Perfect Calc, and Perfect Filer* (all of which came with the Kaypro II, together with *Profitplan* and other software), are produced by Perfect Software Inc., 1400 Shattuck Ave., Berkeley, California, 94709. Since my purchase the software included has been increased and the makers of the Kaypro have a new name: Kaypro Corporation.

[4]Although the configuration of the computer to a printer through Perfect Writer has been agonizing, the problems apparently lie with the user.

[5]A recent review of the Kaypro and its software package, to paraphrase, said that one of the good features is the integration of the software that permits all the *Perfect* brand of software to be used one program with another. One national magazine reported a well-known writer and editor saying that Perfect Writer was so-so. Other computer systems offer similar capabilities, and as for the fine points of difference among them, I cannot comment. Each must judge for himself.

[6]Construction Industry Press, 1105-F Spring Street, Silver Spring, Maryland, 20910, which publishes *CCM, Construction Claims Monthly*, and *CCAN, Construction Computer Applications Newsletter.*

puters in construction. In 1983 within the first four issues of *CCAN* are short articles and bulletins on many topics, general and specific, including:

1. New computer estimating products, including software and hardware;
2. A comprehensive computerized electrical estimating system;
3. Computers used in construction education;
4. Cost-accounting systems for computers;
5. A labor management system for computers;
6. Estimating and job costing by computer;
7. Payroll systems for computers;
8. Critiques of computer software by subscribers (in North America and overseas); and
9. The use of an electronic "mouse" with a computer for measurement.

In addition the newsletter contains material of a more general nature, as may be found in the more topical computer periodicals. The newsletter's annual cost of $60 is a bit steep for individuals, but it is not out of line for newsletters in special fields, and I suspect that most subscriptions are purchased by companies. It is interesting to one who has been in this work for decades to read such a newsletter, but after the initial excitement at seeing all those familiar topics in print, and after reflecting on the content, I cannot say that I found anything really innovative. I read about the Swedish system in the 1960s, and it had an electronic pencil for measurement, even then. This newsletter serves a need, and eventually there will be special magazines on construction applications for computers because they are so numerous. There is a lot to be done, first in rationalizing what it is we ought to be doing in estimating and other aspects of construction management and then in writing innovative computer software for those activities.

The general use of microcomputers and personal computers[7] is expanding at an accelerating rate, and eventually construction will be radically affected by computers. My profession, quantity surveying,[G] has been using computers for construction purposes since the 1960s, but the use of computers by its members is still far from universal. The proliferation of microcomputers and effective software programs, when the latter are abundantly available, will eventually bring us to that state. As far as I can see, nothing fundamental has yet been changed in construction practices by the use of computers. Computers have enabled some of the more enterprising to do the same things faster. Nothing I have read in the recent issues of the mentioned newsletter or anywhere else changes that opinion, but we are on the threshold of great change in construction practices. This is so not because of attitudes in construction but rather because of the general changes that microcomputers and effective software will inevitably bring to all business. But we are still waiting for better and more innovative software.

Estimating seems like a natural activity for computers involving as it does so much information and data and so much number-crunching. But one of the two essential activities of estimating, *measurement of the quantities of work*, is the least affected by computers. The use of the electronic mouse, or an electronic pencil, offers some help, but it can hardly revolutionize taking off.[G] In fact after two decades of computers many quantity surveyors in companies that use computers still find it more efficient to take off quantities in basically the same way as their antecedents. The use of hardware, cheap or expensive, does not fundamentally change what is done by the estimator and what is produced by estimating any more than using a calculator fundamentally changes arithmetic.

There are in this context, it seems, two viewpoints of measurement: (1) those who believe that the measurer is better left alone to measure as in the past and (2) those who advocate radical changes in methods but not necessarily in the fundamental nature of measurement itself. There has been no serious attempt to change the process of measurement radically, but there have been changes in some places in the methods employed—including the use of programmable calculators and computers. There will be no radical change in measurement as long as there is no radical change in the methods of communicating design information. Some will surely say that precious little is communicated by some specifications, and they are right; perhaps that speaks for new methods. Computers have only increased the pace, they have not changed the method of locomotion. Consequently the dyed-in deficiencies of our present systems are still there. The situation remains similar to that of television. In that benighted medium, technology has continued to advance while generally the message has deteriorated. Might not the same thing happen with computers in construction? Shall we find in a few years a microcomputer in every site office—with badly written and inappropriate specifications on their monitors—and still no proper cost accounting? As stated, construction cost accounting cannot be done without frequent

[7]Although the two terms are not strictly interchangeable, generally I shall use the term *microcomputer* to include both.

measurement of the work in progress, and so we return to the importance of measurement, the thing in construction management the least affected by computers.

At present, there exist in construction several applications for computers but, so far as I can see, only to speed up the traditional methods. Not that there is anything wrong with that, provided that the glitter of computerization does not distract us from the need for a renewal in the body of general construction practice.

In addition to the advent of the R. S. Means' computerized systems previously cited, other systems have been around for a while, including the Conceptual Budget Analysis that McGraw-Hill Cost Information Systems advertises. It appears to be a well-advertised but basically unsophisticated elemental analysis of construction costs. An example product illustrated in an advertisement listed only fourteen elements, but it may have been deliberately shortened for the illustration. It seems to offer nothing new, although speed and cheapness are advertised. The Dodge Computer-Aided Estimating Systems, also associated with McGraw-Hill, appears from its advertisements to rival the Means systems, but I have no direct experience with them. The Orr System of Construction Cost Management appears to be more detailed and sophisticated. It is advertised as based on eighteen "functional building systems", three data bases, and eight levels of detail. Seminars in the system's applications are periodically offered around the country. Reportedly the Orr system is used by some federal departments. Illustrative examples of the output of this system show a product such as one might expect from a fairly new system that uses computers. The basic approach is, however, quite traditional. At least one experienced estimator has criticized the limitations of its data base.

Neither of those systems nor anything else that I have seen has yet faced up to and overcome the fundamental problems of estimating and financial management in construction, and no computer program can, as long as it is based on the existing conditions and attitudes in construction. Some of these have been discussed, mainly in Part 1. They include the absence of a North American profession primarily concerned with construction economics and financial management, the absence of general standards of measurement and estimating, and the absence of consumer representation and involvement in construction affairs, including the establishment of needed standards (mainly procedural and contractual) and other standard forms of contracts. (Other deficiencies include inadequate public acountability for spending on public works, unquestioning acceptance of the lowest bid for lump-sum contracts for public and other works, and distorted economics that permit the construction of poorly designed and fast depreciating buildings, especially dwellings.)

There is a lot of talk these days about computers changing the world. Computers cannot change anything on their own. Computers are tools, often effective tools but still only tools that must be manipulated. Put the finest computer hardware (machinery) and software (operating systems and programs on disks, tapes, etc.) together in a room, and nothing will happen until a capable person turns on a switch and uses the computer. Why do we read and hear so much about what computers are going to do expressed in such exclamatory terms so as to imply that we too can have a part in this grand renewal if only we choose to learn to work with the computers? Already we see signs that computers may go the same way as television, with an increasingly more complex and more impressive technology coupled with an increasingly sterile and puerile content. Computer software is already three steps behind hardware, which develops at an accelerating rate, while the development of software for construction appears to be even farther behind. Nevertheless, or even because of this state, it is important that we in construction, as potential users, become aware of what is and is not happening.

I have a brochure for the Orr System of Construction Cost Management (basically, estimating) that shows on one side an illustration of a handwritten estimate sheet described as the "traditional hand-method of estimating." The other side of the brochure shows an illustration of a part of a computer's printout for an estimate. It really is almost unfair to pick holes in advertising because it is so easy, as the hacks who make up this stuff are usually pretty shallow in their tricks to beguile and their techniques are really too obvious. For example, in this brochure the obsolete handwritten estimate sheet is made to appear lacking in content, as there is not even an entry at the top of the sheet in the spaces provided for factual information about a project, while in the "modern" computer printout of part of an estimate 12 lines of preliminary information are included, with such expressions as *historical parameter estimate* and *judgement factors*. (In the army we had a term for that kind of thing, and it was aptly described as being able to baffle brains. The alliteration of the aphorism has a nice ring to it.)

Even the photographs of the estimator on each side are patently contrived. In the one next to the handwritten estimate he is shown head down, surrounded by drawings and by file cabinets symbolically overloaded with notebooks. In the clean world of computers, however, there is nothing except a smiling face and a keyboard. This is especially deceptive: the implication that with the handwritten estimate great effort and support are needed while with the computerized system it is all as slick and easy as water running off a duck's back. Yet—and here is the nub—the actual measurement of work from drawings, and the results of the entry of dimensions and descriptions of items, whether directly on to paper or into a computer by means of a keyboard or electronic devices, or both, are not shown in the illustrations for either the traditional or the computerized estimates because it is fundamentally the same process in both cases. Ironically, however, it appears that the hacks have not done the best possible job here because, so far as actual measurement is concerned, there is some advantage to using a computer, and this point they have not made clear. They could have shown traditional handwritten quantity sheets full of dimensions, not just the total quantity of each item.

My main thrust, however, is not at advertising hacks and their often misleading tricks but at the pretension that computer systems have already revolutionized and remade estimating and that measurement is somehow fundamentally different from what it was in the era B.C. (before computers). It is not so. One day it may be so, but not yet. I am concerned that the needed reformation of estimating and financial management in construction may be overlooked because people may already believe that it is not necessary—because computers have already solved all the problems of the past.

In Britain and some other English-speaking countries where the contract with quantities[G] system of contracting is commonly used, the use of computers in construction dates at least to the early 1960s, if only used by a minority of the professional people involved. In part the reason is the greater emphasis in that system on the written, as distinct from the graphic, means of communicating design information, specifically through bills of quantities[G] (and the detailed measurement that precedes them) that can be produced, checked, and utilized as bidding and contract documents much more efficiently with computer systems. The international journal, *Chartered Quantity Surveyor*,[8] invariably contains advertisements and articles on computer applications. For example, one issue[9] offered articles on:

Microcomputer Applications
Word processing (specifications, communications)
Cash flow
Bending schedules
Bills of quantities
Bookkeeping
Budget reporting
Data processing
Estimating
Feasibility studies
Financial modeling
Insurance
Ordering
Payroll
PERT/CPA/CPM
Stock control
Computer Services
Preparation and analysis of bills of quantities
Interim valuations of work and final accounts
Job costing
Project management
Custom-design of computer software.

An article in the same issue reports on a British quantity surveyor in Abu Dhabi (part of the Middle Eastern garden of delights for enterprising contractors and consultants) who uses a small programmable calculator (one step below a personal computer) with cassette and printer to "take-off dimensions direct *[sic]* onto a keyboard with output to a printer. Dimensions are squared [multiplied together] automatically. . . [The quantity surveyor] considers that with hardware costing only *[$300]* modern technology is within easy each of all quantity surveyors." At the other extreme an adjoining article discusses the uses of 16-bit personal computers (most are still 8-bit machines, which of course is still four times larger than a 2-bit item).

As for appraisal, computers have long been used in similar ways to those of construction, especially in doing mass appraisals and assessments for real-property taxation purposes. The analysis and synthesis by a computer of information and data for assessments collected in the field by a team of which I was a member during 1975 made it possible for a Canadian assessment authority to produce and test the extensive data base and contents of an assessment manual in a relatively short period. Rapid and effective testing of interim results by computer while research was still in progress made for good, quickly obtained final results. As for computers and appraisals based on the cost approach to value, what I have to say about computers and estimates applies just the same to those appraisals.

[8]See the Bibliography.

[9]Dated January, 1983.

Questions and Topics for Discussion

1. Obtain descriptive literature related to computers in construction, and if possible the manual for a computer program, in one of the following fields and write a critique of the program, with special attention to any innovative features that radically change construction practice: (a) planning and scheduling, (b) estimating and cost accounting, or (c) life-cycle costing.

2. Obtain a copy of *INFOWORLD*, "the newsweekly for microcomputer users," and peruse its contents, noting all mention of computer programs specifically described for commercial and business applications, with special attention to any related to construction. Comment on your results.

3. Obtain a copy of *CCAN*, "the construction computer applications newsletter," and peruse its contents. Find a review of a computer program applicable to any aspect of construction and write a critique of the program and of its review.

4. Considering construction management generally, list subjects that you think could be more efficiently handled with microcomputers; give specific outlines for computer programs for at least 10 subjects, indicating briefly their contents and objectives.

5. List and briefly describe as many computer programs especially made for construction purposes (other than *design*) as you can find. Select the one program that you think is the most innovative and write a detailed critique of that program with special attention to any innovation that changes traditional construction procedures.

6. Write an essay of not less than 1,000 words discussing the use of computer data bases and their specific applications to construction management, especially to estimating and cost accounting.

7. Write an essay of not less than 1,000 words discussing the use of computer spreadsheets and their specific applications to construction management.

8. Obtain the manual for one of the following computer programs—Perfect Writer™, Perfect Calc™, WordStar®, Select®, and VisiCalc® (or such other approved program)—and write an essay of not less than 500 words explaining practical applications of the selected program to construction.

9. Obtain a sample of the printed results of one of the construction computer systems of estimating mentioned in the text or some other approved and similar system; write a critique of the results, with specific reference to the handling and presentation of information. Propose specific improvements.

10. Find a construction company that has an in-house computer that is regularly used for the construction business; interview one or more users of the computer and write a report on your findings.

11. Consider the process of measurement for estimating purposes and make carefully considered proposals for increased efficiency and accuracy in measurement by the use of microcomputers, disregarding what you may consider technical limitations in computer hardware and software.

12. Discuss the proposition that electrical contracting companies are in the lead in the construction industry with their computer applications to contracting. Do you think this is true? Why should this be so? Explain your answer.

13. Write an essay of not less than 500 words about the possibilities for the use of microcomputers on construction sites.

14. Suggest one specific way in which innovative use of a microcomputer might radically improve (a) estimating, (b) cost accounting, (c) planning and scheduling, (d) payroll, and (e) cost analysis.

3

Collecting and Handling Information

When I finally began to write this book with a microcomputer and a word-processing program it became my fervent desire to write it without using paper, if that were possible. It was almost but not quite. I have spent my life working in construction and development in one sector or another, and because I seem to have shuffled tons of it during my working life, I am tired of handling paper, except in books written by others. Construction has always meant a lot of paperwork, but it appears that we may be able to get away from it. If one can write a book without paper, doesn't it seem reasonable to go further and build a building without paperwork?

In this chapter we introduce electronic files, and later I shall make some suggestions about collecting information and data for them. Primarily this chapter is intended for the person who is not yet familiar with microcomputers but who realizes that something is happening to change the business world and who may be swinging between two kinds of thinking about computers in business: (1) Perhaps I can manage to avoid getting involved with the darned things and, alternately, (2) I guess that I must face up to them one of these days.

Just before I had substantially finished writing this book, I received from the publisher's parent company[1] a copy of a newly published booklet by Prentice-Hall, the publishers, entitled, "Author's Guide to Electronic Manuscripts." It was newly issued and dated 1983, and I was pleased that I had already decided for myself some six months earlier that I would get a microcomputer for writing and had already substantially written this book. Otherwise I might have felt that I was being pushed into making the change from paper and typewriter to computer and electronic manuscripts, rather like our medieval monk whose quill pen is collected by the abbot as the new printing press arrives.

An Introduction to Personal Computers and to Electronic Files

Superficially, there is nothing extraordinary about the Kaypro computer; in fact, several others look similar to it. If I briefly describe this hardware and its functions, you can get a good general idea about most desktop computers; for that is where it sits, on the top of my desk. It is in two parts: (1) the keyboard, which looks like a flattened typewriter, and (2) the larger part, which looks something like a television set and contains the screen (cathode ray tube, or CRT) and the disk drives. It also contains all the electronic circuits and those mysterious "chips." The screen is a familiar-looking thing, and fundamentally it works the same as a TV screen.

The disk drives are rather like two enclosed high-speed record players, one placed over the other, to hold two disks which are something like the old 45-rpm records except that they are permanently contained and protected in a square, stiff envelope with a circular hole in the middle that matches the hole in the disk, which in turn matches the collett (or hub, as it is popularly called) at the center of the disk drive. Unlike a record, which is comparatively tough, computer disks are much more flexible (therefore they are called *floppy disks*) and also they are much more sensitive, hence the permanent jacket within which the disk revolves at high speed, many, many times faster than the old 45 record. You can read about the technicalities of computers and their disks from other books, such as those listed in the Bibliography. I only want to introduce you to personal computers and tell you something of what they can do for you in construction. *How* they do it can come later. There are two essential parts: (1) the hardware—mainly the keyboard and the "box" containing the screen and the disk drives—and (2) the software—the floppy disks with the operating system and programs on them that make things work—and your own information and data contained in files on the disks. That is why *two* drives are generally needed, one to hold the disk carrying the operating system and the current program (such as for word processing) with which you are working, and one for a second disk to hold your files; the two disks and their contents interact.

Chances are that one of the first obstacles between you and a computer is, you believe, the fact that you cannot type; or that you are not even familiar with a keyboard. Most computer keyboards are almost the same as typewriter keyboards, but do not worry if you cannot type, to begin with you can manage quite well by using just two fingers. Few experienced computer users started as typists. Incidentally most computer keyboards are more comfortable to use than most typewriters. For one thing, with the Kaypro's detachable keyboard (connected to the computer by a coiled telephone-type cord), you can sit in an easy chair with the keyboard on your lap and with the computer on the adjacent table. You just need to be able to read what is on the screen, and it can be set up and used in a construction site office as easily as a typewriter. For another thing the Kaypro computer keyboard feels better and it is quiet.

If you are easily intimidated by computers and all the talk about "losing a month's work by pushing the wrong button," don't be. Anything is possible, but not everything is probable. Besides, once you start to get familiar with a computer, you will not worry about it. It has never really bothered me, and from the outset I made an act of faith by not having a printer to "print out" on paper what I "put in" and to reassure myself that it was really there and really recorded on those floppy disks. At the beginning it was rather like going to a lot of trouble to rehearse and put on an act in a darkened theater, not knowing if anyone were there to see it. As I got responses, and as I saw that each time I typed the words appeared on the screen, and that I could file them and recall them at will, faith turned into conviction. Of course, you *may* lose a few hours work by doing something that you should not; I did, but only once in the last six months of learning and working; and that one must write off to experience.

[1] Prentice-Hall, Inc., Englewood Cliffs, New Jersey, 07632.

You might find it a useful introductory step to buy one of the cheap personal computers that are so popular for games and break the ice while sharing it with the kids, but I did not think that would help me. Writing (word processing) is a lot different from Pac-man. I prefered to jump in the deep end and start right away using this computer for word processing; that is, writing. (One of the irritations about this new computer industry is the jargon; no one speaks plain English.) But it is a matter of personal feelings and choice.

What I am suggesting is this: If you are in construction (or estimating, appraisal, or some other related field) and you haven't yet handled a microcomputer, break the ice and try it. Start to get accustomed to a keyboard and a terminal and to floppy disks and drives. Subscribe to one of the computer magazines published for nontechnical users— they are coming up like weeds and there are as many varieties; some are listed in the Bibliography. In an easy and relaxed way, begin to learn how useful computers can be to you in your work, apart from the dozens of other uses available to you from games (Pac-man is even available for CP/M-operated business-orientated machines like mine) to dealing in stocks and real estate or even for more serious things like estimating. But if you are one of those meticulous people who balance their checkbook, do not buy a computer for that purpose. It is not worth it.

With computers there are filing systems in which you can store information and data of any kind, just as you might store it in paper files and metal cabinets. The difference is that with a computer you can extract, say, seven sheets from seven different files in a few moments and see the information that you needed on your computer's screen. To search manually through paper files to extract and collect one particular kind of information might take 10 times longer. Computers are great for this kind of data filing and extracting, and many people and organizations are using computers solely for this fairly simple but otherwise time-consuming thing. But, of course, you have already seen this kind of computer usage in a number of places, including airlines' offices, for example. The real benefit from computers in construction will come from the use of electronic spreadsheets, a term that we hear with increasing frequency.

Traditionally a *spreadsheet* is the kind of sheet that you make up just before submitting a bid, if you are a general contractor, and on which you enter the names of sub-bidders, the amounts of their subbids, the items included and excluded from their subbids, and the adjustment amounts needed so that you can make proper comparisons among the subbids to select the one that you need. A spreadsheet is also an accounting worksheet, a cash flow projection, or a construction schedule. In fact a spreadsheet is any combination of information or data that requires lines and columns for its arrangement and for the arrangement and interaction of the entries, whatever their purpose. There are dozens of different uses for spreadsheets in all kinds of businesses and professional work, especially in construction. It is with spreadsheets of many kinds that computers are perhaps most able to assist you. In any instance that involves the manipulation of many numbers that are interrelated, such as labor costs in an estimate a cost account (interrelated through a scale of wage rates and fringe benefits), or a cash flow projection that depends on numerous data from many sources, spreadsheets are the means of handling the numbers, and electronic spreadsheets produced by a computer system are the means of handling the numbers much faster and much more effectively.

Computer magazines are full of articles and advertisements for data-base management systems designed for handling information and data on spreadsheets, or worksheets, and for manipulating files and their contents; but in setting aside all of the hype and reading some of the finer print, it seems that few of those are adequate for the large amounts of information and data (which for brevity I simply call data) that are commonly handled in a major construction project, with its many estimates, bids, subbids, sub-sub-bids, micro- and macroschedules, interim reports on production and payments, payrolls involving numerous trades and wage scales, and cash flows that equal the water distribution system of a major city in complexity. Be careful about what software you buy. But it is only a matter of time. The computer systems are getting there, and already the potential storage capacity for small personal computers has increased immensely. The shortfalls are still mainly with software. In fact, you might remember this order of priority for successful computer usage: (1) the operator; (2) the software, and (3) the hardware. Each is essential, but there is an adequate selection of hardware; there is a shortage of good software for many specialized purposes. As for knowledgeable and competent operators—what are you going to do about it?

The other day a charter member of the computer users' group to which I belong[2] (to whom I owe much for my primary computer education) who is involved in computer sales told me of a local businessman who paid several thousands of dollars to a local com-

[2]The Honolulu Personal Computer Users' Group.

puter store, not for computer hardware but for *software*, for a batch of business programs that he could do nothing with because he is a novice with almost no experience in using a computer. Why did he spend so much on something that he could not use? Probably a salesman convinced him that the purchase was exactly what he needed. (The secondhand car business has lost many good salesmen to computer stores.)

Many computer programs, especially those for managing data, require much time in learning how to use them. You might compare a complex and powerful computer program to a high-quality theodolite, a fine instrument of complexity and precision that can be put to both simple and complex uses once you learn how to use it. You can use a theodolite as a simple builder's level, or even as a telescope to look at the stars, but to use it to lay out an accurate curve for a highway requires a full understanding of the instrument and its capabilities. To use a theodolite as a simple level is not an abuse, it is just a waste of a fine instrument. Similarly you do not really need a computer and a spreadsheet program to balance your personal checkbook.

Sooner or later, if you work in construction, you will come up against computers; now is the time to start your education in computers if you have not already. (I trust those readers who are already beyond the introductory stage with computers will bear with me.) I suggest that you start by reading the introductory books on computers listed in the Bibliography. All of these are recommended by many in the field. Then, after purchasing the right software and the right computer for you—and that is the biggest and most difficult step to decide on and take—do some simple word processing and filing (write your own letters; begin by making files for your personal uses), then go on to more and more complex applications so that you grow into it. Fairly soon, you will begin to feel something like affection for your personal computer, and when in the local computer store you hear others talking about the capabilities of their computers, you will want to tell them about your own.

Don't try to matriculate right away; instead begin in primary school and read the set primers. Be tight with your money and do not buy something that you do not really need or cannot learn to use. On the other hand buy something that will fill your needs for a few years. I think the Kaypro (with 64K of RAM and CP/M as the operating system) is the kind of personal computer to start with, or the IBM PC; there are a number on the market. If in doubt, do nothing right away, but do not give up—but do not wait too long as there is nothing gained by delay. Do more reading and more research. Join a personal computer users' group, and listen. Ask all the dumb-sounding questions that come into your mind such as which are the best brands of floppy disks and which brands to avoid. The computer whizzes and hackers do not normally talk about such things, at least not in front of us neophytes, but nevertheless these are important questions. Much computer jargon is so esoteric that you may think that the speakers are wizards, and some of them are; others are just visitors from Oz. At the same time do not be shy about taking advice from somebody who looks like your youngest son; he may be just the computer genius and encyclopedia you were looking for.

Construction has always been outside the mainstream of business management, and it still is, now that computers are a rapidly growing part of that mainstream. A person cannot do much in the management of construction unless that person has a fairly good technical knowledge and understanding of construction work and contracts, unlike most other industries in which it matters little to the manager or the cost accountant whether the product of his company is a collett or a collop. That is not so in construction; consequently it is probable that the truly valuable software for construction must be developed by construction people, those who will step into computer development, who will use and improve the programs that are available for specific construction applications, and who will then go on to create new computer programs that will introduce completely new approaches to the traditional construction fields such as estimating, cost accounting, and construction economics.

Collection of Information for Estimating with Computers

Because most construction companies do not do effective cost accounting or keep up-to-date records of productivity and costs, and because most do not have much of a system for handling information, the easiest way is to buy someone else's system and data and adopt it to your own needs. If, for example, a general construction company purchases Means' General Estimating Program (STAR) to obtain a system within which to develop its own library of cost data and information, the company will get a system based on the 16 standard divisions that can be reviewed in the Means' manual and a large collection of cost data.

Whenever possible, buy the manual for a computer program before you buy the program; compare the products of at least three companies. A good program with a poor manual may not be of much use to you, and a review of a manual gives some idea of the content and capabilities of the program. First, determine what you need; then look for and review the computer systems and programs that offer it; seek out users who can tell you about what you are interested in from their own experience; then finally seek out the right computer hardware to run the software you intend to purchase. Many buyers do exactly the opposite and buy their hardware first, then the software, and then they try to work out what to do with them. That is putting the hardware cart before the software horse.

Hardware is the easier thing to find and choose; software is much more scarce and much more difficult to evaluate. The best way is to see it in operation in another company, if that is possible. Often computer salesmen are quite ignorant of most of the features of both hardware and software, and especially the latter, because they usually deal in a number of brands and do not have the time or inclination to become experts. They appear to be experts to the layman who is unfamiliar with the products and the terminology. Make an effort to learn some of the terms first, read about the hardware that interests you and avoid confusion on your side if not outright deception on the other.

Do not overlook, however, the personal computers such as the Kaypro that come "packaged" with a selection of software. Unless you are buying a fairly expensive business system from a company like IBM or Xerox, and especially if you want a personal computer for your own use and initiation into computers before a larger computer purchase, one of the packaged deals is the best buy. Once you have purchased a system you will want to make it useful by putting into it the information and data that you already have. If you buy something like Means' STAR, you might start by making a general update to bring the data that you have purchased closer in line with your own by applying time and location factors. This is where a computer system begins to show its worth, as you can make such updates easily. The next step is to put your own data into the same format and to compare it with the purchased data. Probably it will be quite different. In other words what you and what Means say about labor productivity and labor costs for specific items of work probably is not the same. It will make it easier if you only deal with productivity expressed in man-hours at the outset. Carefully study what the system's manual says about the subject and compare its basis for data with your own. Many, perhaps most, discrepancies among such data are due to comparing apples with oranges.

Concentrate on those trades of most importance to you. Do not waste time dealing with minor items of work; concentrate your efforts on dealing in important items such as formwork to slabs, if you build reinforced concrete buildings. If you can reconcile the data for two or three such important items in each major section of work, you will be doing well. If you cannot reconcile your data with what you have purchased, reflect on the worth of your own data. If you are confident that the data are valid, consider applying another factor to the purchased data—but only within the appropriate trade or trades—to bring that in line with your own.

When you next make an estimate for a bid, do not depart radically from the costing methods that you have used in the past; using a computer system should not require you to do so. Be conservative at first, and proceed to change slowly. Keep a record on paper of all the steps that you take in setting up the new system with a computer, in integrating your own data with any other new data, and in making updates to data in computer files so that you can, if necessary, backtrack and remake any of the steps already taken. This writing of notes may save much time otherwise spent going around in circles. Later as you start to add newly acquired data from cost accounts to your computer files, and as you gain confidence with the computer system, you will be able to dispense with paper records, but at the outset they are important to a clear view of what you are doing. Always carefully follow the advice of your computer manuals about making duplicate copies of your files on disks, then the risk of loss of data on disks is negligible.

One aspect of cost and productivity information and data that you might consider if you plan to use a computer is the filing of maxima and minima data, the probable maximum and minimum amounts for important items of work (the top 20 percent of items that typically contain 80 percent of a project's total costs). I know of no cost data manuals that publish this kind of information; instead they publish "average" data, and we all know how misleading averages can be. Means publishes ranges of costs and median costs for different classes of buildings (but expressed only as unit costs per square foot of building area). That same approach is also needed for all of the major items of work in each trade; that roughly triples the amount of data to be recorded (for minimum, maximum, and median costs for each item). This kind of information is valuable in estimating costs and also in scheduling work (because scheduling requires earliest and latest times for critical activities).

At first sight the measure of a computer's storage

capacity may seem immense, but most of us underrate the amount of data and information needed to make an estimate of construction costs. Many of the current computer spreadsheets are not big enough for some applications. Some computers do not have enough memory (RAM). There are ways to reduce this problem of storing and handling data: One way is to break down collections of data into smaller sets. Smaller computer files are generally easier and more efficiently handled. Avoid the delusions of accuracy that lead some estimators to calculate every item to the nearest cent. Cents have no proper place in the calculations and extensions in an estimate. Even the total costs of most individual items can be taken to the nearest $10 because the plusses and minusespractically balance. Eliminating cents increases the effective capacity of computer data bases, and rounding off amounts increases efficiency.

As with cost accounting, so in building up a data base—deal first with the most important items, using a coarse sieve to begin with and later increasing the degree of fineness as competency rises. Later as the system is established and consolidated, extend it to include labor items of lesser importance; then extend it to include equipment productivity; lastly include material costs, because for bid estimates it is normal to get quotations for specific materials. Finally a general contracting company may want to include data on major subtrade work, such as masonry, metal work, and finishes. For a construction management company, of course, all major trades, including concrete, formwork, and rebar, are equally important.

Information and Data Sources

Some of the better-known sources of information and data have been cited. Here are some other useful sources of construction cost data and other information affecting construction economics. For conceptual estimates there is the GSA's *Conceptual Estimating Guide* which contains some cost data suitable for conceptual and comparative purposes and is used with the *Capitalized Income Approach to Project Budgeting* already cited. A general source in North America is the American Association of Cost Engineers, but its slant is toward engineering construction rather than buildings. Other sources are, the Canadian Institute of Quantity Surveyors and the newly formed American Institute of Construction Economics.

In the English-speaking world outside North America, probably the single most important source of information on construction economics is the library of The Royal Institution of Chartered Surveyors, London, and the RICS Library Information Service, which is available to subscribing nonmembers. The RICS also has connections through its members with the Building Cost Information Service, a national center for cost information. Much of the specific information may not be of direct interest to subscribers outside Europe and the Middle East, but anybody interested in methodologies in construction economics and contracting will find much of interest from this source. RICS publications are distributed from the same address and include publications on such topics as "Principles of Measurement (International) for Works of Construction" and "Chartered Quantity Surveyors and the Microcomputer." Apart from the cost publications already cited, and those listed in the Bibliography, do not overlook or scorn more prosaic and domestic sources such as the Sears' annual catalog and similar publications in which one can often find and more easily identify certain minor building items not so readily found in the more dedicated cost books.

As a data base expands in scope and content, even a general contracting company may find it useful to collect information and data on a variety of subjects more or less related to construction costs and economics, including:

1. Elemental cost analyses of specific building projects, such as those published in technical journals (for use in elemental estimates, in check estimates, in comparing and validating trade bids and prices, and in life-cycle cost analyses);
2. Unit costs (per square foot, etc.) for buildings published in journals and professional publications (for uses similar to those listed);
3. Price lists published by building suppliers (with varying discounts);
4. Equipment rental schedules;
5. Cost indexes published by cost information services, government departments, and other bodies;
6. Labor costs, including additional burdens for such as travel time, camp costs, board and lodging, union-management agreements, unusual working conditions, and variations in productivity; and
7. Economic factors derived from analyses of completed projects, over and above those found in cost manuals and dealing with such phenomena as the construction market, financing costs, the general economy, the season and climate, unusual locations, government and private sector works, and construc-

tion works performed for specific owners and designers.

The scope and sources of information related to construction are numerous and various, and many have been hardly touched by many who make estimates. Many of the less tangible things that affect costs affect them relatively more than minor differences in the quantity of work, differences that most estimators would not think of overlooking, while at the same time making only the most casual consideration of the chances of an early freeze or a most desultory deliberation on the difference in labor productivity between North and South Carolina, for example. It is not pie in the sky to say that the time has come in which all such things can and should be part of the input into an estimate of construction costs.

Glossary

Construction people often use construction terms loosely—and sometimes incorrectly (such as *plan*)—and, as a result, misunderstandings occur and dispute arise. Certain terms and phrases in the text are shown to be in this Glossary by a G; many of them have an ordinary meaning as well as a special meaning in construction, and that is why they are here. The word *work,* for example, has an ordinary meaning and a special, contractual meaning in construction contracts. A few terms in this Glossary, such as *precontractor,* have been specially coined to describe more precisely a person or thing in construction.

This Glossary originated in another book[1] and was developed further in two later books. It provides a useful means of reviewing the main features of the subject and its background; its primary purpose is, however, clearer expression and better understanding.

Abstract estimate: one made with a summation or abstract of *unit prices* (often totalled and represented by a single unit price) applied to the simple measurements of a building; usually the gross areas at ground level. The easiest, best-known, and often least-effective kind of estimate of *construction costs.*

Activities: see *work activities.*

Advanced purchasing: purchasing of *materials* by an *owner* before awarding contracts for *construction work* to ensure their timely delivery.

Agent of the owner: a construction technician, usually an *architect,* an *engineer,* or a *construction manager,* who represents an *owner* before

[1] *Fundamentals of Construction Estimating and Cost Accounting.*

and during a *construction contract* in which he is named as the owner's agent and advisor (or one of them) and who usually is responsible for or involved in the design of the *work* of that contract. A *quantity surveyor* usually is also an agent of an owner, and there are others. Often an owner has more than one agent.

AIA: The American Institute of Architects.

Alteration work: *new work* done to and in conjunction with *existing work,* other than complete demolition.

Application for payment: made by a *contractor* to a *designer* according to the terms and conditions of the *construction contract* before the certification of payment by the designer to the *owner* and based on the contract's *schedule of values* previously submitted and approved.

Approximate quantities (of work): measured by using rounded-off dimensions and by not making minor adjustments for *voids, wants,* and displacements by other kinds of contiguous *work;* used usually in *conceptual estimates* and *preliminary estimates.*

Architect: a person registered as such; one mentioned in some *standard forms of construction contracts* as the representative and agent of the *owner*; a *designer.*

Area-perimeter ratio: the relationship between a building's gross plan area and the perimeter of that plan expressed as a ratio; for example, for a building 40 feet square and containing 1,600 square feet of area, the area-perimeter ratio is 10.0; for a building 20 feet by 100 feet, and 2,000 square feet of area, the ratio is 8.33. Typical ratios for apartment buildings range from 10 to 30, and for parking garages, from 30 to 150.

Bank measure: (sometimes in-bank measure) the measurement of natural ground to be excavated *in place* and before excavation, because soil increases in bulk once it is excavated.

Bar chart: see *Gantt bar chart.*

Bare costs: the actual *direct costs* before adding any *markup* for *overhead costs* and *profit.* A term usually in need of particular definition.

Basic element: one similar in makeup and usage to a *basic item of work;* an *element* commonly found in a particular class of construction, whose historical costs are therefore useful in making *elemental estimates.* (If necessary an element's *unit costs* may be reduced to those of a basic element for estimating purposes by deducting the theoretical costs of minor atypical features of the element.)

Basic items (of work): *items of work* identified and described in *specifications* and in *cost codes* and commonly found in a particular type of *construction work,* which can therefore be part of a *basic cost code* or a *basic specification* for that type of construction, as opposed to a *particular item of work*; the criterion being an item's practically inevitable occurrence in that type of work (for example, concrete footings in building construction), even though their dimensions and details vary. It is the recurrence of basic items that makes their identification and the use of basic specifications and cost codes possible and effective.

Bid: an offer to do *construction work* for payment, the acceptance of which constitutes a contract between the *precontractor* who made the bid (i.e., the *bidder*) and the *owner* who accepted it. Also known as a *proposal* (in the United States) and a *tender* (in Canada and in most other English-speaking countries). Sometimes called a *general bid* when made by a general contracting company that seeks to become a *contractor* (or a *general contractor*) or a *sub-bid* when made by a specialist contracting company that hopes to become a *subcontractor.* Also known in legal terminology as an *offer.*

Bid bond: a bond provided by a surety company on behalf of a *bidder* to guarantee to an *owner* that the bidder will enter into a *construction contract* with the owner in accordance with his *bid* if it is accepted by the owner. The bid bond is forfeited to the owner if the bidder does not enter into a contract when his bid is accepted or if alternatively the bidder does not pay to the owner an amount (by way of damages) equal to the difference between the amount of the defaulting bidder's bid and that of the next largest bid (assuming there is more than one bid), but only up to the face amount of the bid bond, that is, usually not less than 10 percent of the bid amount.

Bidder: one who makes a *bid.*

Bidding authority: someone who solicits *bids* for *construction work* and provides *bidding documents* for that purpose; usually an *owner* (private or public) or an *owner's agent* such as an *architect,* an *engineer,* or a *construction manager.*

Bidding documents: issued by an *owner* or by a *designer* on an owner's behalf to *bidders;* the same in content as the subsequent *contract documents* (assuming that a contract is made) plus certain other documents needed for bidding but not for a construction contract, such as the *instructions to bidders.*

Bid estimate: one made as a basis for a *bid* for *construction work* and therefore usually more detailed than a *preliminary estimate,* but not necessarily so.

Bills of materials: lists or schedules of *materials* required for *construction work* prepared by a *contractor* or his *subcontractors* after a *construction contract* is made but before and as a means of purchasing. Sometimes prepared by a *designer* as part of the *bidding documents* for a construction contract but not usually in building construction contracts nor as a means to *advanced purchasing.* (Not the same as *bills of quantities.*)

Bills of quantities (BOQ): used in *contracts with quantities* as *bidding documents* and as *contract documents* and containing the terms and conditions (commonly by reference to *standard contract documents*), *specifications,* and accurately measured *quantities of work,* (according to a published *standard method of measurement*) which during bidding are priced by the *bidders* in the calculation of their *bids.* BOQs are usually prepared by an *agent of the owner* such as (in engineering projects) an *engineer* or (in building projects) a *quantity surveyor.* BOQs are roughly equivalent to a *project manual* in North America, except for the insertion, in BOQs, of measured *quantities of (the) work* and "cash columns" for their pricing by *bidders;* bills are usually identified by the names of the *construction trades* represented, hence the use of the plural *bills.* BOQs are not unlike the *schedules of quantities* and *schedules of unit prices* used in some engineering contracts in North America for *engineering construction work* but are usually more detailed and specific and based on a national standard method of measurement, particularly when intended for *building construction work.*

Bonus: see *liquidated damages, penalty.*

Builder's work: generally *trade work* done in conjunction with and subsidiary to the work of the work of mechanical, electrical, or other mechanical installations such as *building equipment* that require a certain amount of cutting and patching of other trade work and minor related trade work (such as excavation and concrete) done in conjunction.

Building (construction) work: (as distinct from *engineering construction work*) generally custom-designed work for *shelter construction* or other buildings for use by people, including residential, commercial, institutional, governmental, and industrial buildings; typically involving the work of a number of or many trades.

Building density: the relative proportion of actual *work* within a building's inner space, or volume; with particular reference to partition walls, fixtures, and similar work, the precise nature of which depends on the class of building; an important thing to consider when comparing buildings and their costs.

Building element: see *element (of construction).*

Building equipment: machinery and equipment permanently installed in and as part of a building as distinct from *equipment* and *construction equipment.*

CAD (computer-aided design): design that employs computers for graphic representations and for calculations (for structures, energy conservation, mechanical and electrical systems, for example).

Calculated risk: in estimating, a risk that is recognized and for which costs are calculated and included in the estimate either by simply making an allowance based on a best estimate or by measurement and calculation of costs (of, say, certain temporary work) and then by including in the estimate a proportion of those calculated costs based on the degree of probability of the risk materializing.

Cash allowances: an amount of money specified to be allowed by all *bidders* in their *bids* for specific parts of the *work* of a *project* for which the *owner* or an *agent of the owner* is unable or unwilling at the time of preparing the *bidding documents* to provide sufficient *design information* to enable the bidders to estimate the costs of that specific part or parts; which specified amount is the agent's best estimate of the costs. (The *contract sum* is subsequently adjusted according to the amount of variation between the actual amount instructed (by the agent of the owner) to be expended on the specified parts of the work and the specified amount of the cash allowance.) Also known as a *prime cost sum (P.C. sum)* or a *provisional sum,* especially outside the United States.

Cash discount: a discount from a written price given by a *supplier* to a customer for payment for the *materials* supplied by a stipulated date. (See also *trade discount* and *volume discount*)

Certificates of payment: periodically made by the *agent of the owner* following approval of an *application for payment,* according to the terms and conditions of the *construction contract,* to certify to the *owner* that a specified amount is due for payment to the *contractor.* A normal prerequisite for payment to a *contractor.*

Certified cost engineer: a member of the American Association of Cost Engineers certified by that body for his experience, ability, and knowledge. (See also *cost engineer.*)

Change in (the) work (of a construction contract): contractually either the subject of a *change order* or of a *minor change in the work (of a construction contract),* as defined by the contract, such as the AIA Document A201. Known also (in AIA Document A201) as a *modification* (a broader term) and in some contracts as a *variation.*

Change order: an order issued by an *agent of the owner* according to the terms and conditions of a *construction contract* to *the contractor* to make a specific *change in the work* that may result in a change in the scope of the contract's *work,* the *contract sum,* or the *contract time,* depending on the change order's purpose and substance.

Chartered quantity surveyor: a member of the Royal Institution of Chartered Surveyors (RICS), qualified in *quantity surveying* and entitled to use this title and the designation ARICS (the general appellation) or FRICS (Fellow). (See also *quantity surveyor.*)

Clerk of (the) works: an inspector and monitor of *construction work* employed by an *owner,* usually under the direction and control of an *agent of the owner,* sometimes according to a construction contract's terms and conditions. Generally an obsolete term used primarily outside North America.

CM: initials representing *Construction Management* or *construction manager.*

CM arrangement (of contracts): one in which an *owner* has several separate *construction contracts* with several *specialist (trade) contractors* for the *work* of a *project.*

CM Project: a *project* administered by a *construction manager,* usually with a *CM arrangement of contracts.*

Component: (as distinct from unassembled *material,* such as yard lumber) a portion of *construction work* that is classed as *material* but usually composed (prefabricated) elsewhere and off the site, often in a factory; sometimes, but not necessarily, made according to a special design for a particular *project.* (See also *products.*)

Conceptual estimate: one made from rudimentary *design information* such as a schedule of space requirements, preliminary design sketches, and outline specifications; usually made by a *quantity surveyor* or other *cost consultant*; often a *preliminary estimate*; sometimes a *bid estimate.*

Conceptual estimator: a convenient but improbable term for an estimator who makes a *conceptual estimate.*

Construction: the design and production of *construction work.*

Construction contracts: see *contract with quantities, cost-plus-fee contract, maximum cost-plus-fee contract, stipulated-sum contract,* and *unit-price contract.*

Construction costs: see *costs of construction work.*

Construction equipment: mobile machinery used in performing *construction work,* such as a bulldozer or a crane. (See also *plant, equipment,* and *building equipment.*)

Construction management: in the ordinary and traditional sense, normal management of *construction work,* usually by a *contractor* with *subcontractors*; but in a more recent and specialized sense, *Construction Management (CM)—capitalized* to distinguish it from the other—involves a contractual arrangement in which an *owner* employs a *construction manager* as an agent and has a number of separate *construction contracts* (instead of one) for the several parts of the *work* of a *project,* all arranged, managed, and coordinated by the owner's construction manager, often working with other *agents of the owner.* (See also *development management* and *project management.*)

Construction management contract: one between an *owner* or his *project manager* and a *construction manager* for the management of a *CM project,* usually for an agreed fee.

Construction manager: a person (corporate or individual) appointed by an *owner* (or by an owner's *project manager*) as an *agent of the owner* to work with other *agents of the owner* in designing *construction work,* preparing *bidding documents* and *contract documents,* arranging *construction contracts,* and managing the several *contractors* to ensure that all of the *work* of the *CM project* is completed within the scheduled time and budget, according to the terms and conditions of a *construction management contract.*

Construction materials: see *materials.*

Construction work: see *work.*

Consumable items: those that are used, worn, or consumed by doing *work,* especially fuel and lubricants, parts of *equipment* such as fan-belts and sparkplugs, and parts of *tools* such as saw blades.

Contact-area (of forms): the area of formwork actually in contact with the concrete; commonly used as the primary basis of measurement and pricing of formwork.

Contingency allowance (sum): an amount included in an estimate by the estimator to provide for specific or nonspecific contingencies that may arise during the *work* (e.g., a snowfall requiring clearing or the discovery of rock below ground surface requiring blasting and removal). (See also *cash allowance.*)

Contract amount: see *contract sum.*

Contract documents: *specifications,* drawings, contract agreement, terms and conditions, and other such documents prepared by *agents of an owner,* initially as *bidding documents,* that describe and illustrate the *work* of a *construction contract* and how it shall be performed and paid for, properly containing nothing that was not part of the initial *bidding documents* except by mutual consent of the *owner* and the *contractor,* as listed in the contract's written agreement and as modified by any subsequent *modification* (which is also a contract document).

Contract manager: one who contracts to manage a *project* and to advise and to represent the *owner.* One hired by a *contractor* to do the same. Not a common title in North America.

Contract sum: the total amount of money paid by an *owner* to the *contractor,* usually in monthly installments, for *construction work* done according to the *construction contract* between them. In *stipulated-sum contracts* the contract sum is explicit but subject to modification; in other contracts it may be only implicit and not finally determined until the work is completed; also called in some contracts the *contract amount.*

Contract time: the period stipulated in a *construction contract* for the *substantial completion* of the *work* or the actual time taken for completion if no contract time is stipulated.

Contract with quantities: a *construction contract* in which the *contract documents* (and the *bidding documents*) include *bills of quantities* prepared by a *quantity surveyor* employed as an agent by the *owner.*

Contractor: the party (of the second part) to a *construction contract* who does the *work* or part of it.

Contractor's estimates: one made for purposes of bidding or negotiation for *construction work.* (See also *conceptual estimate, bid estimate, preliminary estimate.*)

Cost accounting: that part of ordinary *construction management* by which the actual *costs of construction work* are segregated and attributed to specific *items of work,* or groups of items, and to specific items of *job overhead costs,* after which the cost *information* is analyzed and the results are combined with other *data* for use in *planning and scheduling,* in *cost control,* and in *estimating* the costs of other *projects.*

Cost-benefit analysis: a technique for appraising investments in projects that attempts to identify and evaluate all their costs and benefits; not only the purely monetary but also the social benefits; as distinct from other analysis, such as *feasibility studies,* that deal only in purely business terms. A cost-benefit analysis tries to measure in monetary terms not only the economic effects of a project, but this is not always easily done. (How does one measure the aesthetic pleasure that may be derived from developing a park, for example?)

Cost code: figures, letters, or words arranged in a systematic code for the representation of *items of work* and *job overhead items*; used for speed, ease, and convenience in estimating, cost accounting, and other construction management functions. The Masterformat provides a numerical code, for example. (See also *basic item.*)

Cost consultant: a general term for a *quantity surveyor, construction economist, cost engineer,* or other professional person who provides expert advice on construction costs and economics.

Cost control: the part of *construction management* that seeks to ensure during both design and production that construction costs incurred in a *project* do not exceed, in an owner's case, the budgeted amount, and in a contractor's case, the estimated costs.

Cost engineer: one with education and experience in construction and construction economics who uses his knowledge and skills in *cost estimates, cost control, con-*

struction management, and other areas of construction. (See also *certified cost engineer* and *quantity surveyor.*)

Cost estimate: see *estimate (of construction costs).*

Cost information and data: see *information* and *data.*

Cost planning: the technique of estimating and accounting for the *costs of (construction) work* during the *design phase (of construction)* and of selecting *materials* and methods and types of *construction contracts* to ensure the completion of the *work* of a *project* within the financial budget and time schedule.

Cost-plus-fee contract: one in which the *owner* agrees to pay the *contractor* (usually each month) all the actual, *direct costs (of construction work)* (the *reimbursable costs of work*), plus a fee—either a lump-sum fee paid in installments pro rata the actual direct costs or a percentage of the actual direct costs——to allow for the *indirect costs* (the *nonreimbursable costs*) *of the work.* Generally this kind of contract is made because an *owner* or *designer* is unable to provide *bidders* with more *design information* about the required *work* and because the owner is therefore unable to obtain at an acceptable price another kind of construction contract (such as *lump-sum contract*) with less risk for himself.

Costs in use: all of the costs incurred by an *owner* as a result of his ownership of a building (or other development) *over and above* the initial *costs of the (construction) work* and the costs of the land, its development, and the *development costs,* including depreciation, maintenance, taxes, insurances, the costs of financing, vacancies (of rental properties), management, and all other operating costs.

Costs of (construction) work: all of the *direct costs* and *indirect costs (of construction) work*; generally classified as *labor costs, material costs, plant and equipment costs, job overhead costs, operating overhead costs,* and *profit.*

Cover bid: one that is not intended as a genuine *bid* but is submitted in order to satisfy appearances.

CPM: see *critical path method.*

Crashing: speeding *activities* in a construction schedule in order to complete the *work* in a shorter time. The objective is to "crash" only those activities necessary for an economically optimum solution, since crashing costs more.

Critical items of work: those that determine the minimal duration of a *project*; those on the *critical path.*

Critical path: the longest irreducible sequence of *work activities* and *events* that determines the minimal duration of a *project.*

Critical path method (CPM): a method of planning and scheduling using a list of *work activities* that are graphically represented in a *critical path network (diagram)* in order to discover the *critical path.*

Critical path network (diagram): a graphic means of determining and demonstrating the *critical path* of *activities* in a *project.* A network consists of arrows representing *work activities* and nodes (circles) representing *events,* or vice versa.

Cube (item): an *item of work* in an estimate whose quantity is expressed as a *volume.* See also, *super, run,* and *number.*

Custom (items, design): especially designed; not standard.

Cut-and-fill line: a line on a site plan joining those imaginary points at which neither *cut* (excavation) nor *fill* (placing of fill material) occurs, because at those points the ground's existing elevations are equal to the desired finished elevations; consequently cut-and-fill lines separate areas of cut from areas of fill.

Data: (as distinct from *information*) information from several sources analyzed and synthesized to give reference levels of productivity and costs and other things as bases and guides for estimating costs, planning *work,* and other construction management activities.

Density (of a building): see *building density.*

Depreciation: the lessening in value of tangible assets due to any cause; usually physical aging, use, wear and tear, and obsolescence.

Design information: that information about a *project* provided to *bidders* and to *contractors* by the *owner* and *designer* and others (agents of the owner); as distinct from *experiential information* together with which the design information ideally comprises all of the information needed to perform the construction contracts of the project.

Design phase (of construction): the earlier phase in which the *work* is designed, usually followed by the *production phase.* The two phases are sequential in traditional construction, but in *CM projects* the two phases may more or less overlap, thus enabling *fast-track construction.*

Designer: a party to a contract to provide professional design services to an *owner* (the other party); usually an *architect* or an *engineer*; sometimes a *construction manager.*Sometimes a person (corporate or individual) who performs the design function as part of a *package deal, turnkey project,* or a *development project.*

Designer's consultant: a party to a contract to provide specialized design services to a *designer* for such as building structures, mechanical and electrical services, acoustic treatment, and such. Also, one who provides cost consulting or management services to a designer.

Designer's estimate: an estimate made for an *owner* by his agent or employee before receiving *bids* or before entering into negotiations for *construction work.* Also called a *conceptual estimate, preliminary estimate,* budget estimate, or approximate estimate.

Developer: a person (corporate or individual) who develops land through construction and who therefore becomes an *owner* to this end; one who seeks a *profit* from development of land, either by selling a development or by holding it to reap a return on the investment.

Development costs: all the costs from land assembly and aquisition to the fees paid to real estate agents for selling a developed property of which the *construction costs* may be a lesser part; often divided into "hard" and "soft" development costs, the former going for tangible things and the latter for intangibles such as advertising and fees.

Development management: professional services wider in scope and at a higher level of responsibility than either *construction management* or *project management* and embracing both of these. The complete management of an investment in real property, including (if necessary) land assembly and aquisition, *feasiblity studies, design, estimates of construction costs, value engineering, cost control,* production of *construction work,* and possibly property management.

Development management project: one in which the *owner* is represented by a *development manager.*

Development manager: a corporation that provides *development managment* services for a fee.

Direct costs (of construction work): generally classified as *labor costs, material costs, plant and equipment costs,* and *job overhead costs,* all of which are directly attributable to a specific *project.* (See also *indirect costs [of construction work].*

Direct labor costs: those paid directly by a *contractor* to a worker. (See also *indirect labor costs.*)

Discount: see *cash discount, trade discount,* and *volume discount.*

Division (of work): one of the 17 standard divisions of *construction work* in the Masterformat. Divisions in *specifications* are divided into nonstandard *sections* of work by the specification writer, according to the nature and extent of the work, to facilitate the production of specifications and *bidding.* The 17 standard divisions are:

Division 0—Contract
Division 1—General Requirements
Division 2—Sitework
Division 3—Concrete
Division 4—Masonry
Division 5—Metals
Division 6—Wood and Plastics
Division 7—Thermal and Moisture Protection
Division 8—Openings (Windows and Doors)
Division 9—Finishes
Division 10—Specialties
Division 11—Equipment
Division 13—Special Construction
Division 14—Conveying Systems
Division 15—Mechanical
Division 16—Electrical.

Drawings and Specifications: an imprecise but popular term for *bidding documents* and *contract documents.*

Duodecimals: a number system based on twelfths used in calculating quantities from dimensions of work given in feet and inches; now generally obsolete but still of value in manual calculations.

Element (of construction): a part of a structure that always performs the same function in any structure; for *estimating* purposes, a structure may consist of any number of elements, depending on design, function, estimating limitations, and expediency. A practical *minimum* number of elements in a building is three, consisting of (1) those *items of work* mostly related to a building's area, (2) those items mostly related to a building's perimeter, and (3) the remainder. For most *elemental estimates* more elements are preferable. All building elements consist of at least one *item of work* and usually more; some consist of dozens of items. There is a standard list of building elements in the Appendix.

Elemental cost analysis: the allocation of *costs of work* to *elements.*

Elemental (cost) estimate: one made by estimating the costs of *elements*; often a *conceptual estimate*; sometimes a *bid estimate.*

Employer: in many English-speaking countries outside North America, an *owner* in a *construction contract.*

Engineer: (in this text) a professional engineer registered to work as such in one of the construction-related disciplines, including soils, structural, civil, mechanical, and electrical engineering.

Engineering (construction) work: (as distinct from *building construction work*) generally work other than that done in buildings for shelter, and including highways, heavy construction (dams and such), and *industrial construction* (refineries, plants, and such).

Entire contract: one in which the entire work of the contract must be performed before any payment for the work is made.

Equipment: mobile machinery used in the performance of *construction work* such as a bulldozer; sometimes called *construction equipment* to distinguish it from machinery and equipment installed in a building. (See also *building equipment, machinery, plant,* and *plant and equipment costs.*)

Equipment costs: see *plant and equipment costs.*

Estimate (of construction costs): at least the calculation of the construction costs of a *project*; more fully, the measurements and resultant *quantities of work,* the itemized *general requirements,* and the estimated *costs of (construction) work.* (See also *bid estimate, conceptual estimate, contractor's estimate, designer's estimate, elemental estimate,* and *preliminary estimate.*)

Estimating: the process of making an estimate consisting of two main parts: (1) *measurement* and (2) *pricing.* It is reckoned that $M + P = E_c$; that the effort (E) required to produce an accurate and effective estimate is more or less a constant (E_c) for any particular class of estimate, and that, therefore, if the effort put into *measurement* is less than it might be, then the effort required for *pricing* must be commensurately greater, and vice versa. The effort required for *pricing* may be made in part before making a particular estimate by analyzing and synthesizing historical cost (information and data) for future estimates.

Estimator: (in this text) generally, one who makes estimates, usually for a contracting company, a CM company, a *cost consultant,* or an *owner*; also a *cost consultant,* a *cost engineer,* or a *quantity surveyor.*

Event: in *CPM* an event is the finishing of one or more *work activities* or the starting of one or more other work activities, usually dependent on the finishing of the former activities; or more usually both the starting and finishing. In one format of CPM, events are indicated by nodes or circles in the diagram to which work activities represented by arrows are diagramatically connected; in another format of CPM the symbolism is reversed.

Existing work: that which already exists on a particular site when a construction contract is made and which has some physical connection or other relationship with the *work* of the construction contract in question. (See also *alteration work* and *new work.*)

Experiential information: that information about a *project* already known by *bidders* and *contractors* from their experience; as distinct from *design information,* which is unique and particular and together with which the experiential information comprises all of the information needed to perform the contracts of the project.

Extra-over (item of work): part of, or a particular feature of, an *item of work* in an estimate, measured and priced separately from the main item to simplify measurement and pricing and to increase effectively the numbers of *basic items* and the amount of *information* related to basic items.

Fair cutting: cutting of masonry and other *trade work* that will remain exposed and is therefore required to be executed neatly in a workmanlike manner.

Fast-track construction: that in which design and production overlap, so as to reduce construction time. (See also *phased construction.*)

Feasibility study: a study of the economic feasibility of a *project,* usually based on a *conceptual estimate, development costs,* and the estimated *costs in use.* (See also, *cost-benefit analysis.*)

Final completion (of work): total completion of work and total performance of the *construction contract,* except possibly for the subsequent making good of any defects in the *work.* (See also *substantial completion [performance] of work.*)

Fixture: a removable item installed and fixed in place in a building in such a way (more or less permanently) that legally it is considered part of the *improvements,* such as a plumbing fixture or a venetian blind.

Fringe benefits: a popular term for *indirect labor costs.*

Gantt bar chart: a diagrammatic chart used in *scheduling,* attributed to Henry L. Gantt and Frederick W. Taylor, in which *activities* are represented by horizontal bars, the lengths and positions of which indicate the activities' relative order, their duration, and the dates of starting and finishing. In such a chart, usually the horizontal axis shows the weeks, days, and dates, while the vertical axis shows the titles of activities. Probably the most used type of schedule diagram in construction.

General bid: one by a *bidder* who seeks to become the *contractor* (or *general contractor*) for a *project*; as distinct from a *sub-bid.*

General condition items: a popular but imprecise term for those items that usually give rise to *job overhead costs* and are often cited by the written general conditions of a construction contract, hence the name (e.g., a *performance bond*). However, many job overhead costs are not mentioned in contract conditions, and some general conditions do not affect construction costs. (See also *general requirements.*)

General contractor: a popular but noncontractual term for a *contractor* who has *subcontractors.* (In the past the term referred to a contractor employing workmen of several different trades who undertook to build with his own forces and financial resources a complete or almost complete building; hence the term is generally obsolete). (See also *primary contractor, prime contractor,* and *specialist [trade] contractor.*)

General requirements: those temporary services and facilities and other items provided by a *contractor* that give rise to *job overhead costs* and that, because of their general nature, are related to a *project* as a whole rather than to *specific items of work.* Properly, specified in construction documents in *Division 1, General Requirements,* of the *Masterformat.* (Some of these items are commonly referred to in the general conditions of contracts, but Masterformat recommends that general conditions be limited to contractual-legal requirements. Confusion and error may be avoided by references in both parts of a contract.)

GSA: General Services Administration of the United States government; *GSA/PBS:* General Services Administration, Public Buildings Service.

Heavy construction: see *building construction work* and *engineering construction work.*

Holdback (money): (also *retainage*) those portions of cash amounts due to *contractors* for completed work held back (retained) by the *owner* as security and not paid until later, usually at *substantial completion (performance)* or at *final completion of work,* according to the terms and conditions of the contract and to any relevant laws (such as a Mechanics' Lien Act).

Idle time: for *construction equipment,* that time spent on a construction site while not in operation.

Improvement: a legal term referring to anything erected on and affixed to or growing on land, such as buildings, road, fences, services, and trees, which legally (but subject to interpretation by local statute) are seen as part of the land.

Independent contractor: an individual who has a contract to provide certain services for payment, as distinct from an employee. (Not only related to construction).

Indirect costs (of construction work): (as distinct from *direct costs,* which can be entirely attributed to specific *projects*) those *costs of (construction) work* generally classified as *operating overhead costs* and *profit* but which cannot all be absolutely attributed to specific projects. Their magnitude is in part determined by the competitive market.

Indirect labor costs: those paid by an employer on an employee's behalf for such things as insurance, Social Security (retirement pensions), and vacations. (See also *direct labor costs.*)

Industrial construction: See *engineering construction.*

Industrial engineering: primarily concerned with the efficiency of industrial processes through time and motion study, work study, planning and scheduling, and work controls.

Information: (as distinct from *data*) knowledge about some specific event or thing; here, about particular *projects* and their *work;* see also *design information* and *experiential information.*

Instructions to bidders: one of the *bidding documents* that provides *bidders* with instructions on how and where to make an acceptable *bid;* one that should not contain anything that would make it necessary to become subsequently part of the *contract documents.*

Interim payments: those made periodically (usually monthly) to a *contractor* by the *owner* during construction according to the terms and conditions of the *construction contract* on the basis of the total proportional value of the work completed to date, less any required *holdback,* less the total of all previous payments. (In this way the calculation of each interim payment is done anew each time and any errors are not carried forward from payment to payment.)

Item of work: a portion of *construction work* that by its nature can be observed, identified, and separated from the rest of the work for purposes of *estimating, cost accounting, construction management,* and similar purposes. *Direct costs* can be allocated to specific items of work. (For example, a masonry wall is composed of one item of work only; the masonry units and the mortar joints are inseparable and constitute one item.) Items of work usually involve only the work of a single trade. (See also *basic item* and *particular item.*)

Job: a popular term for a *project* or for part of one.

Job overhead costs: those *direct costs (of construction work)* that because of their general nature cannot be allocated to specific *items of work,* only to a specific *project;* as distinct from *operating overhead costs.* (For example, the costs of temporary facilities on a site, or of a building permit for a project.)

Labor: (as distinct from *supervision*) workers the costs of whose work can be allocated to specific *items of work* as a *direct cost;* also their work activities.

Labor and materials payment bond: see *payment bond.*

Labor costs: that part of the total *costs (of construction work)* expended on labor and dependent on the *labor rates* paid and on productivity, including both *direct* and *indirect labor costs.* The *costs of work* other than *material costs, plant and equipment costs, overhead costs,* and *profit.*

Labor item (of work): an *item of work* the *direct costs* of which include mainly *labor costs* and only minor *material costs* if any. (See also *labors.*)

Labor only item: see *labor item.*

Labor rate: the total costs for labor paid by an employee including all *direct* and *indirect labor costs* for a specific period of time (and in a specific place) divided by the total number of hours worked during that period.

Labors: minor *items of work* consisting almost entirely of *labor* done for or to or in conjunction with major items (e.g., vertical cutting of brick masonry work where it abuts up to a concrete wall). (See also *labor item.*)

Laps (in construction work): additional *material* required by and incorporated into *work* because of particular dimensions of the work and of the material used and the need for joining by overlapping, tongue-and-groove, shiplap, or other such feature.[2]

Lien: a legal charge against real property for *materials, work,* or services supplied for that property.[3]

Liquidated damages: (as distinct from damages awarded by a court of law, or from a *penalty* for late completion) settled and agreed to (liquidated) damages included in a construction contract's agreement and payable in the event of late completion of the *work* (*MCC*).

Long-lead procurement: the same as *advanced purchasing.*

Lowest bidder: the *bidder* who has submitted the lowest valid *bid.*

Lump-sum: a fixed and stipulated sum; as in *lump-sum contract* and *lump-sum fee.*

Lump-sum contract: a popular term for a *stipulated-sum contract.*

Lump-sum fee: as distinct from a percentage fee, usually in a *cost-plus-fee contract.*

Lump-sum items: those for which the costs are expressed as a total amount without detailed calculation by way of measured quantities and unit prices.

Machinery: one class of *material* incorporated into *construction work.* (See also *equipment.*)

Market value: that which would be equalled by a selling price arrived at by a seller and a buyer acting prudently and at arm's length in an open market.

Markup: a popular term for the inclusion in an estimate of the *operating overhead costs* and the *profit,* often (either singly or jointly) as a percentage of the *direct costs* to arrive at the total estimated costs as a basis for a *bid* or for negotiations leading to a *construction contract.*

Masterformat: the title in North America of a published system of identification, specification, data filing, and cost accounting based on 17 standard *divisions* made up of non-standard *section* of *construction work.*

Material costs: those expended on *materials,* including the costs of taxes, delivery, handling, storage, and the costs of *laps* and *waste.* The *costs of work* other than the *labor costs, plant and equipment costs, overhead costs,* and *profit.*

Materialman: an old, sexist term for a *supplier.*

Materials: every tangible thing required to be installed and permanently incorporated in *construction*

[2]See Collier's *Fundamentals of Construction Estimating and Cost Accounting* for a fuller explanation of *laps* and other such estimating details. Other instances are noted in this Glossary by the initials *FCE.*

[3]See Collier's *Managing Construction Contracts* for a fuller explanation. Other instances are noted in this Glossary by the initials *MCC.*

work according to a *construction contract,* including *products, components, building equipment,* and *machinery.*

Materials supplier: see *supplier.*

Maximum cost-plus-fee contract: one fundamentally the same as a *cost-plus-fee contract* except that in the former the *contractor* agrees to complete the *work* and perform the contract at a total cost not greater than the stipulated *maximum cost* in the contract agreement. Often this kind of contract contains a "sharing clause" whereby the contractor shares in any "savings" made by completing the work at a total cost less than the stipulated maximum cost. (Similarly sometimes the contractor bears a share of any "loss," according to the contract.)

Methods engineering: primarily concerned with the study of industrial and *industrial construction* methods.

Minor change in (the) work (of a construction contract): contractually (in AIA standard contracts) the subject of an order by an *architect* with authority under the construction contract to order minor changes in the work that do not involve an adjustment to the *contract sum* or the *contract time* and that are not inconsistent with the intent of the contract. (See also *change in [the] work [of a construction contract]* and *change order.*)

Modification: defined in the AIA Document A201 as a written amendment to a *construction contract* signed by both parties (*owner, contractor*), a *change order,* a written interpretation of a construction contract issued by the *architect,* or a written order for a *minor change.*

Net, in place: a term used in reference to the recommended method of measuring *construction work* as set down in (outside the United States) nationally published-*standard methods of measurement (for construction work),* namely, to measure work by the dimensions indicated or, if none are indicated (such as for excavations), according to specific dimensional allowances as laid down, and to make all necessary allowances for *shrinkage and swell* and *waste,* as required, in an item's *unit price* and *not* its measured *quantities of work.*

New work: *construction work* as defined, as distinct from *existing work.* (These terms are significant in renovations.)

Nominated subcontractor/supplier: one selected and named (nominated) by the *owner* or *agent of the owner* in specifications or subsequently in the designer's instructions to the *contractor,* with whom the contractor is reasonably required to enter into a *subcontract/supply contract,* provided the contractor raises no reasonable objection. A nominated subcontractor/supplier is paid for his work and materials from *prime cost sums* or *provisional sums (cash allowances)* stipulated in the specifications. The purpose of such nomination is to enable an owner and his agents to delay design decisions about work covered by such sums (allowances), and which are to be performed or supplied by nominated subcontractors or suppliers, and to give them greater control over a project's work than otherwise in a *stipulated-sum contract.* This practice is common in English-speaking countries outside North America

Nonreimbursable costs (of work): those so defined in a *cost-plus-fee contract* and covered by the contract's fee; usually the *indirect costs (of construction work)* as defined. (See also *reimbursable costs [of work].*)

Notice of change: a written notice without real contractual significance issued to a *contractor* by an *owner* (usually through his agent) giving notice only of an intention to order a *change in the work (of a construction contract);* intended to initiate the necessary negotiations for the change without making a commitment.

Number (item): one for which the quantity is expressed as a number (i.e., by enumeration). (See also *cube, super,* and *run.*)

Offer: see *bid.*

Office overhead costs: a popular but imprecise term synonymous with *operating overhead costs* that refers to the costs incurred at a contracting company's permanent office (as distinct from those incurred at a temporary site office); sometimes called head-office overhead costs.

Operating costs of equipment: see *owning and operating costs.*

Operating overhead costs: those costs of operating a construction contracting company that, because of their general nature or because they are indistinct or unrelated, cannot be accurately allocated to specific *projects* as they are incurred and instead are charged to projects usually as a percentage of the *direct costs* by a *markup;* with *profit,* the *indirect costs (of construction work).*

Other contractor: a contractual term to describe one who is not a party to the contract in question but who also has a *construction contract* with the same *owner* for *work* at the same site, therefore creating a need for recognition and cooperation among the owner's several contractors at that site.

Overhead costs: a general term that includes both *job overhead costs* and *operating overhead costs,* better replaced by a more specific term.

Overhead and profit: a general and imprecise term. See *overhead costs* and *profit.*

Owner: the first party to a *construction contract* who pays the *contractor* (the second party) for the *construction work;* also the one who owns rights to land on which the work of a contract is done and who therefore owns the work; the client of a *designer, construction manager, project manager,* or *development manager.*

Owner-designer: a convenient term when referring to an *owner* in a *construction contract* who is contractually represented by a *designer* who at times may act unilaterally on the owner's behalf.

Owner-developer: an *owner* who develops land in order to make a profit from the development.

Owner's agent: see *agent of the owner.*

Owning and operating costs: total *plant and equipment costs. Owning costs* that are incurred simply by ownership of plant or equipment and primarily consist of the costs of depreciation, maintenance, and investment, and *operating costs* that are incurred by actually using and operating plant or equipment (over and above the owning costs) and consisting primarily of the costs of the operator, fuel, lubricants, consumable items, repairs, mobilization, and demobilization.

Package deal: also a "design and construct contract;" one that includes necessary design services for and in addition to the *construction work* of a *project*; sometimes a negotiated contract; sometimes a *maximum cost-plus-fee contract*; sometimes including land aquisition. Some try to make a distinction between a *package deal* and *turnkey project* (sometimes called a *design and manage contract*), but there are many variations and the terminology is loose and unprecise.

Particular item (of work): (as distinct from a *basic item*) one that requires an original and unique description in *specifications* and a *cost code* because it is not a *basic item* and not part of standard documentation (e.g., standard specifications).

Payment bond: one by which a surety company guarantees that the *contractor* named in the bond will properly pay all legal debts arising from the *construction work.*

P.C. Sum: see *prime cost sum.*

Penalty (for late completion): a monetary penalty contractually payable by a *contractor* to an *owner* for late completion of *construction work*, often not upheld by the courts *(MCC).*

Performance bond: one by which a surety company guarantees on behalf of the *contractor* to the *owner* the proper performance of the *construction contract (MCC).*

Performance specification: one in which the performance of *work* is specified, rather than the specific *materials* and methods for performing the work; as distinct from a *prescriptive specification.* (Many items are specified by a mixture of both kinds of specification.)

PERT: see *program evaluation* and *review technique.*

Phased construction: that in which the design and the production of *work* more or less overlap, thus shortening the time of a *project*; this is achieved by creating more than one *construction contract* for a project; commonly practiced in *CM projects*; sometimes popularly called *fast-track construction.*

Physical depreciation: that caused by aging, usage, and wear and tear. (See also *depreciation.*)

Planning and scheduling: *planning* includes the listing of all *activities* in a logical order; *scheduling* includes calculating the activities' durations and the duration of the entire *project* and putting dates to their starting times and finishing times. (See also *critical path method.*)

Plan room: a location (often in the offices of a contractors' organization) at which *drawings and specifications* as *bidding documents* are displayed for the use of *bidders.*

Plant: (as distinct from mobile *construction equipment*) static machinery more or less planted and fixed in place (perhaps only for the duration of a project), such as a concrete batching plant.

Plant and equipment costs: all of the *owning and operating costs* of plant and equipment; the *costs of (construction) work* other than *labor costs, material costs, job overhead costs, operating overhead costs* and *profit*, and excepting the costs of *small tools*, which are similar in nature to plant and equipment costs but usually of less significance.

Points: one point is one percent of a mortgage amount charged as a fee or premium by a mortgagee to a mortgagor.

Precontractor: (as distinct from a *bidder* or a *contractor*) a useful term for one who is actively negotiating with an *owner* or his agent for *work* or one who has submitted a *bid* that is under consideration but has not yet been accepted and who may be negotiating with the owner over details in the bid or over possible changes in the work.

Preliminary estimate: one made during a project's *design phase.* (See also *conceptual estimate.*)

Prescriptive specification: (as distinct from a *performance specification*) one in which the materials and methods are prescribed. (Many items are specified by a mixture of both kinds of specification.)

Primary bid: one submitted by a *primary bidder* for a *primary contract.*

Primary bidder: one who seeks to become a *primary contractor.*

Primary contract: one between an *owner* and a *primary contractor.* Also a *prime contract.*

Primary contractor: a less common but more precise term for a *general contractor.* Also *prime contractor.*

Prime contract: see *primary contract.*

Prime contractor: see *primary contractor.*

Prime cost: the original costs of *materials* supplied, or *work* done, without a *markup* for the *primary contractor.* (See also *prime cost sum.*)

Prime cost sum: a term used outside North America for a *cash allowance.* (See also *provisional sum.*)

Primordial agreement: the tacit or partly formed agreement that usually exists at the outset and during the early stages of negotiations between an *owner* and a *precontractor*, or their agents or representatives, for a *construction contract*; probably expressed, at least in part, by a large part of the current national *standard forms* of *construction contract* with which the negotiators are familiar and containing such things as terms and conditions dealing with methods of payment for work, the provision of bonds and insurances and the supervision of work; those terms and conditions established by custom and usage that are already generally understood and accepted.

Procurement: in construction the procuring and purchasing of *materials* and other needs.

Production phase (of construction): in traditional construction the latter phase following the *design phase*; in *CM projects* the two phases more or less overlap, as in *phased construction.*

Production rates: unit rates of production; the total amount of *work* done in a given period of time divided by the number of units of time (hours, days); (e.g., placing concrete in continuous footings at ground level at a rate of 0.5 man-hours per cubic yard.)

Products: (as distinct from basic *materials* such as yard lumber and bags of cement) construction materials produced and finished away from a construction site that are less complex than *components* and *machinery.*

Profit: the excess of income over expenditure for *materials, labor, small tools, plant and equipment,* and *overhead costs*; for an *owner*, a *cost of the construction work*; for a *contractor*, a reason for doing work and taking risks.

Profitability: the net return on a business investment often measured by a percentage obtained by dividing *profit* (times 100) by the business's tangible net worth.

Program evaluation and review technique (PERT): a method of planning and scheduling when little historical information is available on which to base the estimation of activity and project durations; not widely used in ordinary construction projects. (See also *critical path method.*)

Project: an undertaking of which the *work* of a *construction contract* may be the whole or a part. In the latter case the several parts of a project may be performed either by *other contractors* (in the traditional mode) or by *specialist (trade) contractors* (as in a *CM project*).

Project management: in an ordinary, looser sense it is practically synonymous with *Construction Management.* In a more precise sense it refers to an arrangement in which an *owner* employs a *project manager* (either as an agent, or as an *independent contractor*, for a fee, or as an employee) as his representative. It involves services at a higher level of responsibility than Construction Management and often includes the appointment and supervision of a *construction manager* and a *designer*, but it is not as comprehensive as *development management.* (See also *project manager.*)

Project manager: a person (corporate or individual) appointed by an *owner* to act as his representative in a *project* and to do more or less those things that the owner would otherwise have to do, including (possibly) hiring the *designer, construction manager, quantity surveyor,* and other agents and employees, making payments for services and for work, and making *changes in the work* and decisions about other related things.

Project manual: the written parts of *bidding documents* that become, for the most part, *contract documents,* as distinct from the drawings; often contained in more than one volume, and often inaccurately referred to as specifications, which are only one part of a project manual.

Proposal: see *bid.*

Provisional sum: similar to a *prime cost sum (cash allowance)* but included in a contract for work that may or may not be needed (i.e., provisional work) such as work in a substructure, or work required by a contingency, hence also *contingency sum.*

Quantities of items of work: measured quantities of *construction work*, not simply of *materials*, in *estimates* and in *bills of quantities.*

Quantity surveying: translating construction drawings into *bills of quantities*, including specifying the *work* and the *general requirements*, measuring and calculating *quantities of work* and their costs, and making cost estimates. (See also *quantity surveyor* and *taking off quantities.*)

Quantity surveyor: one who surveys drawings and measures *quantities of work* and generally does *quantity surveying*; one who prepares *bills of quantities* and other *bidding documents* and *contract documents*, arranges for *bids* and for their review, advises on the selection of *contractors*, administers the financial aspects of construction contracts and projects, negotiates and agrees with contractors the value of *changes in (the) work*, checks *applications for payment*, settles construction accounts, and generally acts as an agent (or an employee) of an *owner* or of a *contractor.* Most professional quantity surveyors also undertake *feasibility studies, conceptual estimates,* make surveys, prepare technical reports, and advise on construction contracts and construction economics. They generally perform the functions of *cost engineers* and *construction managers* (and to some limited extent those of *designers* and appraisers) in all fields of construction but especially in *building construction.* (See also *chartered quantity surveyor* and *cost engineer.*)

Raking cutting (and waste): cutting *material* during installation at an angle (or rake) to the lines of the building or to the lines of the joints in the work, thus creating additional costs of labor and material (through *waste*).

Reflected dimensions and quantities: those of *work* taken from those of other work already measured (e.g., the measured quantities of interior plaster or of gypsum drywall used for painting work).

Reimbursable costs of work: those so defined in a *cost-plus-fee contract*; generally the *direct costs (of construction work).* (See also *nonreimbursable costs of work.*)

Replacement costs: the costs of replacing a building (or part of one) with another building (or part) equal in quality and function to the original but not necessarily identical: the usual basis for assessing the value of a building by the cost approach. (See also *reproduction costs.*)

Replacement value: value of *improvements* based on *replacement costs.*

Reproduction costs: the costs of exactly reproducing an existing building, or a part. (See also *replacement costs.*)

Retainage: see *holdback.*

R.I.C.S: the Royal Institution of Chartered Surveyors.

Rough cutting: see *fair cutting,* its opposite.

Run (item): one whose quantity is expressed as a linear measurement. (See also *cube, super,* and *number.*)

Schedule of quantities: similar to a *bill of quantities,* but whereas the latter is usually only for the work of one trade, and bills of quantities for all trades are part of the *bidding documents* and *contract documents* for a *contract with quantities,* a schedule of quantities often refers to a list of *items of work,* the unit prices for which are submitted with a *bid* for a contract, often a *unit-price contract.* Schedules of quantities are more common in North America than bills of quantities, and also they are found mostly in contracts for *engineering construction work.*

Schedule of unit prices: similar to a *schedule of quantities* but often without quantities, as in a *lump-sum contract* in which the *unit prices* are intended to be used for the valuation of *changes in the work.*

Schedule of values (SOV): a breakdown (analysis) of a *contract sum,* usually required of a *contractor* in a *stipulated-sum contract* before submission of the first *application for payment* for checking and approval by an *agent of the owner,* after which the SOV is the basis for all future applications for payment in that contract. A SOV usually shows the various *sections of work*; the names of the *contractor, contractors,* or *subcontractors* responsible; and the value (total costs to the *owner*) of each *section* or part section, the total of which equals the *contract sum. Cash allowances* are usually shown separately.

Scheduling: see *planning and scheduling.*

Section (of a specification): a distinct part of a *division* with its own title and references that are usually numerical-alphabetical and consist of a standard division number (from 0 to 16) followed by a nonstandard alphabetic letter that varies according to a project's scope as determined by a specification writer; usually the work in a section of a specification should be:

1. Recognizable as a distinct entity and part of a project's work and consisting of one or more related *items of work.*
2. Done by only one *trade contractor.*
3. The subject of not more than one trade's *sub-bid* so that the scope of all sub-bids contain one or more whole and identifiable sections of work so that no sub-bid need contain only part of the work specified in one section.

The extent to which these requirements are met varies because specification writers cannot determine exactly how a project's work shall be divided among a contractor's subcontractors (unless they are *nominated subcontractors*); therefore specification writers should create as many valid sections of work as possible in order to facilitate bidding.

Services: water, gas, electrical, drainage, sewerage, and other lines carrying supplies to and wastes from buildings and other structures, and also installed within structures; usually classified as either private (within a site's boundaries) or public services.

Shelter construction: generally *construction work* in buildings to house people; the construction of dwellings of all kinds.

Shop drawings: those made for production purposes by other than a *designer.*

Shrinkage and swell: the increase (*swell*) and the decrease (*shrinkage*) in the volume of excavated and imported fill materials that occur when excavation and filling are done. Soils increase in bulk when excavated; fill materials decrease in bulk when placed and consolidated. Proportions of shrinkage and swell vary greatly and depend on the physical nature and moisture content of the material and the manner in which material is handled and treated. Indicative proportions intended as illustrative guidelines only are:

Sand and gravel	Swell 5–20%	Shrinkage 10–15%
Loamy soil	Swell 15–25%	Shrinkage 15–20%
Ordinary soil	Swell 20–30%	Shrinkage 20–25%
Heavy clay	Swell 25–40%	Shrinkage 25–30%
Solid rock	Swell 50–75%	Shrinkage 0%

One cubic yard of ordinary soil in *bank measure* becomes, say, 1.25 cubic yards when excavated and 1.05 cubic yards when backfilled and compacted. (Such figures are only illustrative; actual values should be established by experience for specific local materials with differing moisture contents in different locations.)

Small tools: generally hand tools and small items of *equipment* (using the term loosely) the costs of which are individually and relatively small and which therefore are normally accounted for and estimated by approximate methods, such as a percentage of *labor costs* established by experience; as distinct from *plant* and *construction equipment,* the individual costs of which are relatively large; although the difference is solely one of degree and the principles involved are the same in both cases. (See also *small tools' costs.*)

Small tools' costs: see *small tools;* the costs vary greatly among projects and for different trades and different costing methods are needed accordingly; for plumbing work, the costs may be relatively high (depending in part on a contracting company's obligations in management-labor agreements for the supply of tools), while for another trade the costs of small tools may be practically nonexistent. The answers are found by *cost accounting.*

Source building: for estimating purposes, that from which information is obtained. (See also *subject building.*)

Specialist contracting company: see *specialist (trade) contractor.*

Specialist (trade) contract: one between an *owner* and a *specialist (trade) contractor;* a *primary contract* in a *CM project.*

Specialist (trade) contractor: a contracting company with a contract to do *trade work* in a *project;* usually the work of only one or two trades, hence the term *trade* (or sometimes *subtrade*) in the title; often referred to as a *subcontractor* whether there is in fact a *subcontract.* Only a subcontractor has a subcontract, and a company doing trade work directly for an *owner* is a *contractor,* but not a *primary contractor* or a *general contractor.*

Specifications: the major part of a *project manual* apart from the *bidding documents,* contract agreement, and the conditions of the contract usually found therein; the written descriptions of *work* that complement the construction drawings. (See also *performance specification* and *prescriptive specification.*)

Standard form (of construction contract): one published or endorsed by one or more professional bodies, such as of *architects, engineers, contractors,* or an association or board made up of representatives of such bodies; usually consisting of a contract agreement and general conditions of the contract. (See the Bibliography.)

Standard method of measurement (for construction work): a document published by an institute of quantity surveyors (or other professional construction body), with or (in some cases) without the cooperation of contracting organizations, or published by a committee of representatives from such bodies, which sets out agreed methods of measurement for construction work, trade by trade, for purposes of standardization and to facilitate and increase the proper understanding and use of *bills of quan-*

tities, estimates (of construction costs), schedules of quantities, and *schedules of unit prices* based upon the standard methods *(FCE)*.

Stipulated sum: that stipulated in a *bid* for a stipulated-sum contract for which the *bidder* offers to do the *work* and perform the contract, which stipulated sum becomes the *contract sum* if the offer is accepted and a contract is made.

Stipulated-sum contract: see *stipulated sum.*

Sub-bid: an offer to a *contractor* or to a *precontractor* to do *trade work,* the acceptance of which constitutes a *subcontract* between the *specialist (trade) contractor* (now a *subcontractor*) who made the sub-bid and the contractor who accepted it; a contract subsidary to a *primary contract* between an *owner* and a contractor. (See also *bid.*)

Sub-bidder: one who submits a *sub-bid.*

Subcontract: see *sub-bid* and *subcontractor.*

Subcontractor: one who does *trade work* in a contract subsidary to a *primary contract;* one defined as such by a primary contract, as distinct from a *supplier.*

Subject building: the one under study; the one the costs of which are to be estimated.

Substantial completion (performance) of work: completion (performance) such that the *work* is substantially ready and usable for the purpose for which it was constructed, even though some minor items may yet remain to be completed that do not significantly detract from its readiness for use; something less than total completion or *final completion;* completion so certified by the *designer* (or other with contractual authority) as substantial completion according to the terms and conditions of the *construction contract.* Substantial completion is usually more contractually significant than final completion since certain major contractual and other legal matters depend on it.

Sub-sub-bid: an *offer* to a *subcontractor* (or to one expecting to become a subcontractor) to do *trade work,* the acceptance of which constitutes a *sub-subcontract* between the *trade contractor* who made the offer and the one who accepted it; a *trade contract* subsidary to a *subcontract.* (See also *sub-bid.*)

Sub-subcontract: see *sub-sub-bid* and *sub-subcontractor.*

Sub-subcontractor: one who does *trade work* in a contract subsidary to a *subcontract;* one defined in a subcontract as a *subcontractor* to the contracting party of a subcontract.

Subtrade: a construction trade the work of which is usually performed by a *specialist (trade) contractor* and so-called because the work is traditionally performed by a *subcontractor;* as an apparent contraction of subtrade contractor, often loosely used in referring to a specialist (trade) contractor. In *CM projects* this work of subtrades is mostly performed by *contractors.* (See also *trade contractor* and *trade work.*)

Subtrade contractor: (as distinct from a *primary contractor*) loosely used in refering to a *specialist (trade) contractor,* because such persons often are *subcontractors.*

Subtrade work: see *subtrade, subtrade contractor,* and *trade work.*

Super: a popular contraction of *superintendent.*

Super (item): one whose quantity is expressed as a *superficial* measurement (of area). (See also *cube, run,* and *number.*)

Superintendent: one appointed to represent a *contractor* on a site, usually according to the requirements of a *construction contract,* who receives instructions (from an *agent of the owner*) and generally manages the *work* for the contractor; one in charge of *labor,* including foremen, on a site *(MCC)*. See also *supervisor.*

Supervision: a *contractor's* supervisory and managerial staff on a site, as distinct from *labor;* a distinction made primarily for purposes of *estimating* and *cost accounting.*

Supervisor: one appointed to represent a *construction manager* on a site (as distinct from a *superintendent,* which many supervisors originally were) *(MCC).*

Supplier: one who supplies *materials* for *construction work* and who is not a *subcontractor* as defined in the *primary contract.* (See also *subcontractor* and *supply contract.*)

Supply bond: one given to guarantee proper and timely delivery of ordered *materials.*

Supply contract: one between a *supplier* and an *owner, contractor, subcontractor,* or *sub-subcontractor,* for the supply of *materials.* (The nature of the materials supplied is often the contractual point on which turns the distinction between a supplier and a subcontractor.)

Swell: see *shrinkage and swell.*

Takeoff (of quantities): the result of *taking off (quantities).*

Taking off (quantities): measuring *construction work* from drawings. (See also *quantity surveying.*)

Target figure (contract): practically the same as a *maximum cost-plus-fee contract;* one in which a target figure, or maximum amount, is stipulated, beyond which the *owner* is not obliged to pay, unless the contract provides otherwise; for example, the owner may be obliged to pay only a specific proportion of the costs over and above the target figure. There are many possible variations.[4]

Tender: see *bid.*

Terminal units: fixtures serviced by a mechanical, electrical system, or both, including plumbing fixtures, air registers, diffusers, and light fixtures; the units that terminate a branch of a service system.

Third party: one who is not a party to the contract in question.

Tools, plant, and equipment costs: see *plant and equipment costs.*

Total completion (of a contract): see *final completion* and *substantial completion.*

[4]For more information see Collier's *Construction Contracts.*

Trade contract: one for the *work* of one trade made with a *specialist (trade) contractor;* either a *contract* or a *subcontract.*

Trade contracting company: one that does *trade work.*

Trade contractor: see *specialist (trade) contractor.*

Trade discount: one allowed by a *supplier* to a customer who is a contracting company; the amount of the discount is deducted from a list price or from an already discounted price (e.g., list less 30 percent, less 5 percent). (See also *cash discount and volume discount.*)

Trade quantities: quantities of *trade work.*

Trade work: that done by one particular trade according to local custom, trade union agreement, or both; the *work* of a *trade contract.*

Turnkey project; (also, design and management project) one in which the contract between *owner* and *contractor* makes provisions for such things as land aquisition, design of the work, and production, so that an owner can in fact simply pay, turn the key, and walk into a completed building. Similar to a *package deal* and a *development project.*

Uniform Construction Index (UCI): the original title form and publication of *Masterformat.*

Unit cost: the unit cost of an *item of work* is an average cost per unit calculated by dividing total costs of the item by the measured (net) quantity (the number of units). As applicable and needed, unit costs may include *material costs, labor costs, plant and equipment costs, job overhead costs, operating overhead costs,* and *profit.* Therefore, the content of unit costs must, to avoid error and misunderstanding, always be made clear. An item may have separate unit costs to include different costs (e.g., in estimates it is usual to have at least separate unit costs for labor); in *contracts with quantities* unit costs may include all *direct costs* and some or all *indirect costs.* All costs vary, therefore unit costs for similar items in different projects will be different. Nevertheless, unit costs are the only means whereby costs are conveniently compared, and they are the primary source of *unit prices.*

Unit price: similar to a *unit cost* but usually consisting of all *direct costs* and some or all *indirect costs,* as in a *bill of quantities* or a *schedule of unit prices;* usually based on historical unit costs that are based on actual costs.

Unit-price contract: one in which *unit prices* for the *items of work* are stipulated by the *bidders* and related to the (usually) approximate quantities of the items listed. The actual quantities of work done are later measured and priced at the unit-price contract's unit prices as the basis for payment. (See also *unit cost* and *unit price.*)

Unit rate: see *production rates.*

Use and waste: usage (of plant, equipment, forms) and the resultant wear and tear, or *depreciation* that gives rise to costs.

Value engineering: the comparison and economic evaluation of alternative construction methods to produce a required result; an aspect of construction economics.

Variation (to a construction contract): see *change order* and *modification.*

Void: in measuring, a deduction made for an opening or for a minor area (e.g., a stairwell) in a major area. (see also *want.*)

Volume discount: a discount, or rebate, allowed by a *supplier* to a regular customer because of a certain minimum volume of business transacted between them over a certain period. (See also *cash discount* and *trade discount.*)

Want: in measurement, a deduction made for an overmeasurement deliberately made to simplify measurement of an irregularly shaped area. (See also *void.*)

Warranty period: the specified period (usually one year) immediately following *substantial completion* during which a *contractor* undertakes to correct *work* found not in accordance with the *construction contract.*

Waste: construction *material* that is extra to the actual net quantity required by the *work,* as indicated by a *contract,* but that is nevertheless required by or used in performing the work or is somehow lost as a result of doing the work and therefore contributes to the *material costs.* (Some waste is usually unavoidable, as in cutting a sheet. Other waste may be caused by careless design and production practices that overlook dimensions of products and the work. Waste is subject to individual efficiency and other contingencies and is therefore essentially variable. It is usually more easily ascertained by *cost accounting* if it is first estimated and stated in an item's description and thereby kept separate from the net quantity of work and by allowing for waste in the unit price.) (See also *laps* and *shrinkage and swell.*)

Weasel clause: a pejorative term for those clauses in documents that unreasonably pass on to a *contractor* risks that could be better dealt with by more equitable contractual means; a clause that is an attempt to weasel out of what otherwise would be a responsibility.

Work: the substance of a *construction contract;* consisting of *labor, materials,* and the use of *small tools, plant, equipment,* and all other services and things required of a *contractor* by a *construction contract* for which the *owner* pays. (See also *alteration work, costs of (construction) work, existing work, item of work,* and *new work.*)

Work activities: a term used in *planning and scheduling* to indicate the discrete parts into which total work is divided for this purpose. In some projects work activities may be relatively large and may even consist of all the work of one trade; this is common in *Gantt bar charts* for smaller projects. In large engineering projects, in which there may be large quantities of a relatively small number of *items of work,* each item of work may be divided for scheduling purposes into a number of separate work activities (e.g., laying a pipeline may be divided into one-mile lengths). In other projects some work activities may be fractions of items while others are groups of items; it is a matter of expediency.

Work study: the systematic analysis of *work* with the intention of achieving greater efficiency from *labor* and *supervision;* it involves detailed examination of work in progress and its analysis *(MCC).*

Wrap-up insurance: a method of insuring *construction work* of a *project* by combining all required insurances for all purposes in one insurance policy, usually arranged and paid for by the *owner*. The objectives are, to ensure that proper and adequate insurances exist for the work and all involved in it, and to reduce the costs of insurance partly by eliminating overlapping policies and by reducing the costs of writing and administering the insurance. Wrapup insurance is complex and requires the services of an insurance consultant-broker.

Appendix

1. SUBSTRUCTURE

(a) Normal Foundations

Includes: Excavation, concrete or masonry for wall and column footings, column caps, grade beams, base plates, anchor bolts, foundation walls to the top of the lowest floor construction and weeping tile..
Excludes: Lowest floor construction, excavation for basements, basement walls and waterproofing, caissons and piles.
Measure the gross area in square metres to the outside face of perimeter walls or to the outer extent of foundations if there are column or similar foundations isolated beyond the perimeter walls. This area will usually be the gross area of the lowest floor and is sometimes known as the footprint of the building.

(b) Basement

Includes: The mass excavation and backfill required to construct a basement.
Excludes: Basement walls and waterproofing, and any work included in Element 1(a) Normal Foundations.
Measure the cube of the basement in cubic metres to the outside face of the perimeter walls and from the underside of the lowest floor construction to grade level.

(c) Special Foundations

Includes: Caissons, piling, extra cost of excavating in rock, special shoring, unusual dewatering and other special foundation conditions.
Measure rock excavation in cubic metres; shoring in square metres; and caissons and piling in metres keeping different types and sizes separate. Unusual dewatering is included as a lump sum allowance.

2. STRUCTURE

(a) Lowest Floor Construction

Includes: The floor slab, whether a slab on grade or a suspended slab over a crawl space (in the latter case it will include the columns and beams supporting the slab), fill under the slab, waterproofing and skim coat or vapour barrier to the slab, small sump pits, construction and expansion joints.
Excludes: Final finish to the slab and machinery or equipment bases.
Measure the gross area of the lowest floor slab in square metres to the outside face of the perimeter walls. (Note: This area is frequently the same as that for Element 1(a) Normal Foundations.)

(b) Upper Floor Construction

Includes: Columns, beams, slabs, floor joists, sub-floors and fire-proofing for all suspended floors.
Excludes: Floor and ceiling finishes; and columns, beams, slabs, etc. for roof construction.
Measure the gross area of suspended floors in square metres to the outside face of the perimeter walls, keeping each type of construction separate. Make no deduction for openings. (Note: The gross area of Upper Floor Construction plus the gross area of Element 2(a) Lowest Floor Construction should equal the total gross floor area of the building.)

(c) Roof Construction

Includes: Columns, beams, slabs, rafters, purlins, trusses, roof boarding, etc. for roof construction.
Excludes: Roof finish, insulation, cant strips, flashings, roof lights, roof drains, eavestroughs and rainwater leaders.
Measure the gross area of the roof in square metres to the outside face of the perimeter walls, or if roof projects beyond the walls, to the

farthest extent of the roof, keeping each type of construction separate. The area of sloping roofs are to be measured on the slope.

3. EXTERIOR CLADDING

(a) Roof Finish

Includes: Roof finish, insulation, cant strips, flashings, fascias, eaves soffits, barge boards, parapet walls above the roof level, copings and roof lights.
Measure the gross area of the roof finish as for Element 2(c) Roof Construction.

(b) Walls Below Ground Floor

Includes: Perimeter basement walls of reinforced concrete, brick, block, etc., including waterproofing, insulation and integral pilasters.
Excludes: All finishes (plaster, paint, panelling, etc.) to the interior face of the wall.
Measure the area of the walls in square metres from the top of the lowest floor construction to ground level. Keep different types of wall construction separate.

(c) Walls Above Ground Floor

Includes: Exterior walls of wood, stone, precast concrete, marble, face brick, curtain wall, etc.; brick, block, tile or concrete backup; insulation, vapour barriers, damp-proof courses, construction and expansion joints as well as applied finishes to the exterior face of the walls.
Excludes: All finishes (plaster, paint, panelling, etc.) to the interior face of the wall and parapet walls above the roof level.
Measure the gross area of the exterior walls in square metres along the general face of the walls from the top of the walls below ground floor, if any, or from ground level to the top of the roof finish if the roof is flat, or to the top of the wall if the roof is sloping. Ignore any extra perimeter caused by projecting isolated columns or recessed windows, etc., and deduct for windows and doors. Keep different types of wall construction separate.

(d) Windows

Includes: Steel, wood and aluminum sash; glazing; louvres; hardware; mullions; transoms; sills; stools; flyscreens; storm windows; lintels; precast concrete, stone, etc. surrounds; damp-proof courses; caulking; and all necessary painting or finishing.
Excludes: Curtain walls.
Measure the area of windows in square metres, keeping each type separate. (Note: Record for future reference the percentage of area of the windows to the gross area of Walls Above Ground Floor.)

(e) Exterior Doors and Screens

Includes: Wood, hollow metal, kalamein, steel, bronze and aluminum doors; roll-up or sliding shutters; store fronts; glazed entrance screens; revolving doors; decorative screens; metal and wood frames; subframes; sills; lintels; damp-proof courses; concrete, stone, etc. surrounds; caulking; electric or hand-operated opening devices; hardware; and all necessary painting and finishing.
Measure all glazed screens, store fronts and doors in square metres keeping each type separate.

(f) Balconies and Projections

Includes: Any item which, because of its existence, increases the area and cost of exterior cladding.
Examples are:

(1) Overhangs: If a floor projects beyond the floor below include the exposed soffit and insulation (but not the floor slab).
(2) Balconies: Include projecting balconies in their entirety. For recessed balconies include the cost of balcony railings, soffit finishes and insulation, but not the building wall, floor slabs or floor finishes. (Floor finishes are included in Element 3(a) Roof Finish)
(3) Canopies: Include projecting canopies attached to the building in their entirety.
(4) Sunshades: Include precast concrete, stone, wood, etc. sunshades which stand out beyond the face of the building.

General Comments on Exterior Cladding

(a) Record total cost of Elements 3(b) – (f) inclusive, and express as a price per square metre of the total area of vertical exterior cladding. This rate is very useful in comparing the costs of various sytems of enclosing space and can also be helpful in very early cost estimates when little or no information is available. Because of the wide range of materials available to enclose space this rate will vary considerably and must be used carefully in making early estimates. Record the ratio between the total area of vertical exterior cladding and total gross floor area as an indication of design efficiency.
(b) As it is sometimes difficult to distinguish between walls, doors, windows and screens, as in the case of curtain walls, certain Elements may have to be combined to avoid entirely arbitrary divisions.

4. INTERIOR PARTITIONS AND DOORS

(a) Permanent Partitions and Doors

Includes: Wood studs; metal studs; clay tile, concrete block, poured concrete or other solid walls or partitions; toilet partitions; and all interior doors located in permanent partitions, including hardware and all necessary painting or finishing of doors and frames.
Excludes: All finishes (plaster, gypsum board, paint, panelling, etc.) to walls or partitions.

Measure the gross area of all partitions in square metres and keep separate in different thicknesses and types of construction. Measure doors in square metres keeping each type separate.

(b) Movable Partitions and Doors

Includes: All movable, glazed or unglazed partitions, regardless of material, including doors; and folding or demountable partitions.
Measure movable partitions and doors in square metres keeping each type separate.

(c) Glazed Partitions and Doors

Includes: All permanent interior glazed partitions and doors (such as those forming vestibules at main entrances) glazed internal store fronts, and borrowed lights.
Measure glazed partitions and doors in square metres keeping each type separate.

General Comments on Interior Partitions and Doors

Record and express the total cost of Element 4 (a) - (c) inclusive as a price per square metre of all interior partitions and doors. This unit price is very useful for comparative purposes when considering the cost of this Element in various buildings. However, a wide range is possible because of the large number of materials available and the number of combinations of systems, and when using this unit rate to estimate projects in the very early stages great care must be taken to assess the many factors which are involved.

5. VERTICAL MOVEMENT

(a) Stairs

Includes: Treads, risers, stringers, landings, supporting framework, balustrades, handrails, soffits, steps, ladders, painting and other finishes, and floor framing around stair openings.
Excludes: Enclosing walls.
Measure the total length of risers in metres keeping each type of stair separate.

(b) Elevators and Escalators

Includes: Passenger and freight elevators, hoists, dumbwaiters and conveyors, including entrances, cabs and guide rails; escalator treads and balustrades, etc.; machinery and controls; and general contractor's work to provide bases, pits, fixings and floor framing around openings, etc.
Excludes: Hoistway enclosure walls and machine room power supply.
It is advisable to obtain quotations from manufacturers for this equipment and add general contractor's work required therefor. Record costs per elevator for different sizes, number of floors, speeds and degree of automatic control; and per escalator for complete installation, kept separate according to width, height, speed, etc., for

calculating preliminary estimates for future projects. Because this is very specialized work this informaion should not be relied on too heavily except where very similar installations are being considered.

6. INTERIOR FINISHES

(a) Floor Finishes

Includes: All floor finishes including toppings, steel trowel or wood float finish together with bases, curbs, mat sinkages, frames and mats, construction joint cover plates, etc.
Measure floor finishes in square metres keeping each type separate.

(b) Ceiling Finishes

Includes: All ceiling finishes, cornices, bulkheads, etc.
Excludes: Special illuminated or heated ceilings.
Measure ceiling finishes and bulkheads in square metres keeping each type separate. Measure cornices in metres keeping each type separate. (Note: In the case of sloping, stepped or special ceilings measure the area on plan rather than the actual area for the quantity to be shown in a Cost Analysis.)

(c) Wall Finishes

Includes: All applied interior finishes to exterior walls and partitions, including grounds and strapping.
Excludes: Self-finished walls, the finish for which is included with the wall.
Measure wall finishes in square metres keeping each type separate.

General Comments on Interior Finishes:

Measuring interior finishes, particularly base and wall finishes, can be very time-consuming, and frequently their cost amounts to only a small percentage of the total construction cost. To save time the following is suggested as a quick method of measuring wall finishes for a preliminary estimate. Take once times the area of Elements 3(b) and 3(c) and twice times the area of Element 4(a). When added together this gives a gross area of interior wall surfaces, but since this usually exceeds the total area to be finished because of unfinished spaces and areas above ceilings, a percentage must be applied to reduce it to the total finished area. The percentage will vary with the type of building and each time a detailed estimate is made the percentage of finished wall surfaces to gross wall surfaces should be calculated for future use. As a general rule this usually works out to between 70 and 75 percent. As a further refinement to aid in calculating interior finishes when doing preliminary estimates, the percentage of each type of floor, ceiling and wall finish to the total finished area should be calculated and recorded when doing detailed estimates.

7. FITTINGS AND EQUIPMENT

(a) Fittings and Fixtures

Includes: All fittings and furnishings normally supplied under the construction contract.

Excludes: Equipment described in Element 7 (b)
Information is not usually available on fittings and fixtures at the time of preparing a preliminary estimate. If the cost of this element is recorded as a cost per square metre of the gross floor area whenever a detailed estimate is prepared, this cost can be used for preliminary estimates.

(b) Equipment

Includes: All equipment supplied and/or installed under the construction contract such as kitchen, laboratory, gymnasium, swimming pool, laundry, x-ray and sterilizing equipment, etc.
Excludes: Mechanical and electrical roughing in.
Obtain quotations from specialists who normally supply and/or install these items. Keep records of details and costs of special equipment. Costs can be recorded on a square metre basis for use in preliminary estimates as described for Element 7(a).

8. SERVICES
(a) Electrical

Includes: All the work inside the building normally included in the Electrical Division of the Specification.
Excludes: Electrical work outside the building.
Record and use the costs of the various electrical systems from detailed estimates as follows:

i) Transformers and main switchboard: Lump sum costs to be used.
ii) Lighting and power distribution: This includes distribution panels for light and power and all associated conduit, wiring and outlet boxes, and should be recorded as a cost per square metre of the gross floor area.
(iii) Lighting fixtures: Enumerate light fixtures if possible, otherwise record the cost per square metre of the gross floor area.
(iv) Branch wiring: Cost per square metre of the gross floor area.
(v) Special services: This includes systems such as empty conduit for telephones, T.V. and radio systems, lightning conductors, fire alarms, under-floor ducts, clocks, P.A., emergency lighting equipment, stage lighting, doctors' and nurses' call, intercom, central dictating, automatic electric door locks and closers, car heaters, snow melting, telautograph, photofloods, etc. The extent of these special services must be known and the cost of each system estimated by obtaining quotations from suppliers or by reference to previous detailed estimates.

(b) Plumbing and Drains

Includes: All the plumbing and drainage work inside the building normally included in the Plumbing Division of the Specification.
Excludes: Outside drainage.
Record and use the costs of the plumbing systems from detailed estimates as follows:

(i) Fixtures, Services and Wastes: This includes all sanitary fixtures, floor and roof drains, firehose cabinets, piping, hangers, insulation, fittings, valves, traps, etc. Enumerate plumbing fixtures if possible, otherwise record the cost per square metre of the gross floor area.
(ii) Special Services: This includes systems such as liquid soap, compressed air, oxygen, vacuum, gas, etc. Record as lump sums or as the cost per square metre of the gross floor area for each system.
(iii) Sprinklers: Cost per square metre of sprinkled area.

(c) Heating, Ventilation and Air Conditioning

Includes: All the heating, ventilation and air conditioning work normally included in the Mechanical Division of the Specification plus the cost of any heat storage tanks, fireplaces and chimneys. Record and use the costs of the mechanical systems from detailed estimates as follows:

(i) Boiler plant: This includes boilers, boiler controls, fuel storage and supply, pumps, motors, etc. Record the cost per square metre of the gross floor area, or per horsepower of the boilers.
(ii) Refrigeration plant: This includes the cooling tower, equipment, controls, motors, bases, anti-vibration pads, etc. Record the cost per square metre of the gross floor area, or per ton of refrigeration.
(iii) Distribution systems: This includes all supply and return piping, insulation, hangers, fittings, valves, radiators, convectors, etc. required to radiate or absorb heat. When possible enumerate radiators, convectors, etc., otherwise record the cost per square metre of the gross floor area.
(iv) Ventilation system: This includes all ductwork, fans, coils, dampers, grilles, sound and acoustic insulation, hangers, etc. Record the cost per square metre of the gross floor area being ventilated.
(v) Controls: Cost per square metre of the gross floor area.
(vi) Chimney: Cost per metre of height of the chimney.

9. SITE DEVELOPMENT

(a) General

Includes: Roads, paths, parking areas, curbs, bumpers, grading, seeding, sodding, planting and landscaping, retaining walls, canopies and covered ways [other than those included in Element 3(f)], pools, fencing and gates, permanent signs, outside steps, balustrades, paved terraces, etc.
Measure each type of area separately in square metres.
Measure curbs, bumpers, fences and balustrades in metres.

(b) Services

Includes: Incoming electrical services, connection to the electric main, water mains, septic tanks and tile beds, sewers, site drainage, manholes, catch basins and fountains.

Record the cost per square metre of the site for this work.

(c) Alterations

Includes: Any minor works of alteration to an existing building caused by erecting the new building next to it.
Excludes: Major alterations in an existing building.
Measure alterations in detail recording all structural removals in cubic metres, walls and partitions or finishes to be removed or replaced in square metres.
Enumerate and describe all openings to be cut or filled in, doors and windows to be removed or installed, fittings and fixtures to be removed or replaced. Note that since the gross floor area is that of the new addition only, alterations are expressed as a cost per square metre of the new addition and not of the area of alterations.

(d) Demolition

Includes: All demolition of existing buildings to make way for new construction.
Demolition of existing buildings can only be done on a basis of known costs of similar jobs or by consulting a specialist in this type of work.

10. OVERHEAD AND PROFIT

(a) Site Overhead

Includes: The contractor's indirect and site expenses such as:

(i) Access to the Site — temporary roads and traffic control.

(ii) Site Accommodation — temporary offices, storage sheds, toilets, hoardings, signs, telephone, stationery and site office heating.

(iii) Site Protection — barricades, ladders, scaffolding, guard rails, fire precautions and safety precautions.

(iv) Temporary Services — water and hydro.

(v) Clean-Up — cleaning glass, cleaning site, cleaning building and removing rubbish.

(vi) Labour Expenses — overtime, lost and reporting time, strike delays, lodging and travel, and wage increases.

(vii) Supervision — superintendent, engineer, watchman and other site staff.

(viii) Insurance and Bonds — fire insurance, public liability and property damage insurance, performance bond, and bid bond.

(ix) Equipment — elevator and hoist, crane, trucks, buggies and vibrators, pumps, fuel and oil, and small tools.

(x) Winter Conditions — winter heat, winter concrete, premium for winter work, clearing ice and snow, tarpaulins, straw, and enclosures etc.

(xi) Miscellaneous — building permit, land survey, layout and preparation, attendance on architect, attendance on subcontractors, critical path, rough hardware, testing and samples, photographs, cutting and patching, and financing.
For preliminary estimates assess this as a percentage of the direct costs, that is, the work in Elements 1(a) - 9(d) inclusive.

(b) Head Office Overhead and Profit

11. CONTINGENCIES

(a) Design Contingency

The design contingency is intended to provide an allowance so that necessary design changes can be made as the design is developed. It may therefore be as high as ten percent of the estimated construction cost — Elements 1(a) - 10 (b) inclusive — in the preliminary stages, reducing to zero for the final estimate before tenders are called.

(b) Escalation Contingency

The escalation contingency is intended to allow for increases in costs between the date the estimate is prepared and the date tenders are called. (Note: The escalation normally carried by the contractor in his tender for increases in costs between the date of his tender and the time the work is carried out should be included in the estimate and should not be allowed for here.)

(c) Post-Contract Contingency

The post-contract contingency is intended to provide an allowance so that changes can be made during construction. The amount of this contingency is left to the discretion of the architect and his client.

Bibliography

Construction Costs and Estimating

The Building Estimator's Reference Book, Chicago, Il., Frank R. Walker Publishing Company. A classic that has gone through many editions, it contains much factual information about construction and many examples of cost analyses.

Collier, Keith, *Fundamentals of Construction Estimating and Cost Accounting*, Englewood Cliffs, N.J.: Prentice-Hall, Inc., 1974. The predecessor to this book on conceptual estimating, this earlier book is about bid estimating; it includes examples of detailed measurement from three of the same buildings in this book: the residence, the warehouse, and the frame apartment block. This enables useful comparisons among estimating methods and their results.

Means, *Building Construction Cost Data*, (annually) Kingston, MA.: Robert Snow Means Company, Inc. One of the best sources of data that can be converted into useful information for both estimator and appraiser. The companion publications cited below are of similar quality. Refered to in this text; see Index.

Means, *Mechanical and Electrical Cost Data*. Very detailed information that enables one to make estimates of M & E work following the tree format described in this book.

Means, *Square Foot Costs*. Useful to owners, contracting companies, appraisers and others seeking quick approximate estimates. Best used in conjunction with more detailed cost data such as is found in the companion manuals cited above.

Standard Methods of Measurement

Method of Measurement of Construction Works, Toronto, Ontario: The Canadian Institute of Quantity Surveyors. In metric mode.

Principles of Measurement (International) for Works of Construction, London, England: The Royal Institution of Chartered Surveyors.

Standard Method of Measurement of Building Works, London, England: The Royal Institution of Chartered Surveyors and the National Federation of Building Trade Employers. Several editions have been published; commentaries and interpretations are also available. Similar standards are published in some other Commonwealth and European countries.

Appraisal Manuals

Boeckh Building Valuation Manual, Volumes I, II, & III, Milwaukee, WI.: Boeckh Division, The American Appraisal Company. Refered to in this text, see Index.

Means, *Square Foot Costs*, cited previously, is also an appraisal manual as well as a source of data for making quick estimates of construction costs. It includes examples of residential, commercial, industrial, and institutional buildings.

Computers

Construction Computer Applications Newsletter, Silver Spring, MD.: Construction Industry Press. Published monthly, this newsletter should grow into an informative magazine with an increased circulation.

Frankel, Steven, *The Compleat Kaypro*, Reston, VA.: Reston Publishing Company, 1984. A comprehensive guide to using Kaypro computers, covering both hardware and software, and a useful guide to those thinking of buying a Kaypro or any other personal computer for business purposes.

INFOWORLD, The Newsweekly for Microcomputer Users, Menlo Park, CA.: A readable source of computer news and reviews of both hardware and software and a useful guide for the computer customer.

McWilliams, Peter A., *Word Processing on the Kaypro*, Los Angeles, CA.: Prelude Press, 1983. Essentially the same as his bestseller, *The Word Processing Book*, but with additional material about the Kaypro computer added. A companion book deals with the IBM PC. All are non-technical and easy to read. Nevertheless, to read one of these books is for the computer neophyte a useful introduction.

PC, The Independent Guide to IBM Personal Computers, New York, NY.: Readable and useful, even if you do not own an IBM computer.

PROFILES, The Magazine for Kaypro Users, Solana Beach, CA.: Slimmer than the last publication, it nevertheless contains much valuable information for the Kaypro user and is well worth the money.

Standard Contract Documents

American Standard Contract Documents, published by the American Institute of Architects, 1735 New York Avenue, N.W., Washington, D.C. 20006.

British Standard Contract Documents, published by RIBA Publications Limited (for the Joint Contracts Tribunal), 66 Portland Place, London, England.

Canadian Standard Contract Documents, published by the Canadian Construction Documents Committee, 85 Albert St., Ottawa, Ontario, K1P 6A4.

A Miscellany

Capitalized Approach to Budget Estimating, Washington, D.C.: General Services Administration, Public Buildings Service (GSA/PBS)., 1881. Refered to in this text; see Index.

Collier, Keith, *Construction Contracts*, Reston, VA.: Reston Publishing Company, 1979. Examines the basis of and the differences among the several kinds of construction contracts, in part by comparisons among the standard forms of contracts used in the United States, Canada, and Britain. One of the few books on construction contracts.

__________ , *Managing Construction Contracts*, Reston, VA.: Reston Publishing Company, 1982. Views this large subject from the standpoint of construction contracts, rather than from that of the soft sciences. As a sequel to the author's book on construction contracts it deals in particular with that relatively new approach to building called Construction Management (as defined in the Glossary).

Conceptual Estimating Guide, Washington, D.C.: GSA/PBS.

The GSA System for Construction Management, Washington, D.C.: GSA/PBS, 1975.

Hardie, Glenn M., *Construction Contracts and Specification*, Reston, VA.: Reston Publishing Company, 1981. Part 1 deals with the basics of contracts; Part 2 with specifications; and Part 3 with practical applications, including specifications for the same warehouse building dealt with in Collier's book on fundamentals of estimating and in this book on conceptual estimating.

Masterformat, published jointly by the Construction Specifications Institute, Washington, D.C., and Construction Specifications Canada, Toronto, Ontario. A standard format for specifications, databases, and cost codes in construction, now based on 17 standard divisions of work and information (originally 16 divisions), as described in this book's Glossary.

Index

C

D

E

F

G

H

I

V

W